Ciarcia's Circuit Cellar Volume VII

Volume VII

Ciarcia's Circuit Cellar

by Steve Ciarcia

McGRAW-HILL PUBLISHING COMPANY

New York St. Louis San Francisco Auckland Bogotá Caracas Hamburg
Lisbon London Madrid Mexico Milan Montreal New Delhi Oklahoma City
Paris San Juan São Paulo Singapore Sydney Tokyo Toronto

Library of Congress Cataloging-in-Publication Data
(Revised for vol. 7)

Ciarcia, Steve.
Ciarcia's Circuit cellar.

Articles written by the author for a Byte magazine column, Ciarcia's Circuit cellar, which began in Nov. 1977.
Vol. has imprint: New York : McGraw-Hill.
Includes bibliographies and indexes.
1. Microcomputers—Collected works. I. Circuit cellar.
TK 7888.3 C58 621.391'6 78-20920
ISBN 0-931-71807-4 (v. 1)

ISBN 0-07-010963-X (v. 2)

ISBN 0-07-010965-6 (v. 3)

ISBN 0-07-010966-4 (v. 4)

ISBN 0-07-010967-2 (v. 5)

ISBN 0-07-010968-0 (v. 6)

ISBN 0-07-010969-9 (v. 7)

1234567890 SEM/SEM 8965432109

ISBN 0-07-010969-9

The editors for this book were Daniel Gonneau and Fred Bernardi, and the production supervisor was Richard Ausburn.

Printed and bound by Semline, Inc.

This book is dedicated to Ed Nisley.
His support and friendship were instrumental
in the success of many projects.

ABOUT STEVE CIARCIA

If you're writing the input text for a 6K-byte speech-to-text program, Ciarcia is spelled "see-are-see-ah." Otherwise he needs no introduction to computer buffs.

A regular contributor to BYTE since 1976, Ciarcia has been repeatedly voted that magazine's most popular columnist. The reason? In the words of *Science Books and Films:* "His technical breadth and depth of knowledge... displayed without pretense yet with a sense of exuberance and energy... is an inspiration. He's brilliant!"

When he's not providing expert advice to our major computer companies, or cooking up ingenious projects in his famous cellar, or writing his state-of-the-art books and articles, Ciarcia can be found in his kitchen creating such specialties as chicken cacciatore and shrimp provençale!

Contents

Preface

Over the last 12 years, I've developed a very special relationship with the readers of *Circuit Cellar*. Because the projects I write about work in the real world, and because I make no bones about being a working engineer, *Circuit Cellar* readers and I tend to regard one another as colleagues. I don't think that I would want to continue writing if I had to keep the traditional journalist's arms-length from the readers. For over 11 years, the projects in *Circuit Cellar* have served as jumping off points for correspondence, conversations, and collaboration.

This close relationship with the readers has given me some advantages that most writers can't share. For one thing, after thousands of letters and telephone calls (and thousands more messages on the Circuit Cellar BBS) I think I know what most readers are interested in knowing about microcomputer applications. I've heard about the problems and pleasures of individual situations, the processors in use, the limits of time and money that constrain most people, and the frustrations that come from being asked to provide too much solution from too few resources. I've also come to respect the people who read *Circuit Cellar*. The vast majority are hardworking, creative individuals who are anxious to learn better ways to solve problems, and eager to share the knowledge they've gained from hands-on experience. I'm sure that this combination of knowledge and respect is a large part of what made *Circuit Cellar* a success.

I'm indulging in this nostalgia because this is the last volume of *Circuit Cellar* columns from BYTE magazine. The December 1988 issue of BYTE saw the final installment of *Ciarcia's Circuit Cellar*. After nearly 100 projects I decided that it was time to do something different. My something different is CIRCUIT CELLAR INK, a magazine dedicated to the kind of hands-on articles I've always enjoyed. If you want to know more about the magazine, just write me at 4 Park Street, Vernon, CT 06066, and I'll send you some information.

The projects in this book run the gamut from the simple to the complex. There are projects for development systems, human brainwave monitoring, video digitizing, and multitasking process control. The final *Circuit Cellar* project is a tutorial on multiprocessor computers that show you how to build your own 64-processor computer for generating Mandelbrot sets. As always, these projects have been built and tested so that you can build and use them with confidence. I've always figured that there are plenty of opportunities for disappointment and disaster in the world: what you read in *Circuit Cellar* shouldn't be among them.

I've enjoyed the last dozen years. We've learned a lot, built a lot, and had a lot of fun together. And I'm looking forward to the next dozen years. In the computer industry there's no danger of running out of projects to build or applications to design. I'll head back to the Circuit Cellar to keep making yesterday's dreams into today's hardware.

—Steve Ciarcia

1

BUILD A GRAY-SCALE VIDEO DIGITIZER

PART 1: DISPLAY/RECEIVER

An imaging system with remarkable features for the price

Video technology has always interested me. One look at all the monitors, TVs, and displays around my house suggests that it goes deeper than interest. Freudian views aside, this is not the first time I have covered video technology in a Circuit Cellar project. In previous articles, I have described high- and low-resolution video display systems and even a low-cost digital camera. However, the one project I've always wanted to do has eluded me. Until now, I have had to hold off on the presentation of a cost-effective general-purpose high-performance gray-scale video-digitizing "frame grabber."

I'll explain all this later, but the key terms for the moment are "gray scale" and "frame grabber." Such terms usually indicate commercial units costing thousands of dollars.

While some video digitizers are designed as peripherals for specific computers, virtually all digitizers endent and involve significant trade-offs in performance to maintain low cost. Generally, their digitizing speed is significantly less than the rate necessary to capture a video image as it is transmitted in real time (1/30 to 1/60 second). Instead, they must repeatedly sample many sequential video frames. Digitizers like these—sequential field scanning digitizers—can deal only with stationary objects in front of a camera and can take as long as 30 seconds to scan and record an image. Such digitizers are useless if you are working with moving objects.

Another factor to consider is how a digitizer represents the intensity of each pixel. Most low-cost digitizers meet the minimum video display capabilities of their host computer and digitize each element only as black or white. Some allow a limited gray scale. Higher-performance digitizers offer 64 or more levels of gray scale as well as a high digitizing speed.

Fortunately, both static memory and video integrated circuit technology have progressed to the point where I can finally offer a project that attempts to meet the level of perfection I have outlined. In parts 1 and 2 of Chapter 1, I will describe a complete digital video system. You can use it independently as a video camera digitizer and display (to implement a video telephone, for example), or you can connect it to any personal computer for tasks like image processing, character recognition, and desktop publishing graphics.

ImageWise consists of two separate boards: a digitizer/transmitter and a display/receiver. Each board can be used independently, or they can be connected to form a complete digitizer/transmitter/receiver system (see photo 1).

In contrast to other digitizers, ImageWise is a true frame grabber that takes only 1/60 second to capture an image. It accepts the video signals from devices like a standard TV camera (either monochrome or color), VCR, laserdisc player, and camcorder, and it then stores the picture as 244 lines of 256 pixels with 64 levels of gray scale—256 by 244 by 6 bits (see photo 2). The ImageWise digitizer/transmitter board converts the stored video image to RS-232 serial data that can be transmitted to any computer or to the ImageWise display/receiver board. Transmission rates are selectable from 300 bits per second up to 57.6k bps.

The ImageWise display/receiver board has a serial RS-232 input and a composite video output. It receives serial data directly from the digitizer/transmitter board or transmitted from a file downloaded from your computer and converts this data back into a picture on a composite video input black-and-white monitor (adding a pair of modems lets you send the images over telephone lines). The displayed image is an interlaced 256 by 244 by 6-bit gray-scale picture. The following specifications for the display/receiver board sum it up better:

- Resolution: The three selectable resolutions are 256 by 244, 128 by 122, and 64 by 61. All resolutions support 64 levels of gray scale (each picture element is represented by 6 bits). Note that, regardless of resolution, the system displays all pictures as interlaced full-screen images. Lower-resolution images are composed of larger pixel blocks.
- Video output: 75-ohm, 1.5-volt peak-to-peak composite video.
- Serial input: RS-232, 8 bits, 1 stop bit, no parity. Transmission rate is selectable from 300 bps to 28.8k bps.
- Hardware: 8031 microprocessor, Telmos 1852 video D/A converter, 64K bytes of static video RAM.

The specifications of the digitizer/transmitter board are

- Resolution: Same as above.
- Video input: 1-V peak-to-peak, black and white or color, 75-ohm termination.
- Serial output: RS-232, 1 start bit, no parity. Transmission rate is selectable from 300 bps to 57.6k bps.
- Hardware: 8031 microprocessor, RCA CA3306 6-bit flash A/D converter, 64K bytes of static RAM.

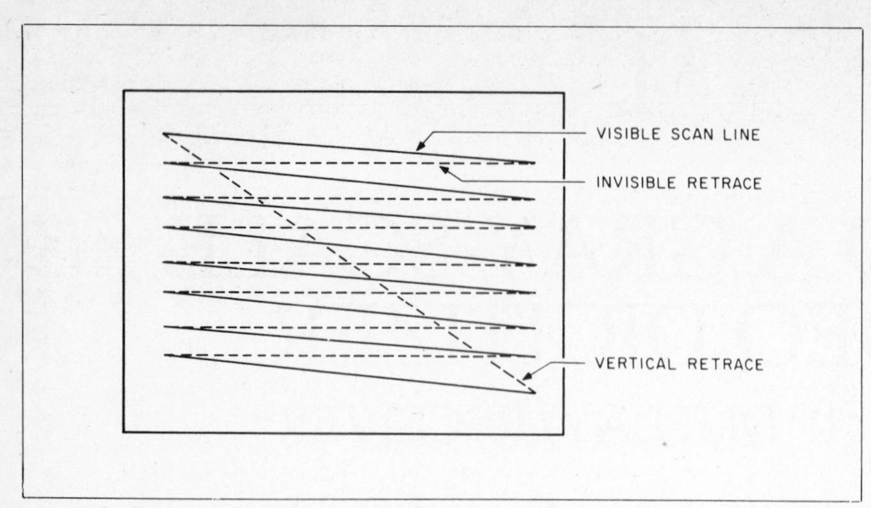

Figure 1: *The path of the electron beam during a video scan.*

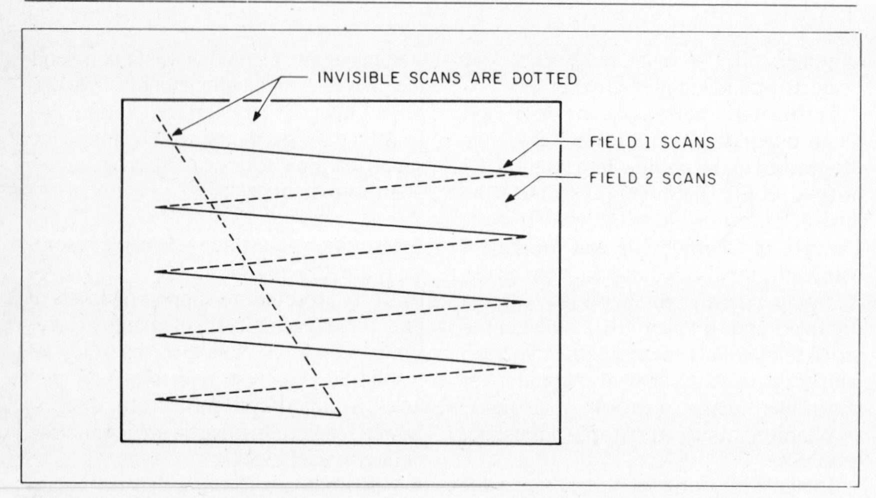

Figure 2: *The path of the electron beam for an interlaced video scan.*

Photo 1: *The ImageWise video digitizer system consists of two boards that can be used together or independently. The display/receiver board, shown on the right, reassembles and displays images that are flash-digitized and serially transmitted by the digitizer/transmitter board on the left.*

The ImageWise video system is designed to be intelligent as well as functional. You can control many of its features—such as digitizer resolution and picture update—remotely from the receiver or another computer. In addition, the system incorporates various compression techniques, including run-length encoding, to considerably reduce image-transmission time (essential with slow modems).

ImageWise is no small project. Consequently, I will present it in two parts, beginning with the display/receiver section. (If you built the digitizer/transmitter section first, you would have no way of displaying a gray-scale picture other than converting it to a dot-dithered black-and-white-only image on your computer—and what would that prove?) You can check out the receiver using its internal test patterns and by downloading picture files from a computer or my bulletin-board system. It's much easier to verify that the transmitter is sending the correct data after you have a working receiver to show any problems.

Before I delve into the hardware and software, however, I think it's a good idea to review what goes into a "standard" TV signal. With that in mind, it will be easier to see how the ImageWise transmitter digitizes the video and the receiver reconstructs it. As you'll discover, there is a lot more to video than just another pretty picture.

Today's Class: TV Basics 101

Although there may be a few folks tucked away in odd corners of the country who don't have a TV set, I think it's safe to say that everyone who reads BYTE has at least seen a TV picture at one time or another. While most TV is color TV these days, I'll describe only monochrome (black-and-white) TV signals because that's what ImageWise uses. The circuitry required to digitize and reconstruct color TV signals is considerably more complex than seemed reasonable for this project. Fortunately, the color video standards include monochrome as a subset, so we can use color cameras and monitors as well.

Figure 1 shows a simplified diagram of the process used to build an image on a TV monitor. An electron beam is scanned horizontally across the screen, starting from the upper left corner. It is moved downward after each line, and the result is a set of lines scanned left to right and top to bottom filling the screen. After scanning the last line, the beam is returned to the upper left corner to begin scanning the next screen.

The faceplate of the screen is covered with a phosphor that glows when struck by the electron beam and continues to glow even after the beam passes on. Because the entire screen is scanned rapidly enough to get the beam back to each spot before the phosphor glow fades out, the entire screen seems to be illuminated at once.

One key difference between a TV display and most computer CRT displays is that the electron beam in a TV set can take on a wide variety of intensities, ranging from completely off (black) through shades of gray to completely on (white). A computer display may allow only black, one shade of gray, and white. We'll see what this difference means a little later on. Most composite video input amber or green monitors also have some gray-scale (or should I say green-scale?) capability.

As you might expect, the actual details are a bit more involved. A TV screen is scanned twice for each image, with the two sets of scan lines interlaced on the screen (computer displays are generally noninterlaced). This allows the whole screen to be scanned in half the time a noninterlaced scan would take, without reducing the number of lines in the image. Figure 2 shows how interlaced scanning paints lines on the screen. Each vertical scan is called a *field*, and two matched fields make up a *frame*.

One field is completed in 1/60 second. There are 262.5 lines in each field (see figure 2 to locate the half lines), so each line must be scanned in 63.5 microseconds (1/60 second divided by 262.5). Figure 3 illustrates the signal voltage sent to the display for a single scan line, along with the allowable times for each part of the waveform. (Note: By strict definition, ImageWise is a field rather than frame grabber, since it digitizes only the first 244-line field of an interlaced frame. However, since the term frame grabber has come to mean a digitizer with the speed to digitize within the time period of a video frame, I shall continue to use the term frame grabber.)

The horizontal sync pulse occurs every 63.5 μs and tells the monitor to end the current line, return to the left edge, and begin another line. Surrounding each sync pulse is a blanking voltage that ensures that the track of the retrace will not be displayed on the screen. After allowing time for retrace and blanking, about 52 μs are left for the actual video picture on the horizontal scan line.

A vertical sync signal tells the monitor when to end the current field, return to the top of the screen, and begin the next field. Because the two fields in each frame are offset by exactly half a line, the timings at the end of each field are slightly different. Figure 4 diagrams the analog voltage required to display a complete field. To keep the size of the diagram down, only the first and last lines of active video are shown. The horizontal scale is distorted so that you can see the details.

You should note that the horizontal

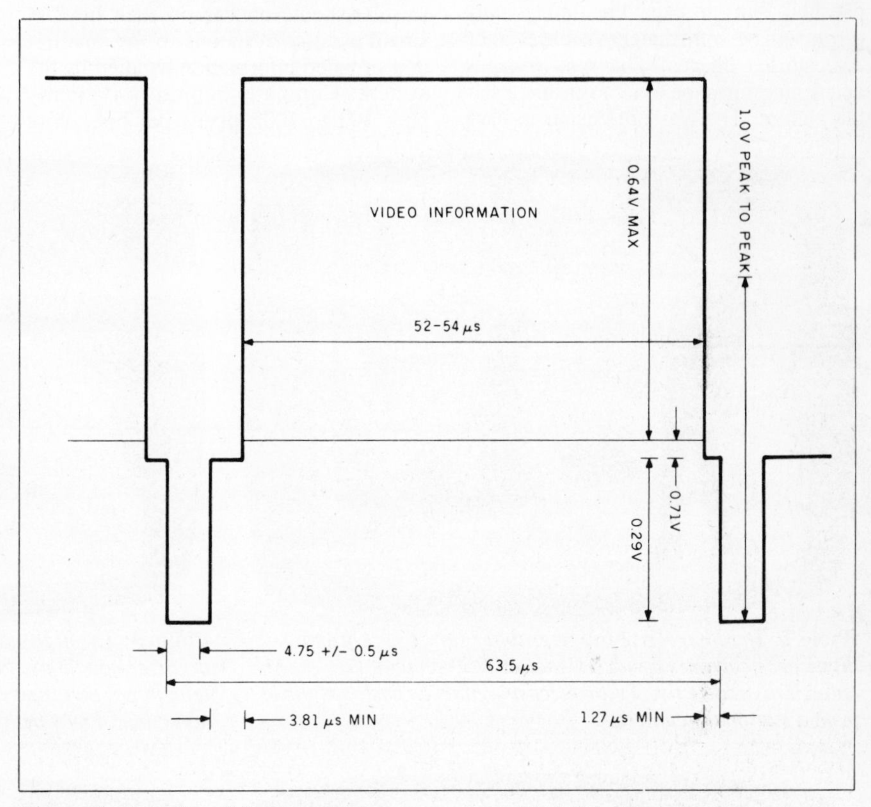

Figure 3: *A profile of the signal voltage sent to the display during a single scan line. Note the bracketing 63.5-microsecond horizontal sync pulses.*

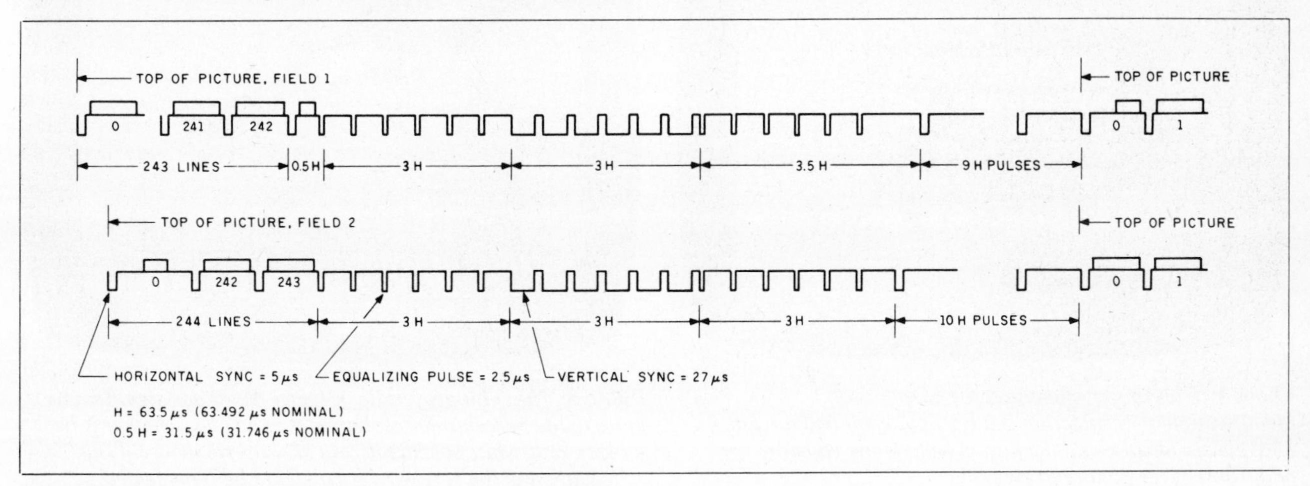

Figure 4: *Signal voltage timing diagram for a complete field.*

sync pulses do not stop during the vertical sync and retrace. In fact, to make sure that the monitor switches smoothly across the half line, sync pulses occur every 31.75 μs during most of the vertical sync period.

A blanking voltage surrounds the vertical sync pulse itself to ensure that the vertical retrace is not visible on the screen. Each field contains 244.5 visible lines, which you can verify by examining the timing diagram in figure 4.

Although video monitors can tolerate small variations in the number of displayed lines or the exact line timings, any errors will be immediately visible as jitter or distorted images. The sync voltages and timings must be exact to ensure a stable picture. The worst offense is to have timings that vary "just a little" from field to field; this will cause the picture to jitter annoyingly.

Variations on a Theme

Now that you're acquainted with standard video, I can explain some of the basic design criteria for ImageWise. As with any project, there are trade-offs between "the ultimate system" and "the one that got built." I will try to explain why I made the decisions I did.

The most basic question was one of resolution: How many picture elements (pixels) should appear on each line? A single pixel corresponds to the smallest unit of video information handled by the system. Computer monitors typically have 300 to 1000 pixels per line. With about 50 μs available in each line to display those pixels, a 1000-pixel line requires a new pixel every 50 nanoseconds. Since typical dynamic RAMs have a cycle time of 300 ns, allowing a few nanoseconds for the other circuitry would require about eight banks of RAM to ensure that a pixel was ready every 50 ns. Using 50-ns static RAMs, while feasible, would be very expensive.

I decided to look at it the other way: How many pixels will fit on a line with affordable hardware? We find that 32K-byte static RAMs are increasingly affordable, and they are considerably faster than DRAMs. Since even the "slow" ones have a cycle time of about 130 ns, one pixel every 200 ns is reasonable. That allows 256 pixels in 51.2 μs. Be-

Photo 2: *The ImageWise frame-grabber (or "freeze-frame digitizer") captures a video signal in 1/60 second; fast enough to digitize live TV broadcast signals as easily as those from a stationary camera.*

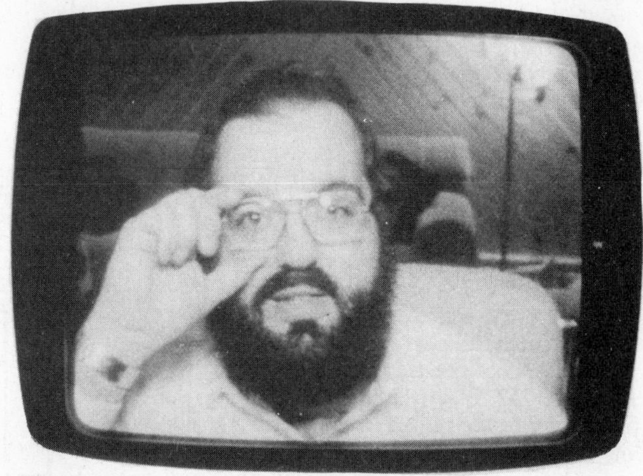

Photo 3: *The high-quality gray-scale images of the ImageWise digitizer can be used in security, pattern recognition, a video telephone, and image database applications. (Phone me sometime; I may be looking back at you through the camera.)*

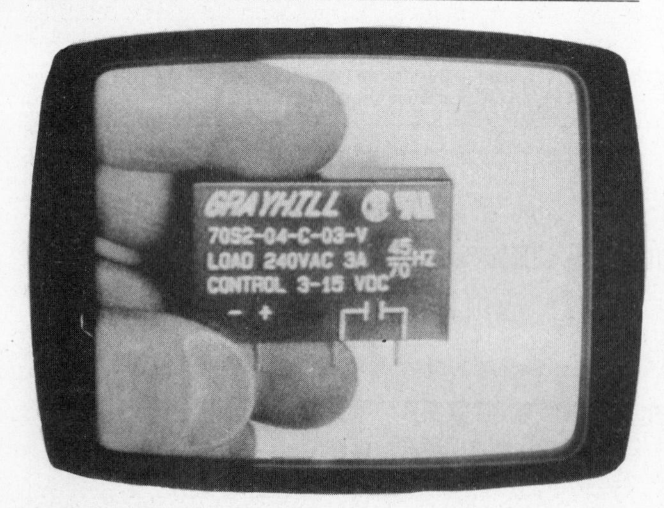

Photo 4: *Teletransfer of pictures for purposes of identification or verification is a legitimate application for ImageWise. Simply hold the part in front of the camera and transmit the picture to everyone.*

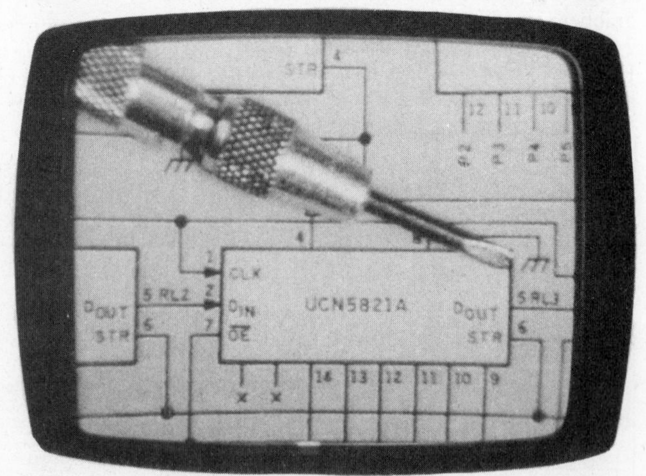

Photo 5: *The 6-bit gray scale of ImageWise adds significantly more to the perceptible resolution of an image, whether it be a black-and-white schematic or a silver screwdriver. (The schematic and the screwdriver could not be represented accurately without gray scale.)*

cause 256 is a "magic" number, I knew I was on the right track.

As I mentioned earlier, each field has about 244 visible lines. Therefore, a 64K-byte buffer could hold one field with some room left over. Two fields could be contained in 128K bytes. With two fields, however, the vertical resolution (488) is twice the horizontal resolution (256). This seemed excessive. Fortunately, because both fields often contain redundant information, I decided to keep the amount of RAM within reasonable bounds and digitize only a single field. But how would a 256- by 244-pixel picture look compared to the original?

All my experience with 320 by 200 computer displays suggested that I might not like the results and be forced to go back to expensive plan A. However, seeing is believing, so I figured I'd build it and decide then. (Often it is easier and faster to build a prototype and take a look at the results than to argue about what might be.)

Photo 3 shows the quality of the image I got with a 256 by 244 display. (So much for my prior experience with computer displays!) There's a good reason why I was wrong, and if you're as surprised as I was, here's the explanation.

You see jagged diagonal lines or "jaggies" on low-resolution computer displays because each pixel can have only a few levels of brightness. The jaggies can be reduced only by increasing the number of pixels on each line. Depending upon the subject material, resolutions of 640 pixels per line and 350 to 400 lines per screen are required to see noticeable improvement.

But there is another way to reduce the jaggies: If each pixel can take on many levels of brightness, the sharpness of the edges can be reduced. ImageWise uses 6 bits to represent each pixel, allowing 64 shades of gray. Real-world images don't have crisp, computer-generated edges, so each pixel tends to shade into the adjoining ones. The effect is a rather smooth picture that has more "effective" resolution than you'd expect (see photos 4, 5, and 6). This is why pictures shown on color displays that incorporate palette D/A converters often look better. Look closely at a line boundary and see if there is some gradual shading.

As an example, you're probably aware that a standard TV does not make a good computer monitor. Trying to fit more than 40 characters on a line results in an unreadable display. However, newspaper headlines displayed on a TV are easily readable even though the characters are very small, simply because each pixel can take on many brightness levels. Watch your TV carefully and see.

Finally, why does ImageWise use 6 bits per pixel and not more if gray scale is such a good idea? Again, it is a cost tradeoff. We have to digitize and determine the gray-scale value of 256 data points in 50 μs, or one every 200 ns. This requires a fast A/D converter called a flash A/D converter. The price of one is directly related to the number of bits it resolves. Eight-bit models are considerably more expensive than 4- or 6-bit chips. The device I ultimately chose was the RCA 3306 6-bit flash A/D converter, which can operate at 12 million to 16 million samples per second (our sample rate is 5 megahertz). I'll talk more about this next month.

Display/Receiver Hardware

The receiver has two main functions: It accepts data from the RS-232 serial port and displays the resulting picture on a monitor. Figure 5 shows the receiver hardware.

As in many recent Circuit Cellar projects, the receiver uses an Intel 8031 single-chip microprocessor to control the rest of the hardware. A 2764 EPROM stores the 8031's program. An Intel 8254 Programmable Interval Timer (PIT) produces the sync pulses. The video field data is held in a pair of 32K-byte static RAMs and is converted to an analog voltage by a specialized video D/A converter. The MC145406 converts RS-232 voltages into TTL levels for the 8031's serial port.

It was tempting to use the 8031 to produce the sync pulses directly, but a little study showed that there was no way to get the precise timings required for a stable picture. The 8254 is connected to produce repetitive pulses, so the 8031 need only program the appropriate values into

Often it is easier and faster to build a prototype and take a look at the results than to argue about what might be.

the 8254's registers when a change is required. The 8254 uses a 500-ns (2-MHz) clock divided from the 10-MHz crystal oscillator. Figure 6 shows the 8254 pulses for a normal video line.

The Telmos 1852 is a specialized video D/A converter that accepts up to 8 video data bits, a blanking input, and a sync input. The analog output conforms to the standard video specifications. Using this D/A converter eliminates a lot of hardware that would otherwise be required to combine the video, blanking, and sync signals to produce the right output voltage with enough power to drive the monitor. The 8031 ensures that the 2 unused bits (the low-order ones) are always 0.

The 16-bit address required by the 64K-byte field buffer is divided into two parts: a high byte supplied by the 8031 and a low byte that can come from either the 8031 or an 8-bit counter. Normally, the counter steps through the 256 pixels on each line, and the 8031 counts out the lines in the high byte. Both bytes are supplied by the 8031 when it reads or writes buffer data.

A pair of LS244s isolate the field buffer's data bus from the 8031's data bus,

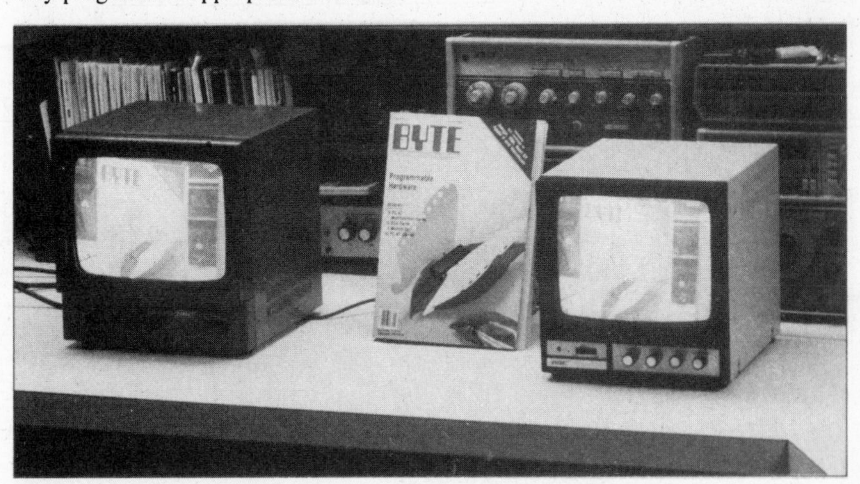

Photo 6: *An in/out comparison. The camera is pointed at the magazine in the middle. The monitor on the left displays the image seen by the camera and the digitizer/transmitter. The monitor on the right shows the digitized image received by the display/receiver board.*

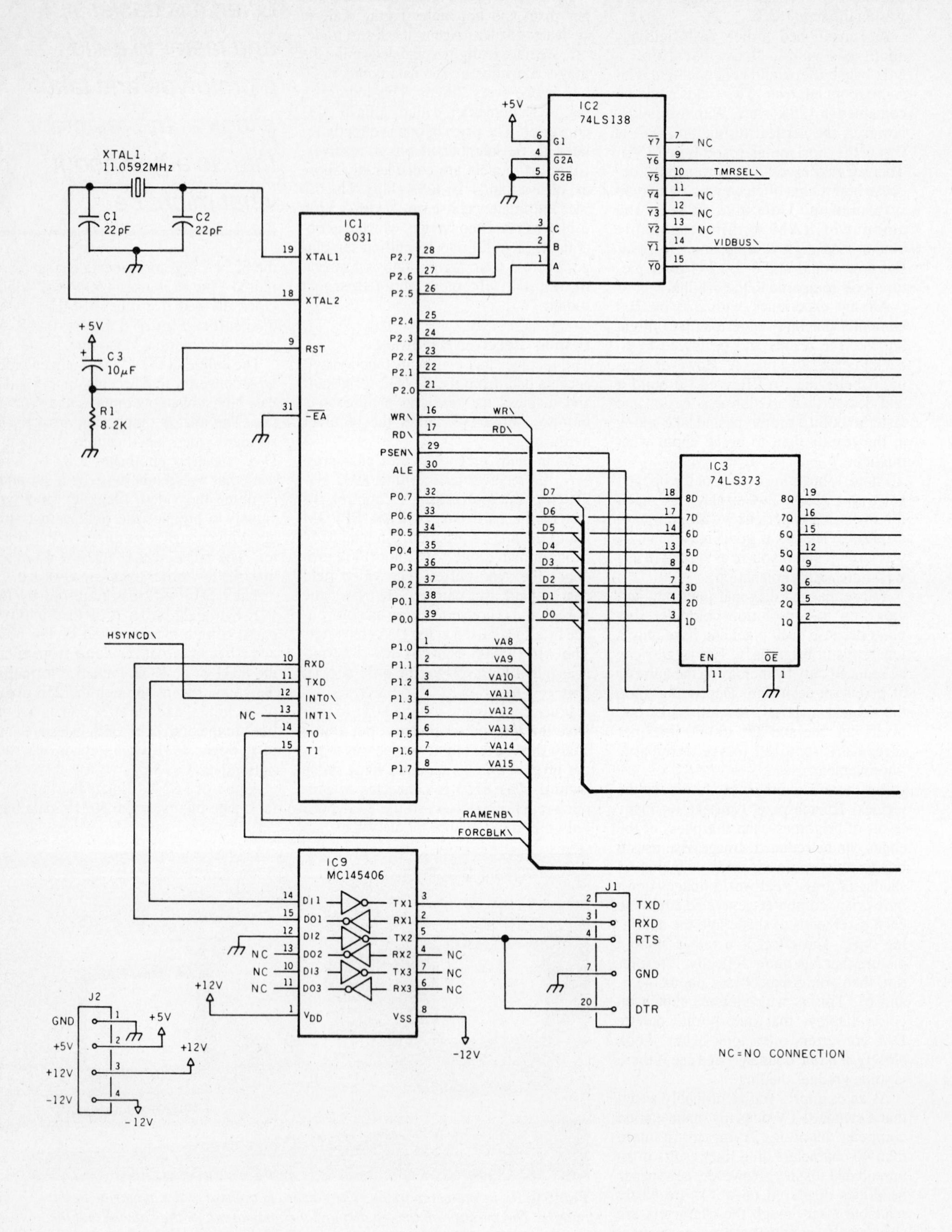

Figure 5: *Schematic diagram of the ImageWise display/receiver hardware.*

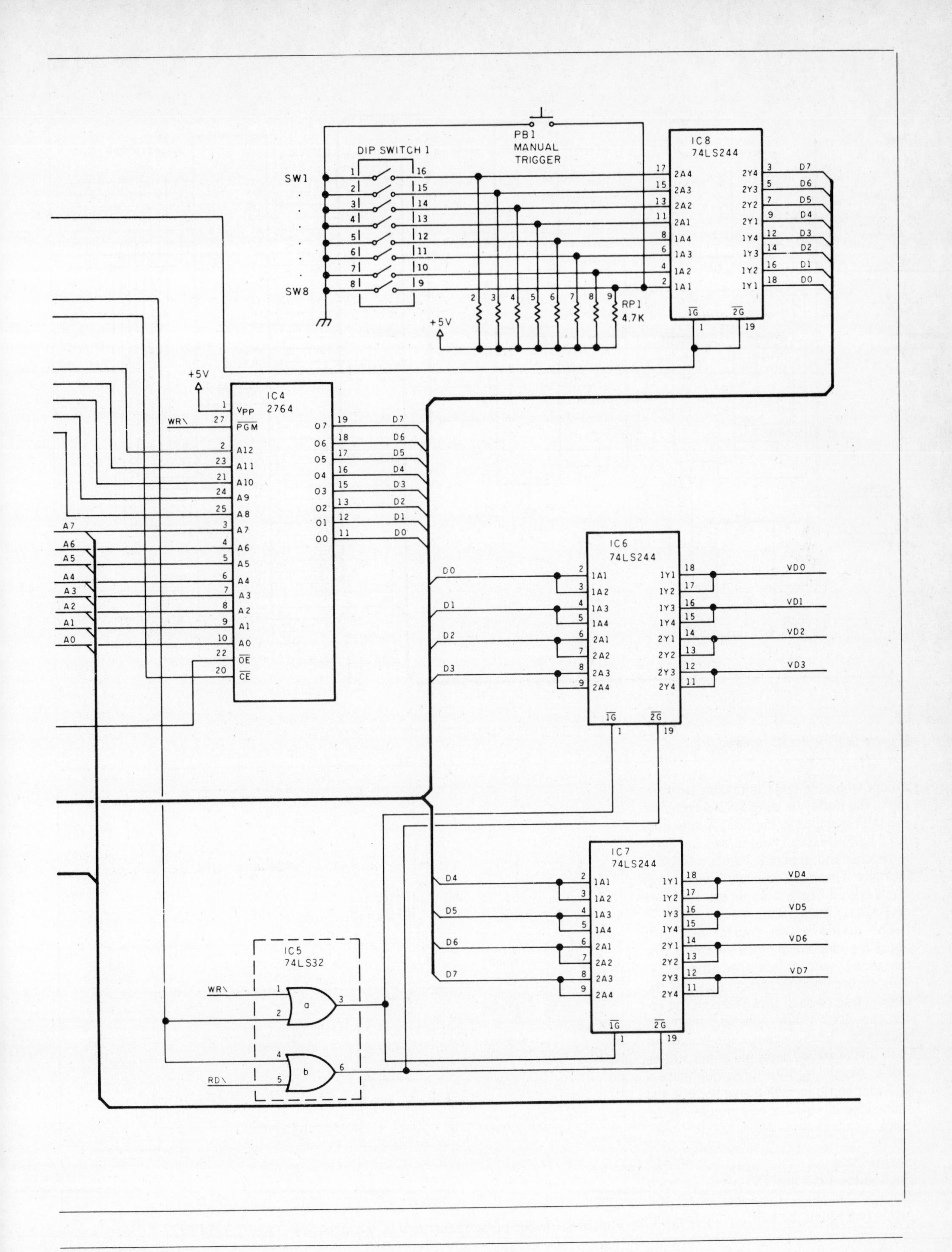
DIP SWITCH 1
PB1
MANUAL
TRIGGER
SW1
SW8
RP1
4.7K
+5V
IC8
74LS244
IC4
2764
IC6
74LS244
IC7
74LS244
IC5
74LS32
WR\
RD\
VD0
VD1
VD2
VD3
VD4
VD5
VD6
VD7

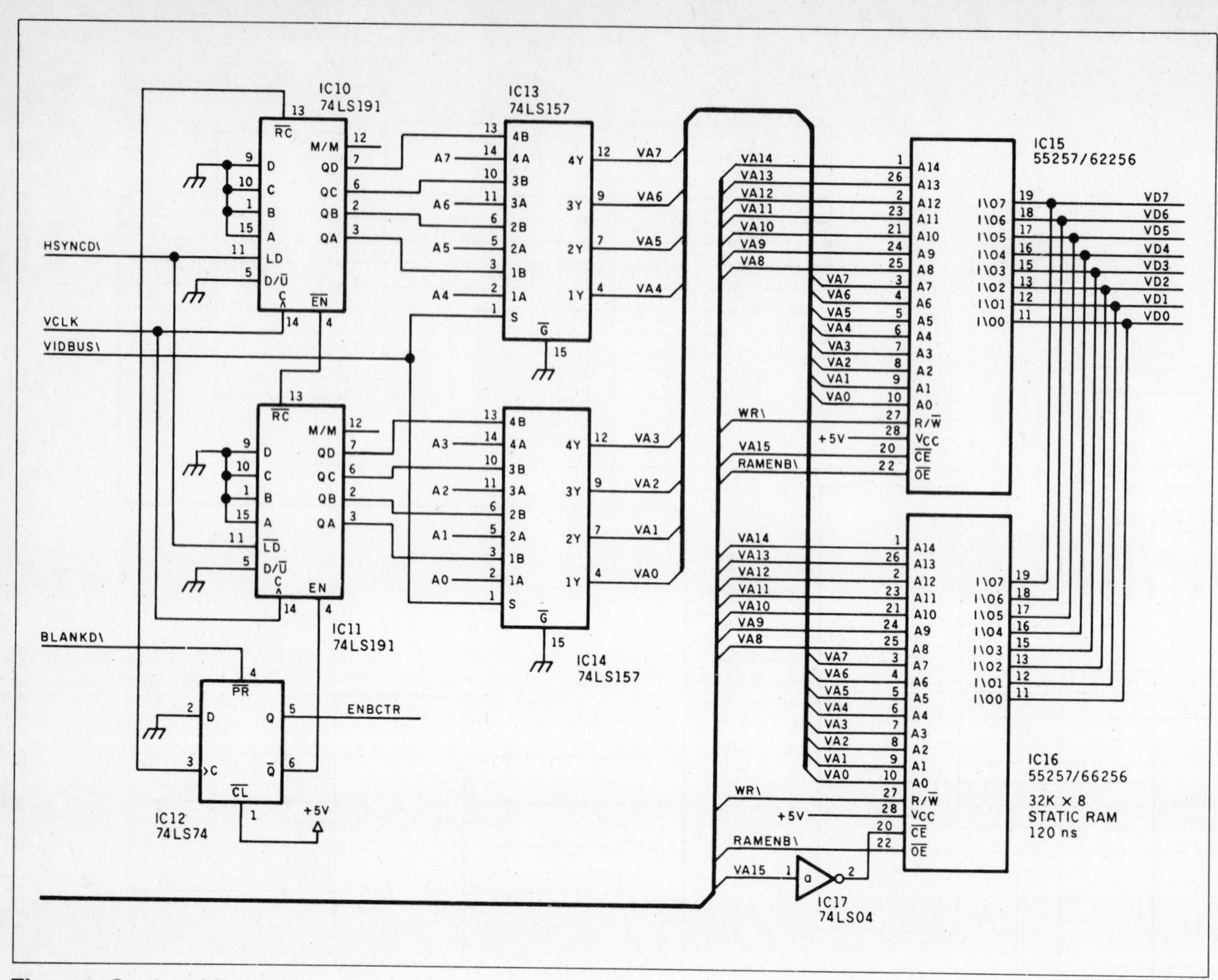

Figure 5: *Continued from page 7.*

except when the 8031 is reading or writing to the buffer. A third LS244 connects the DIP switches to the 8031's data bus. An 11.059-MHz crystal allows the 8031 to receive and generate standard RS-232 bit rates. The video data and sync timings are derived from a separate 10-MHz crystal oscillator circuit.

The divide-by-two counter that produces the video data clock is reset by the horizontal sync pulses from the 8254. This ensures that the pixel clock has the same phase in each line. Without the reset, the clock would alternate phases in successive lines because the length of each line is an odd multiple of the 500-ns clock. Worse, a given line would have a different phase in each frame because the frame length is also an odd multiple of the clock.

Serial Data

Each video field has 256 pixels on each line and 244 lines, for a total of 62,464 pixels (we round the half line up to a full

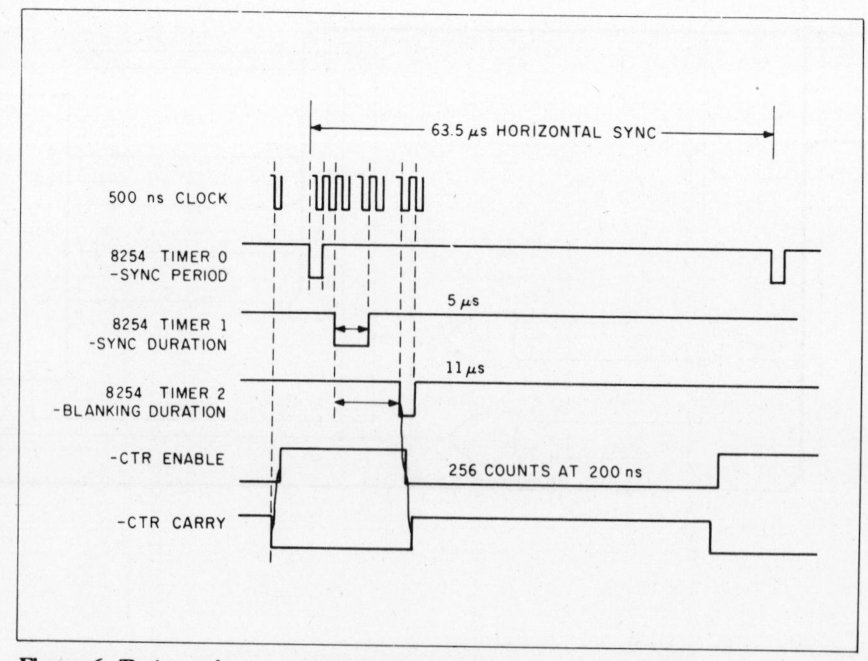

Figure 6: *Timing pulses generated by the display/receiver's 8254 PIT.*

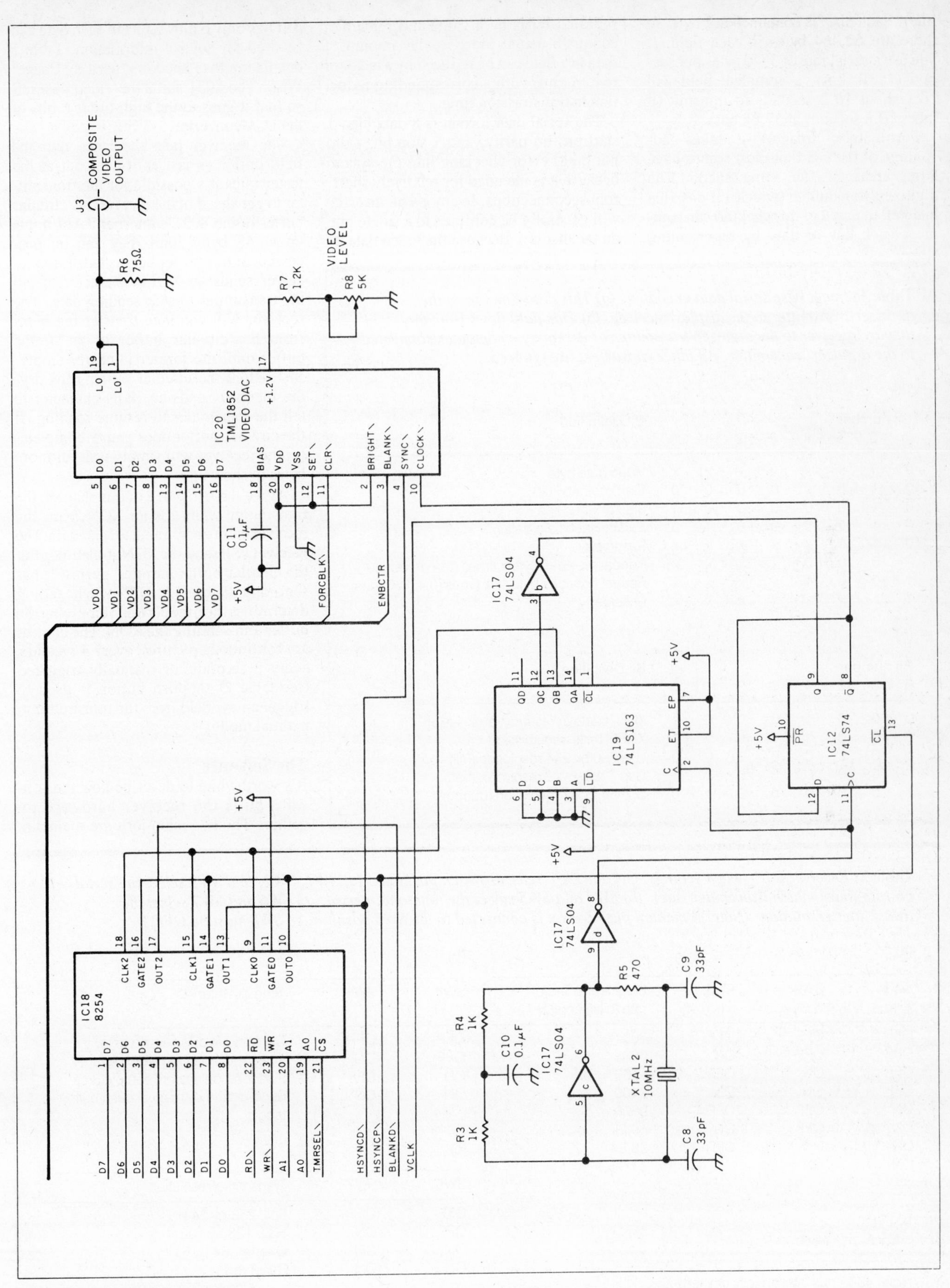

Figure 5: *Continued from page 8.*

line). Each pixel is contained in 1 byte, so there are 62,464 bytes in each field. If you use a serial rate of 3840 bytes per second (38.4k bps), a complete field will take about 16.2 seconds to transmit (it takes 10.8 seconds at 57.6k bps).

Fortunately, ImageWise takes advantage of the fact that most scenes have large areas of the same shade. The digitizer/transmitter (which I'll describe in detail in part 2 of this chapter) can compress each line of data by representing repeated bytes by a value and repetition count. In actual practice, the amount of data in a field can be reduced by a factor of two to four, with a corresponding reduction in transmission time.

The serial data format is 8 data bits, 1 start bit, no parity, and 1 stop bit. I did not build error checking into the system because it is intended for relatively short, robust connections. In any event, an error will generally be confined to a single line on the display. Because the video data itself has only 6 bits, 2 bits in each byte can be used for control information. Table 1 details the byte encoding used by ImageWise. The 8031 shifts the video data left so that it goes to the high-order 6 bits of the D/A converter.

The receiver puts the bytes into the field buffer as fast as it can, but at the faster rates it's possible for the transmitter to get ahead of the receiver. A circular buffer in the 8031's internal RAM holds up to 48 bytes until they can be processed. If this buffer begins to fill, the receiver sends an XOFF character to tell the transmitter to stop sending data. The receiver will continue transferring bytes from the circular buffer to the frame buffer until the former is nearly empty. Just before the circular buffer runs dry, the receiver sends an XON character to tell the transmitter to resume sending. If the circular buffer does empty completely, the receiver will simply wait for more bytes to show up.

As we'll see in part 2 of this chapter, the transmitter waits for an XON from the receiver before beginning to send data. The receiver will send the XON sometime after the circular buffer empties, even if it has sent one before. A DIP-switch setting determines the time between emptying the buffer and sending the XON. The choices are continuous pictures, every 4 seconds, every 8 seconds, or manually triggered (see table 2). A push button is used to trigger a new field from the transmitter in manual mode.

The Software

It's worthwhile to describe how the software pulls the receiver hardware together. The two main jobs are maintain-

Table 1: *ImageWise serial data encoding. (a) This data flows from the digitizer/transmitter to the display/receiver. (b) This data flows from the display/receiver to the digitizer/transmitter or is sent by a computer connected to the digitizer/transmitter. All other characters are ignored.*

(a)

Bit number 7	6	5	4	3	2	1	0	Bit Definition
0	0	x	x	x	x	x	x	Video data byte
0	1	0	0	0	0	0	0	Start of video field
0	1	0	0	0	0	0	1	Start of video line
0	1	0	0	0	0	1	0	End of video field data
0	1	1	x	x	x	x	x	Reserved
1	0	0	0	x	x	x	x	Repeat previous byte x times (0 = 16 reps)
1	0	0	1	x	x	x	x	Repeat previous byte $16x$ times (0 = 256 reps)
1	1	x	x	x	x	x	x	Reserved

(b)

Bit number 7	6	5	4	3	2	1	0	Bit Definition
0	0	0	1	0	0	0	1	XON, starts or restarts transmission
0	0	0	1	0	0	1	1	XOFF, halts transmission
1	0	0	0	0	0	0	0	Use 256 by 244 resolution (full)
1	0	0	0	0	0	0	1	Use 128 by 122 resolution (half)
1	0	0	0	0	0	1	0	Use 64 by 61 resolution (quarter)

Table 2: *Receiver DIP-switch settings. ON and OFF refer to switch positions. (a) SW1, SW2, and SW3 select the serial bit rate (must match transmitter rate). (b) SW4 and SW5 select the time-out interval. (c) SW6 and SW7 select the transmitter resolution. (Note: A manual push button is connected to the SW8 position, so SW8 must be OFF.)*

(a)

SW1	SW2	SW3	Serial transmission rate (bits/second)
OFF	OFF	OFF	300
OFF	OFF	ON	600
OFF	ON	OFF	1200
OFF	ON	ON	2400
ON	OFF	OFF	9600
ON	OFF	ON	19.2k
ON	ON	OFF	28.8k
ON	ON	ON	57.6k

(Note: 4800 bps intentionally omitted.)

(b)

SW4	SW5	Time-out interval
OFF	OFF	Continuous pictures, no delay
OFF	ON	4-second delay between pictures
ON	OFF	8-second delay between pictures
ON	ON	Send picture by manual push-button trigger

(c)

SW6	SW7	Transmitter resolution
OFF	OFF	Full: 256 by 244
OFF	ON	Half: 128 by 122
ON	OFF	Quarter: 64 by 61
ON	ON	Reserved

ing stable video sync and accepting bytes from the serial interface. The code is written in assembly language to maximize the performance of the 8031. Figure 7 is a flowchart of the software's major components.

The 8254 PIT generates the precise sync signals for each line, so the 8031 need only reprogram the PIT when a change is needed. Because changes to the 8254's settings take effect with the next 8254 sync output, the 8031 must make the changes one sync pulse before they're actually needed. All timings are determined by counting sync pulses, which are connected to the 8031's INTO interrupt request pin.

The INTO interrupt handler decrements a counter and checks to see if it's 0. If so, an 8254 change is required; otherwise, the handler simply returns to the mainline code. Each change to the 8254 involves writing a few bytes and reloading the counter to tell how many interrupts will pass before the next change.

Each 8031 instruction takes 1 or 2 μs. At most, only about 50 instructions can be executed per horizontal line. During the vertical retrace interval the sync pulses are only 31.5 μs apart, giving time for only 20 instructions per sync pulse. The interrupt routine must have enough

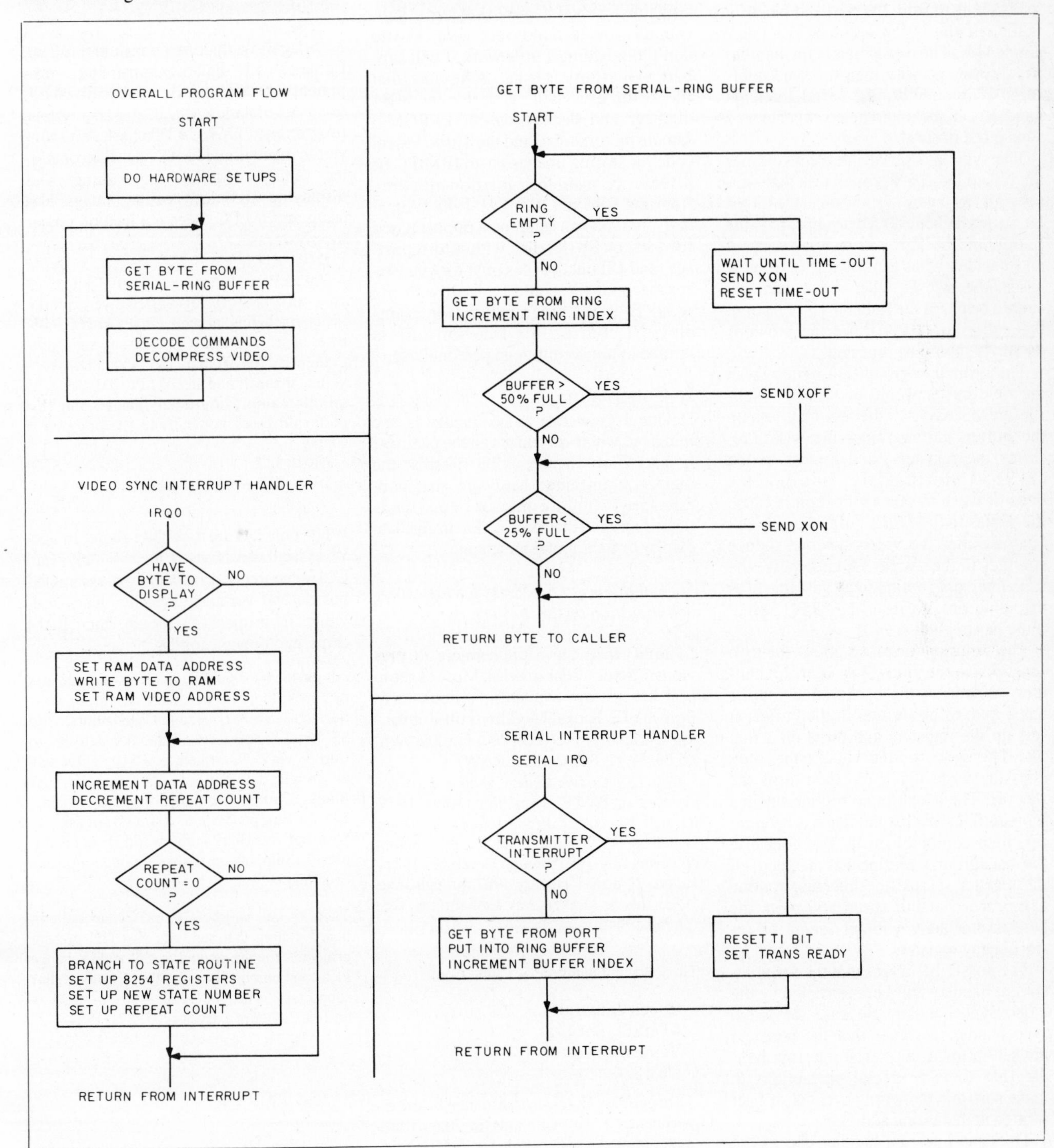

Figure 7: *Flowchart for the ImageWise display/receiver system's software.*

time to get ready for the 8254 loading during the short sync pulses in the vertical retrace interval, so control is passed to the routine two sync pulses before the change is needed. The interrupt routine then uses a polling loop to detect the last sync pulse.

Another interrupt is generated within the 8031 whenever a byte is received on the serial port. This interrupt awakens the serial interrupt handler routine, which reads the byte from the port and places it in the circular buffer in the 8031's internal RAM. The serial interrupt handler has a lower priority than the sync interrupt; consequently, the serial interrupt handler can be interrupted whenever a sync pulse occurs.

The sync and serial interrupt routines are linked by a background task that simply waits for bytes to show up in the circular buffer. Whenever a byte appears in the buffer, the background task takes it out and decides what to do with it. In most cases, the byte is either video data that should be put in the field buffer or a count that tells how many times the previous data byte should be repeated.

The ordinary way of putting a byte in the field buffer would be to have a subroutine that saves all the registers, sets up the buffer address, does the write, restores the registers, and returns to the caller. Unfortunately, this scheme doesn't work in our application because the writes to the frame buffer have to occur just after the video syncs to reduce sparkles in the display. Additionally, the sync interrupt routine must get control at the same time to reset the 8254. Something has to give!

The solution is to combine the two functions in the video sync interrupt handler. Whenever the background routine has a byte to be written in the buffer, it sets up the registers and turns on a flag bit. The sync routine checks the flag, does the write if it's on, then turns the flag off. The background routine sits in a loop until it sees that the flag has been reset, then continues on its way. Because the background routine has handled all the register setups, the interrupt routine can proceed at full speed and write the byte immediately without saving or restoring any registers.

The possibility arises that the serial interrupt handler will be interrupted by the video sync handler. Because the video sync handler assumes that the registers are set up for it, the serial interrupt handler has to take special precautions to make sure that the wrong byte doesn't get written at the wrong address.

The sync interrupt handler checks the switches once every frame (at the end of the second field) to see if anything's changed. If so, it drops what it's doing and runs through the power-on initialization routine again. If a picture is being received when you flip the switches, it will get garbled because the serial port will miss a few characters. The rule of thumb is to change switch settings only when nothing else is happening.

Experimenters

While printed circuit boards and kits are available for the ImageWise system, I encourage you to build your own. If you don't mind doing a little work, I will support your efforts as usual. A hexadecimal file of the executable code for the 8031 digitizer and display system EPROMs, sample picture files, and the Turbo Pascal code for storing images on an IBM PC or SB180 are available for downloading from my BBS at (203) 871-1988. Alternatively, you can send me a preformatted IBM PC or SB180 disk with return postage, and I'll put the files on it for you (the hexadecimal file could be used with my serial EPROM programmer, for example). Of course, this free software is limited to noncommercial personal use.

In Conclusion

Having a gray-scale video display is one thing, but where do you get all the pictures? In part 2 of Chapter 1, I'll describe the digitizer/transmitter hardware that captures images from a camera or TV and sends them to either the receiver for immediate display or a computer for storage.

Special thanks to Ed Nisley for his expert collaboration on this project.

Editor's Note: Steve often refers to previous Circuit Cellar articles. Most of these past articles are available in book form from BYTE Books, McGraw-Hill Publishing Company, P.O. Box 400, Hightstown, NJ 08250, (1–800–2–MCGRAW).

Ciarcia's Circuit Cellar, Volume I covers articles in BYTE from September 1977 through November 1978. *Volume II* covers December 1978 through June 1980. *Volume III* covers July 1980 through December 1981. *Volume IV* covers January 1982 through June 1983. *Volume V* covers July 1983 through December 1984.

The following items are available from

Circuit Cellar, Inc.
4 Park St., Suite 12
Vernon, CT 06066
(203) 875-2751

1. ImageWise digitizer/transmitter board experimenter's kit. Contains digitizer/transmitter printed circuit board, 11.05-MHz crystal, programmed 2764 EPROM with transmitter software, and CA3306 flash A/D converter and manual with complete parts list.
DT01-EXP $99

2. ImageWise display/receiver board experimenter's kit. Contains gray-scale display/receiver printed circuit board, 11.05-MHz crystal, programmed 2764 EPROM with receiver software, Telmos 1852 video D/A converter, manual with complete parts list, and an IBM PC 2.0 disk containing sample digitized images and test patterns.
DR01-EXP $99
DT01-EXP and DR01-EXP
together $179

3. ImageWise digitizer/transmitter full kit. Contains all digitizer/transmitter components, including printed circuit board, 64K bytes of static RAM, IC sockets, crystals, programmed 2764, CA3306 flash A/D converter, manual, and IBM PC 2.0 disk containing utility routines for storing and displaying and downloading image files using an IBM PC. Does not include power supply or case.
DT01-KIT $249

4. ImageWise display/receiver full kit. Contains all gray-scale display/receiver components, including printed circuit board, 64K bytes of static RAM, IC sockets, crystals, programmed 2764, Telmos 1852 video D/A converter, manual, and an IBM PC 2.0 disk containing sample digitized images and test patterns. Does not include case or power supply.
DR01-KIT $249
DT01-KIT and DR01-KIT
together $489

While only kits are described above, ImageWise has been licensed for assembly. Call CCI for source and availability of assembled boards and complete systems, black-and-white TV cameras, 32K-byte static RAM chips, and power supplies.

All payments should be made in U.S. dollars by check, money order, MasterCard, or Visa. Surface delivery (U.S. and Canada only): add $3 for U.S., $6 for Canada. For delivery to Europe via U.S. airmail, add $10. Three-day air freight delivery: add $8 for U.S. (UPS Blue), $25 for Canada (Purolator overnight), $45 for Europe (Federal Express), or $60 for Asia and elsewhere in the world (Federal Express) . Shipping costs are the same for one or two units.

There is an on-line Circuit Cellar bulletin board system that supports past and present projects. You are invited to call and exchange ideas and comments with other Circuit Cellar supporters. The 300/1200/2400-bps BBS is on-line 24 hours a day at (203) 871-1988.

1

BUILD A GRAY-SCALE VIDEO DIGITIZER

PART 2: DIGITIZER/TRANSMITTER

An imaging system with remarkable features for the price

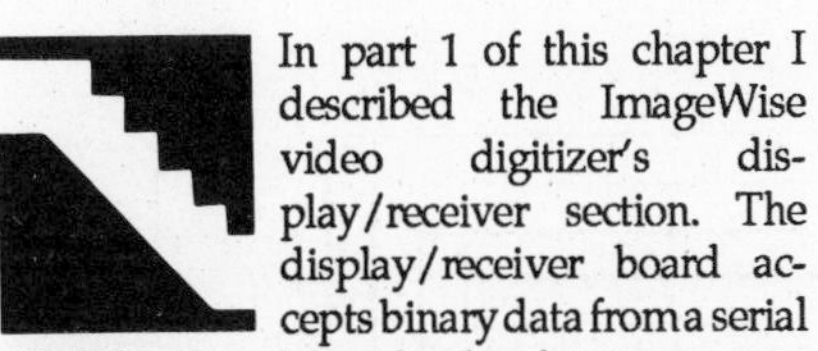

In part 1 of this chapter I described the ImageWise video digitizer's display/receiver section. The display/receiver board accepts binary data from a serial RS-232 port and decodes that data to generate a gray-level display (with 64 levels) on a standard TV monitor. In part 2, I'll complete the project by describing the digitizer/transmitter board (see photo 1) and discussing some possible applications. (Note: Certain portions of this chapter depend heavily on material presented in part 1.)

As I mentioned in part 1, ImageWise is technically a "field" grabber rather than a "frame" grabber. The digitizer/transmitter board makes no distinction between the two fields in a frame: One is as good as the other. The digitizer/transmitter must decide when a new field is starting, wait until the first active line begins, then begin converting the analog video signal into digital bytes that are stored into the field buffer. Because the video can't be "slowed down," all this must happen when the video occurs rather than when the processor is ready.

You might think that you could locate the start of the first active video line by counting horizontal sync pulses after the conclusion of the vertical sync pulse, but it's not that easy. Some cameras do not produce "standard-width" vertical sync pulses, so counting pulses won't work. Instead, I used an internal timer on the 8031 to provide a fixed delay period (DIP-switch-selectable, either 16- or 20-line times). The first horizontal sync pulse after that delay becomes the first active line to be digitized.

Digitizer/Transmitter Hardware

The digitizer/transmitter has two main functions. First, it digitizes the analog video signal; second, it transmits the data to the display/receiver over a serial RS-232 link. Figure 1 shows the digitizer/transmitter circuitry.

The analog circuitry merits a detailed description. Many people who are familiar with the level of integration possible in digital circuitry are appalled at the number of components needed to accomplish even the simplest analog task. The whole mass of circuitry attached to the analog video input performs five functions: termination, clamping, filtering, buffering, and level comparison.

Standard composite video is communicated over coaxial cable with a character-

Photo 1: *The ImageWise digitizer/transmitter in prototype printed circuit form. The digitizer/transmitter board flash-digitizes the video output of a TV camera or other video source (connectors in upper right corner) and converts it to serial data that can be stored and manipulated by a computer or redisplayed on an ImageWise display/receiver board (see Chapter 1, part 1). The flash A/D converter is the 18-pin chip in the upper right corner; the two 28-pin chips on the left are 64K bytes of static RAM.*

istic impedance of 75 ohms. To prevent reflections from the end of the cable, it must be terminated in that impedance. The 75-ohm resistor at the video-input connector of the digitizer/transmitter (J3 and J4) accomplishes this goal. A jumper (JP1) disconnects the terminator if you have a terminated device (perhaps a monitor to watch "live" video) connected to the second parallel connector.

Video signals never seem to exhibit "textbook" profiles. Frequently, inexpensive cameras produce video signals with a DC offset that depends on the scene being viewed. The digitizer incorporates clamping circuitry that forces the tips of the sync pulses to 0 volts. This means that the brightest white will be in the area of 1 to 1.4 V (different cameras give different results).

Both color and black-and-white sig-

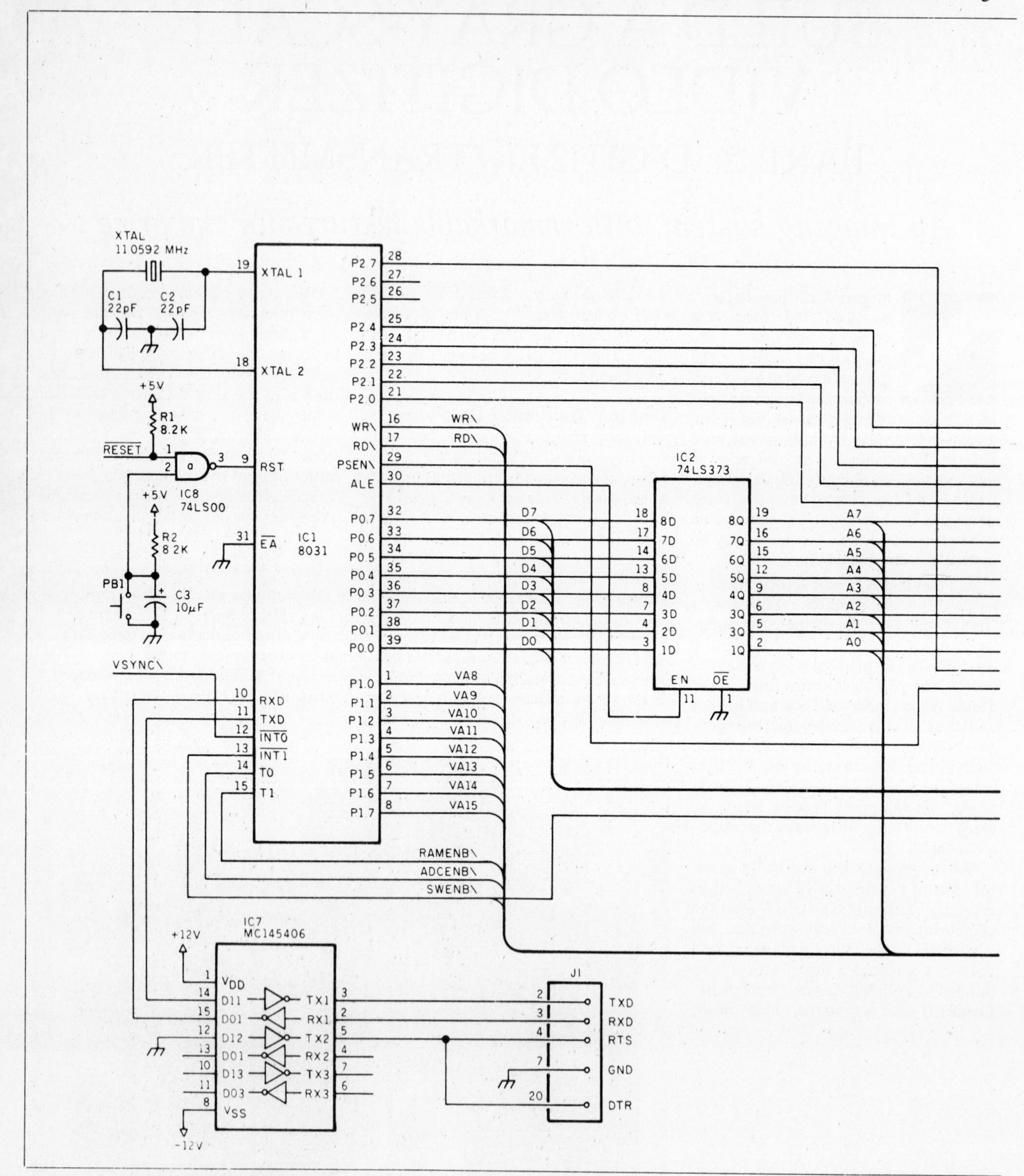

Figure 1: *Schematic for the ImageWise digitizer/transmitter board.*

nals will work with the ImageWise digitizer; however, a color signal contains more information than is necessary. This extra color information is detrimental because it can impart a herringbone pattern on the black-and-white digitized image. I solved this problem by using a filter that removes the frequencies used to encode the color. The remaining signal contains the important intensity voltage that the digitizer measures. You should disconnect this color filter (jumper-selectable JP2) when using a black-and-white camera because the filter doesn't have a sharp cutoff, and it can "soften" the picture by removing some of the finer details. Try it both ways and use the jumper setting that works best for your application.

The 2N4401 transistor serves as a power amplifier to ensure that the re-

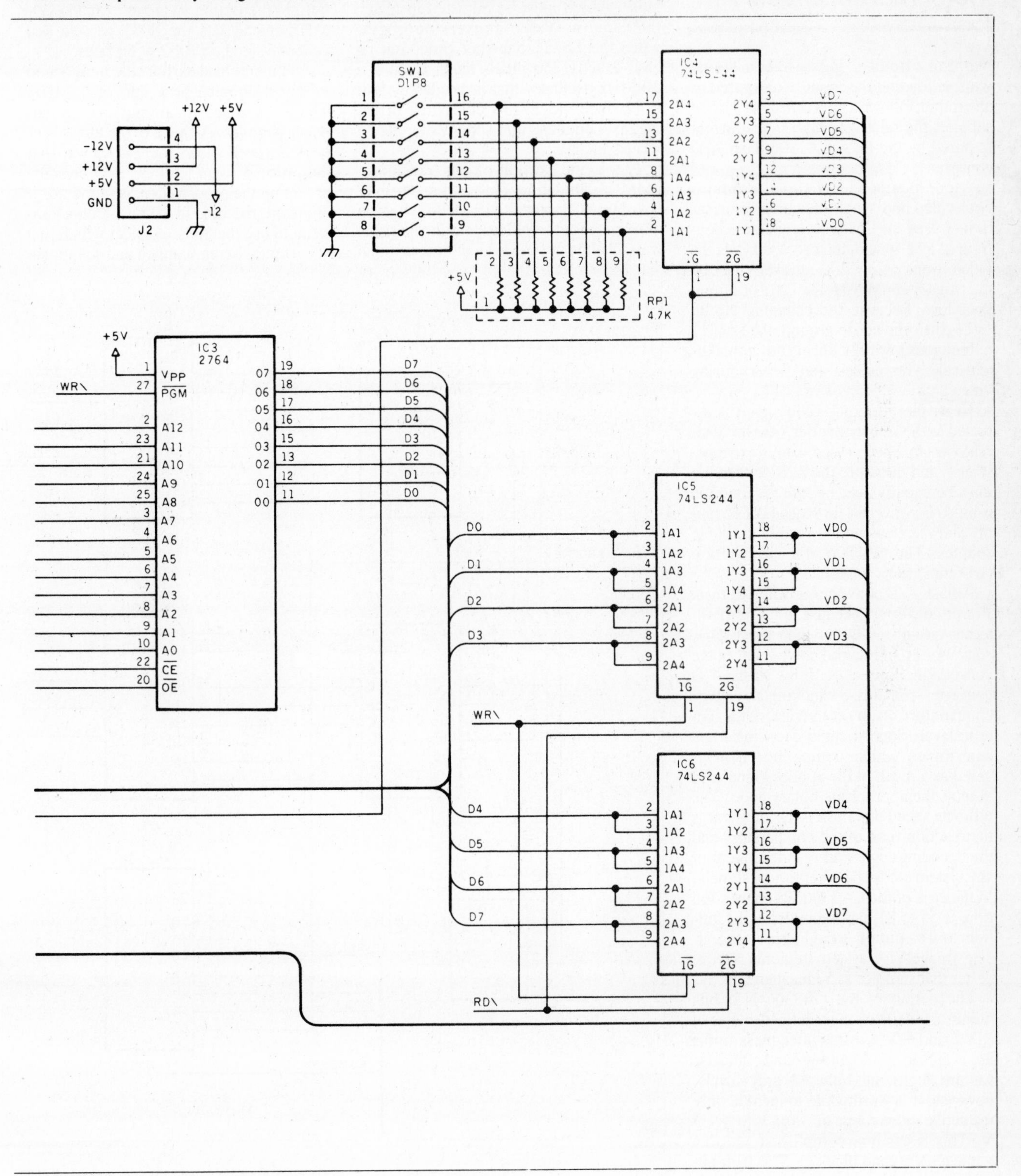

A flash A/D converter differs from slower converters in having more internal circuitry.

maining circuitry gets a clean signal without loading the input. Configured as an emitter follower, this transistor circuit supplies the relatively high input current required by the RCA CA3306 flash A/D converter.

A composite video signal contains both video and synchronization information. I used an LM311 comparator configured as a sync detector. Whenever the video drops below 200 millivolts (set by the resistors on pin 3), the LM311 output goes low. Because the clamping circuit forces the sync tips to ground, the LM311 output goes low only during horizontal or vertical sync pulses and never during video data.

Next, the buffered video signal is directed to the A/D converter. As we've already established, video data is extremely fast, and therefore the A/D conversion must be equally fast. I chose the CA3306 6-bit A/D converter for reasons of economy and the existence of readily available sources. The CA3306 is a special flash A/D converter.

A flash A/D converter is different from slower converters because it has more internal circuitry. Rather than use a D/A converter as an integral component in the conversion process, a 6-bit flash A/D converter contains 64 individual voltage comparators, each set to trigger at a specific level. Sophisticated decoding logic determines which comparator is triggered as a result of the applied signal and outputs the appropriate binary code. The ultimate speed of a flash A/D converter is the reaction time of the comparators and the decoding logic and is independent of any system clocks or other timing signals. In the case of the CA3306, its conversion time is 55 to 83 nanoseconds, or 12 million to 18 million samples per second. The ImageWise system requires a converter that samples at 5 megahertz.

The output of the CA3306 is 6 bits within a relative range defined between $+V^{ref}$ and $-V^{ref}$. Most often these limits are +5 V to +8 V and ground. Because we are digitizing only 64 gray levels, however, it is worthwhile to include only the active video range of +0.2 V to +1.5 V. This is easily accomplished with the white ($+V^{ref}$) and black ($-V^{ref}$) trim potentiometers that adjust the CA3306's conversion thresholds so that a bright white will be digitized as hexadecimal 3F and dark black will be hexadecimal 00. I'll describe the adjustment procedure later.

Finally, the delay trim potentiometer (R21) determines the blanking delay from the start of each horizontal sync pulse, which must be adjusted to match the camera. Video conversion begins when the LS221 one-shot times out. It ends exactly 256 pixels later. IC20 is a 20-MHz oscillator that is divided by four to produce the 200-ns clock that drives the pixel counters and supplies the RAM write signal.

The digital portion of the transmitter board is similar to that of the receiver board (see Chapter 1, part 1). Both boards must process the video data in the same way, albeit in opposite directions, so much of the circuitry can be the same.

Without the 8254 counter/timer used on the receiver board, fewer control lines were needed, so I was able to eliminate the 74LS138 decoder. Two 74LS244s isolate the processor data lines from the video data lines, and a third 74LS244 buffers the option switches. An 11.059-MHz crystal lets the 8031 receive and generate standard RS-232 bit rates.

As I mentioned earlier, the input video must be sampled at a rate of 5 MHz, which translates to around 200 ns between samples. A check of the 8031's execution speed shows there is no way that the processor can read the A/D converter, set up the proper address in memory, and store the byte in 200 ns. I chose instead to use the 8031 to select which line is currently being scanned and set up that

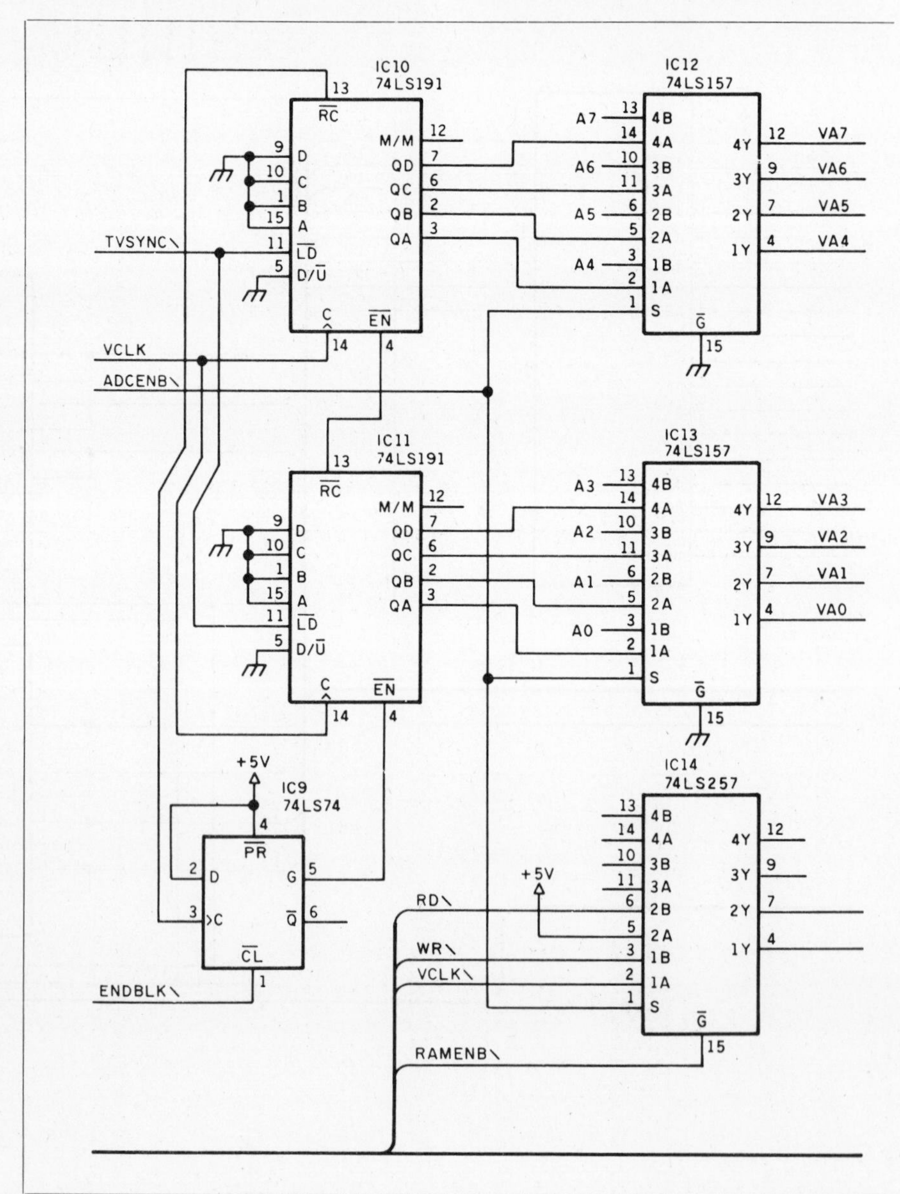

Figure 1: *Continued from page 15.*

address in the upper 8 bits of the memory address. Two 74LS191 4-bit counters provide the lower 8 bits automatically. At the start of each scan line, the counters are cleared, and the 8031 sets up the line number. After each pixel has been digitized and stored, the counters increment the address. Since there are about 66 microseconds between the start of each scan line, the processor has plenty of breathing room.

The 74LS257 ensures that we are always writing to the RAM during digitization. Normal processor reads and writes can be performed at any other time for image transmission or other processing.

The Software Connection

The digitizer/transmitter software is a simple loop that captures a video field in the RAM buffer, then compresses and transmits the data via serial RS-232. A DIP-switch setting determines whether the digitizer/transmitter will transmit continuously or wait for an XON from the display/receiver before starting each field (see table 1).

The software begins video data capture when it detects the first vertical sync pulse in a field, as described above. The program then waits for the vertical-blanking delay (determined by using the 8031's internal timer) before enabling the RAMs and the A/D converter. Next, the program counts sync pulses and increments the RAM line address after each pulse. When the buffer is full, it disables the RAMs and the A/D converter to prevent further buffer writes.

The flowchart shown in figure 2 describes the process used to compress the video data in each line. Notice that a unique sync byte designates the start of the field and the line within the field, as well as the end of the video data.

The software captures a video field, then it compresses and transmits the data.

Getting Started

Assuming that you have an ImageWise display/receiver in good working order from the project in part 1 of this chapter, use the procedure that follows to make your first digitized video connection.

First, connect the camera to the monitor (a coaxial cable without any fancy hardware between) and get a picture that's well-lighted and focused. The adage about "garbage in, garbage out" certainly applies to this operation! Make sure that you've got the monitor terminated in 75 ohms.

Connect the display/receiver to the monitor (remember to disconnect the camera first). Set the DIP switches to 28.8 kilobits per second and no time-out (continuous pictures; see chapter 1, part 1 for a DIP-switch-setting guide for the display/receiver).

Turn the video level (R8) trim potentiometer to midrange and then plug in the power. Do not connect the digitizer/transmitter. The display/receiver will display a diagonal test pattern that includes a gray scale ranging from full white to full black. Adjust the video level trim potentiometer so that the monitor shows the complete range of shades. You may have to tinker with the monitor's hold, brightness, and contrast controls.

Connect the camera to the digitizer/transmitter and install the 75-ohm terminator jumper (JP1). If you're using a color camera, install the color filter jumper (JP2); otherwise, remove it. Set the DIP switches to

28.8k bps
16-line vertical delay
Compression enabled
Ignore +/−1 count changes
Paced mode disabled

Turn the delay (R21), black level (R18), and white level (R14) trim potentiometers to midrange; connect a serial cable between the digitizer/transmitter and display/receiver; and plug in the power. (Note: Set up the serial cable so that pins

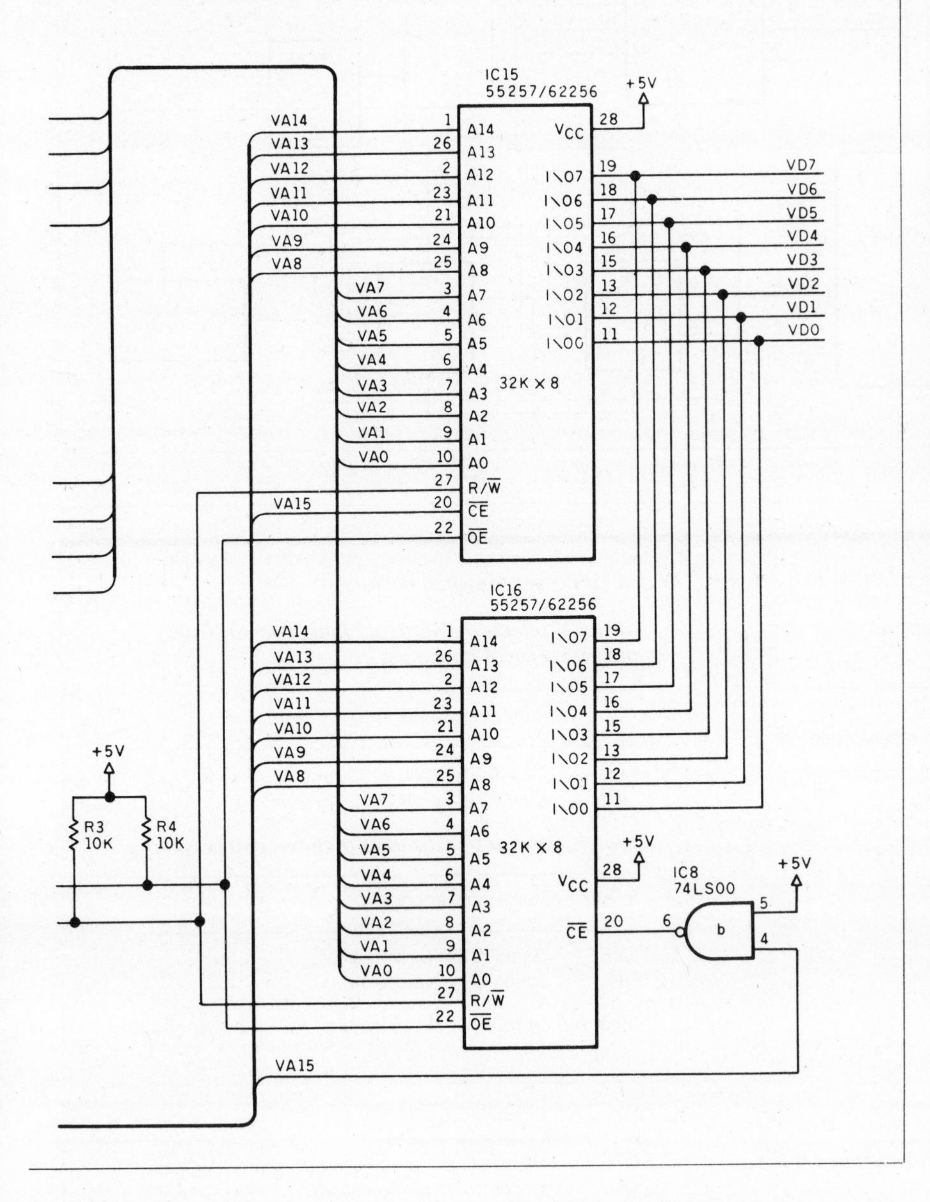

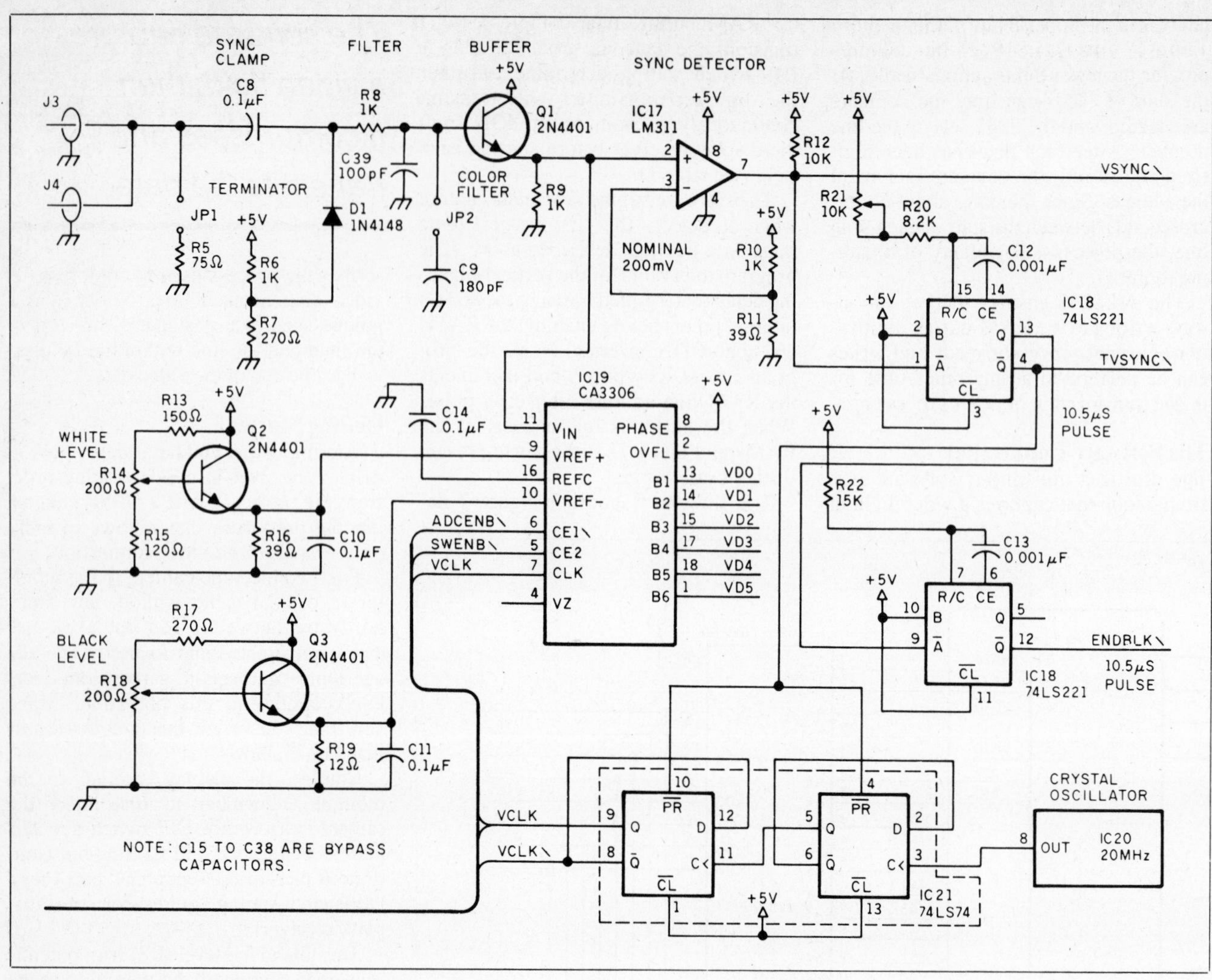

Figure 1: *Continued from page 17.*

Table 1: *ImageWise digitizer/transmitter DIP-switch settings. ON and OFF refer to switch positions.*

Switches 1, 2, and 3 select the serial bit rate (must match the receiver's rate).

1	2	3	Serial bit rate (must match receiver rate)
OFF	OFF	OFF	300 bps
OFF	OFF	ON	600 bps
OFF	ON	OFF	1200 bps
OFF	ON	ON	2400 bps
ON	OFF	OFF	9600 bps
ON	OFF	ON	19.2k bps
ON	ON	OFF	28.8k bps
ON	ON	ON	57.6k bps

(Note: 4800 bps was intentionally omitted.)

(Note: Switches 7 and 8 are not used and must be OFF.)

Switch 4 selects the vertical blanking delay from the first vertical sync pulse.

4	Vertical blanking delay
OFF	16 lines (normal)
ON	20 lines (extended)

Switch 5 enables or disables picture compression.

5	Run-length encoding
OFF	disabled (no compression)
ON	enabled (compression)

Switch 6 enables or disables the +/−1 count change compression.

6	Compress +/−1 count changes
OFF	encode (less compression)
ON	ignore (more compression)

2 and 3 are exchanged and pins 5 and 7 are straight through.) You should see the digitized picture appearing on the monitor, painting from top to bottom. After the entire scene is done, another image will be sent. You'll be able to see a horizontal line marking the descending edge of the new picture overlaying the old one.

You should now adjust the black and white trim potentiometers to get the maximum amount of detail in the picture. If the black level is too high (clockwise), dark areas will have little detail, and the whole picture will be dark. If the white level is too low (counterclockwise), the bright areas will suffer, and the picture will be light. You should have a high-contrast scene in front of the camera to make sure you have the right levels. (On the other hand, don't make the black level too low or the white level too high, because that will reduce the number of digital levels in your scene. For example, if your camera's highest voltage is 1.4 V, it does no good to have the white setting at 2.0 V; that 0.6-V difference contains some digital codes that will never be used.)

Adjust the delay trim potentiometer so that the scene is horizontally centered on the monitor. If your monitor has a great deal of overscan, it won't matter too much what the delay setting is. You can also adjust the monitor's horizontal hold knob slightly.

If a line of trash (there's no better way to describe it) appears at the top of the monitor's display, try setting the digitizer/transmitter DIP switch for 20 lines of vertical delay instead of 16. Some cameras produce a few lines of trash at the beginning of the field, and the digitizer/transmitter faithfully digitizes it. If you don't see anything, or if the trash moves to the bottom, leave the DIP switch set for 16 lines.

If you have a monitor connected to the camera (a viewfinder doesn't count), you must make sure that either the monitor is terminated or the termination jumper is installed, but not both or neither. You should terminate the device at the end of the camera cable, not the one in the middle.

You may want to try the filter jumper to see what effect it has on the scene. If you have a color camera, the filter is probably going to be essential; if you have a monochrome camera, you may simply like a softer picture. Do not confuse the effect of the filter jumper with the output of an unfocused camera; make sure the scene is crisp to start with.

Try flipping the digitizer/transmitter switch that ignores +/−1 count changes

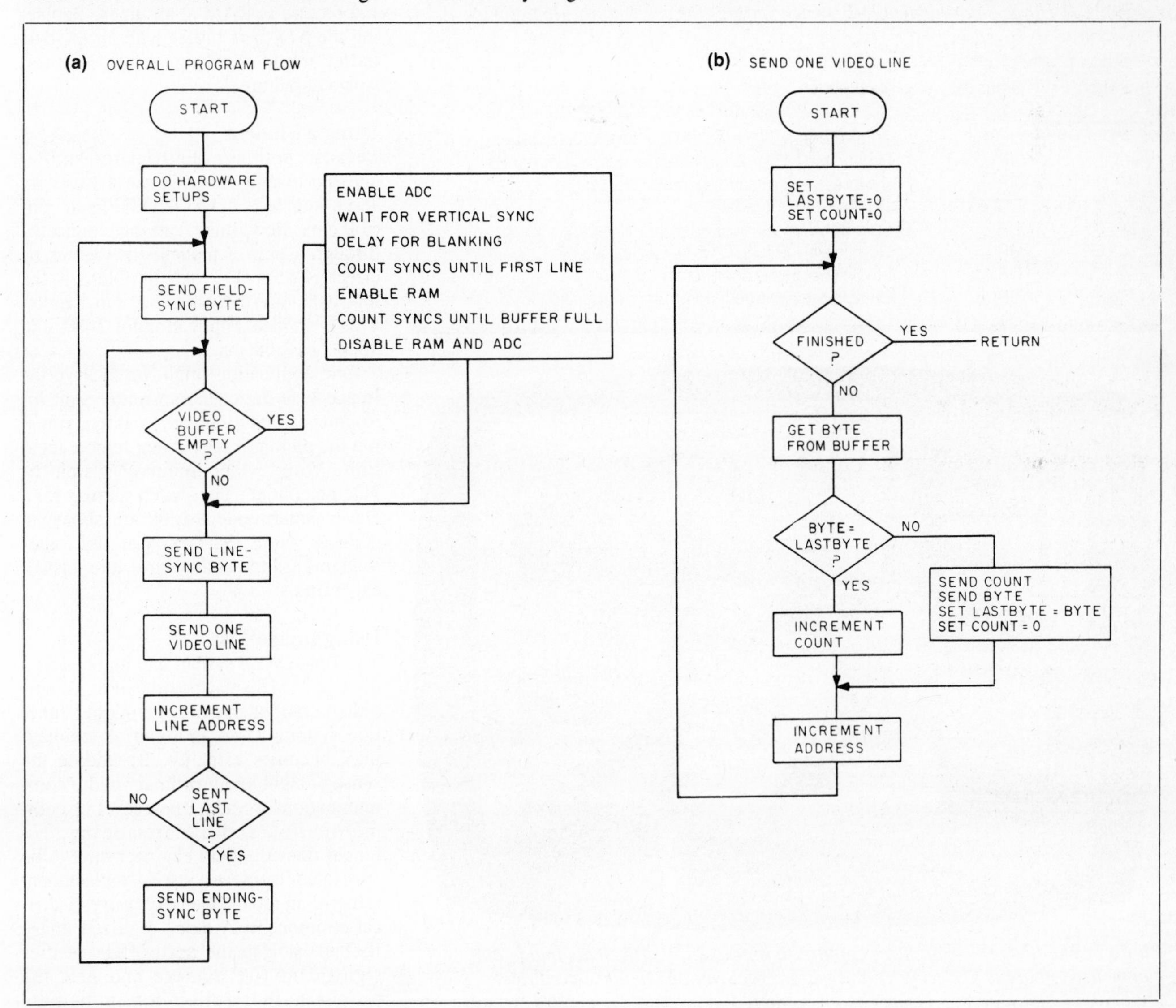

Figure 2: *Flowchart of the digitizer/transmitter software.* **(a)** *This outlines the overall flow of the software.* **(b)** *A flowchart of the routine to transmit a single video line.*

to see what happens to the flat areas in the picture. Disable compression and measure the increase in transmission time.

You can also try changing the bit rates to see which rates work for you. A direct connection can run at 28.8k bps, but for longer wires running through the house, you may need to use a slower rate. At 300 bps you can see the compression working. Remember that the display/receiver will get confused when you change the rate, so you may need to reset the system after changing any DIP-switch settings.

Cheap Buffer Option

I've described how the digitizer/transmitter works with a 64K-byte RAM field buffer made of two 32K-byte static RAM chips. It turns out that you can use an 8K-byte buffer in the digitizer/transmitter. Because the 32K-byte static RAM chips are still rather expensive, I thought it would be worthwhile to reduce the cost for applications that don't need the advantages of the full 64K-byte buffer (Note: The ImageWise kit contains 32K-byte static RAM chips.)

An 8K-byte RAM can hold 32 lines of video (8K bytes divided by 256 bytes per line). The digitizer/transmitter digitizes 32 lines, transmits them, captures another 32 lines, and so on. It keeps track of the last digitized line and starts with the next line for the next group. When it's done, it begins again with the first video line.

Using an 8K-byte RAM is simply a matter of plugging it into the first RAM socket. The 8031 program "feels around" after a reset to determine the RAM size during initialization and uses the proper addresses automatically; no DIP-switch settings are required.

The only difference between using 8K- and 64K-byte RAM buffers occurs when the scene contains moving objects. The 64K-byte buffer can hold a complete frame that is captured in 1/60 second, and it is the only configuration that can legitimately be called a frame grabber. The 8K-byte buffer holds 32 line groups that are digitized several seconds apart, so it's possible to get confused images. If you've ever seen those "panoramic" shots of a line of people with the same person at both ends of the line, you'll recognize the problem right away. But if your application doesn't involve rapid motion, you can save some money on buffer RAMs.

This trick doesn't work in the display/receiver because it must have the entire picture in RAM at all times. Replacing the 64K-byte buffer with an 8K-byte buffer would simply give you 32 lines' worth of picture.

Table 2: *Hayes Smartmodem DIP-switch settings for use with the ImageWise system.*

Switch	Setting	Description
1	DOWN	Smartmodem DTR input on pin 20 is forced active.
2	UP	Don't care, UP for English responses to commands.
3	UP	Suppress responses to commands.
4	DOWN	Do not echo characters.
5	UP	Smartmodem will auto-answer on first ring.
6	UP	Don't care, UP for CD when carrier detected.
7	UP	Single-line phone connection.
8	UP	Disable Smartmodem command recognition.

Photo 2: *The ImageWise digitizer captures a high-quality gray-scale image that can be used in areas like security and pattern recognition. In the surveillance application shown here, an empty room is entered, used, and then exited. To save space in this presentation, all four scenes are displayed on a GT180 (in 16-level gray scale).*

Using a Modem

Because both the digitizer/transmitter and the display/receiver use a two-wire serial interface and XON/XOFFs to control data flow, they can be connected through a pair of modems as well as by direct wiring. The only problem is the low data rate that modems can handle. This is an ideal application for 2400- and 9600-bps modems.

One application might be to have the ImageWise digitizer/transmitter wait for a phone call, auto-answer, and then transmit the picture it sees at that remote location. Voilà, video security system or videotelephone. The switch settings for a Hayes Smartmodem 1200 are shown in table 2. Note that you can use other modems as long as they have auto-answer capabilities.

Using ImageWise

The ImageWise system can be used as a digitizer and display board pair or as individual components of some higher-function system. Used as a pair, "teleimaging" becomes a reality. By adding the sense of sight to our ordinary audio communication, we add a new level of communication and understanding. No longer does the field engineer have to be frustrated trying to justify replacement rather than repair of an expensive electrical component. A quick digitized image flashed back to the head office verifies significant fire damage and gets the proper authorization for immediate action.

The key factor in this new level of communication is the old saying that "seeing is believing." Consider another example: You are a consultant at a customer's site, and some question arises as to the actual wording and the date of a revision note on an important schematic. Rather than waste a day with express delivery, you can call your office and have them transmit a digitized image of the portion of the schematic in question complete with the authorization signature and date as they appear.

ImageWise has an infinite number of stand-alone uses. It can be used to instantly communicate fingerprints and ID photos, monitor traffic at remote intersections, monitor remote security risk zones (see photo 2) via auto-answer modems, aid in conducting company-wide lectures (standard audio teleconference with pictures of the blackboard periodically sent to all locations; see photo 3), or send x-rays and CAT scan pictures to medical personnel for corroborating diagnosis.

Some of these uses might seem ambitious for ImageWise, but similar more expensive units are already being applied in these areas. My immediate application may seem mediocre by comparison. Recently, I've been spending time out of the Circuit Cellar at an office across town. Since I already had a TV camera in my driveway (no windows in the cellar, remember), I simply attached it to the digitizer/transmitter and a 2400-bps auto-answer modem. Now I can call the house and get the latest snapshot or simply leave it on all the time as a real-time display of all the activity around the house (see photo 4). (I have four telephone lines; I can call the Home Control System on its own line if I want to have some real fun with someone like a delivery man. These guys all think my house is haunted.)

You'll note that I have described ImageWise only in terms of a 256- by 256- by 6-bit digitizer. Because I intended to use it with a modem, I felt the need to increase the picture-transmission speed. One way is to reduce the resolution to 128 by 121 or 64 by 61 bits. Even though such resolutions produce grainy images, they are still quite recognizable, especially if they are of familiar faces or geography (the recently advertised Mitsubishi video telephone has a 94- by 94- by 4-bit resolution by comparison). The 64- by 61-bit image is transmitted eight times faster than a 256- by 256-bit image and is suitable for monitoring gross changes in a driveway scene when a car or a person approaches.

When something appears, I can immediately change the DIP switches on the display board for a higher resolution and trigger another picture while the form is still in view. (The frame is grabbed instantly and is independent of the transmission time.) The picture-repeat rate and resolution, remember, are commanded from the receiver and not fixed by the transmitter. The interaction is completely dynamic. My next activity is to connect the ImageWise digitizer to a computer and let it decide what's happening for itself and make all the decisions.

Fortunately, this is as easily said as done. Probably the most significant feature of ImageWise is that it is computer-nonspecific. It is a serial RS-232 I/O device that does not depend on any computer-specific bus. The ImageWise digitizer/transmitter's serial port can be connected to any personal computer. The computer can receive image data and store it on disk or send it to a similarly connected display/receiver board. So far we've written the software for my SB180 and the IBM PC. Others will follow.

Experimenters

While printed circuit boards and kits are available for the ImageWise system, I encourage you to build your own. If you don't mind doing a little work, I will support your efforts as usual. A hexadecimal file of the executable code for the 8031 digitizer and display system EPROMs, sample picture files, and the Turbo Pascal

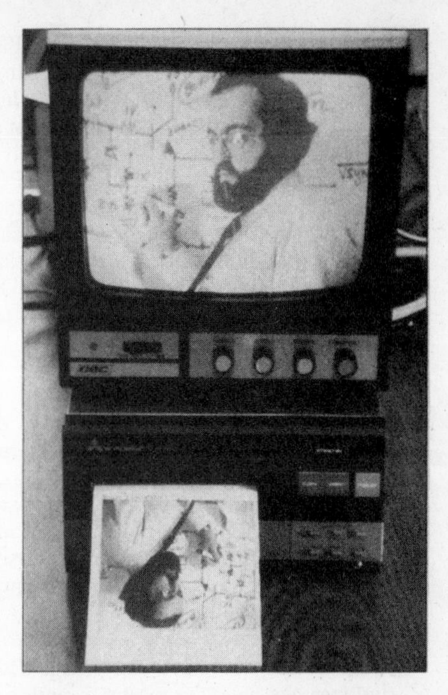

Photo 3: *This is how a picture of me standing in front of a blackboard in Connecticut would be received by an ImageWise display/receiver board in California. A video printer like the Mitsubishi unit shown here can save the current scene while another is being transmitted.*

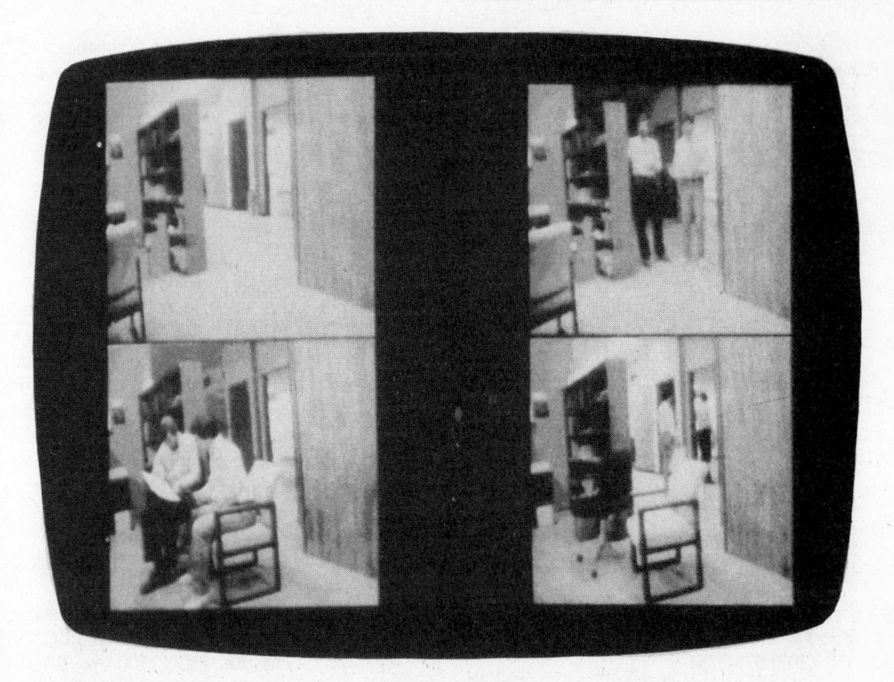

Photo 4: *The equipment counter in the Circuit Cellar where I took most of the photos with the setup you see. The interesting point to realize here is that this picture is completely digitized. A video camera (out of view) is pointed at the two stacked monochrome monitors and the film camera. The monitor on the bottom displays the picture as it is produced by the video camera and input to the digitizer/transmitter board. The display on top shows the output of the display/receiver board after it receives the data from the digitizer. It is the object of view by the film camera. This picture is the screen of the top monitor.*

code for storing images on an IBM PC are available for downloading from my bulletin board at (203) 871-1988 (similar code written for the SB180 in machine language is also available). Alternatively, you can send me a preformatted IBM PC or SB180 disk with return postage, and I'll put all the files on it for you (the hexadecimal file could be used with my serial EPROM programmer, for example). Of course, this free software is limited to noncommercial personal use.

In Conclusion

Once you've got a picture in digital format, you can write programs that perform magic tricks with it (hardware people like to think that way). By manipulating the binary data making up the picture, you can transform it into another picture that may be more meaningful. You can even combine two pictures to find differences—this is called image processing.

Now that we have the ImageWise digitizer, we have the means to perform some real experiments. I know many tutorial articles on image processing have been published, but the true Circuit Cellar creed is to build it yourself. Using ImageWise, in Chapter 2, I'll demonstrate how the basics of picture comparison, enhancement, and other image-processing fundamentals can be done for real.

Special thanks to Ed Nisley for his expert collaboration on this project.

Editor's Note: Steve often refers to previous Circuit Cellar articles. Most of these past articles are available in book form from BYTE Books, McGraw-Hill Publishing Company, P.O. Box 400, Hightstown, NJ 08250, (1-800-2-MCGRAW).

Ciarcia's Circuit Cellar, Volume I covers articles in BYTE from September 1977 through November 1978. *Volume II* covers December 1978 through June 1980. *Volume III* covers July 1980 through December 1981. *Volume IV* covers January 1982 through June 1983. *Volume V* covers July 1983 through December 1984.

The following items are available from

Circuit Cellar, Inc.
4 Park St., Suite 12
Vernon, CT 06066
(203) 875-2751

1. ImageWise digitizer/transmitter board experimenter's kit. Contains digitizer/transmitter printed circuit board, 11.05-MHz crystal, programmed 2764 EPROM with transmitter software, and CA3306 flash A/D converter and manual with complete parts list.
DT01-EXP $99

2. ImageWise display/receiver board experimenter's kit. Contains gray-scale display/receiver printed circuit board, 11.05-MHz crystal, programmed 2764 EPROM with receiver software, Telmos 1852 video D/A converter, manual with complete parts list, and an IBM PC 2.0 disk containing sample digitized images and test patterns.
DR01-EXP $99
DT01-EXP and DR01-EXP together $179

3. ImageWise digitizer/transmitter full kit. Contains all digitizer/transmitter components, including printed circuit board, 64K bytes of static RAM, IC sockets, crystals, programmed 2764, CA3306 flash A/D converter, manual, and IBM PC 2.0 disk containing utility routines for storing and displaying and downloading image files using an IBM PC. Does not include power supply or case.
DT01-KIT $249

4. ImageWise display/receiver full kit. Contains all gray-scale display/receiver components, including printed circuit board, 64K bytes of static RAM, IC sockets, crystals, programmed 2764, Telmos 1852 video D/A converter, manual, and an IBM PC 2.0 disk containing sample digitized images and test patterns. Does not include case or power supply.
DR01-KIT $249
DT01-KIT and DR01-KIT together $489

ImageWise is also available assembled. Call CCI for source and availability of assembled boards and complete systems, black-and-white TV cameras, 32K-byte static RAM chips, and power supplies.

All payments should be made in U.S. dollars by check, money order, MasterCard, or Visa. Surface delivery (U.S. and Canada only): add $3 for U.S., $6 for Canada. For delivery to Europe via U.S. airmail, add $10. Three-day air freight delivery: add $8 for U.S. (UPS Blue), $25 for Canada (Purolator overnight), $45 for Europe (Federal Express), or $60 for Asia and elsewhere in the world (Federal Express). Shipping costs are the same for one or two units.

There is an on-line Circuit Cellar bulletin board system that supports past and present projects. You are invited to call and exchange ideas and comments with other Circuit Cellar supporters. The 300/1200/2400-bps BBS is on-line 24 hours a day at (203) 871-1988.

2

USING THE IMAGE WISE VIDEO DIGITIZER: IMAGE PROCESSING

This digitization and display process is easy to duplicate

While I was writing the second part of the ImageWise project, a BYTE editor sent me copies of the image processing theme articles used in Chapter 4. After I got over my first reaction to the common thread of the articles (it's almost all software—yech!), I realized that, while I was covering the hardware specifics of ImageWise more than adequately, to do real justice to the subject I should include more on using and processing the data created from the digitizer.

Getting from here to there constituted a problem, however. While many people can read and instantly visualize the image transformations described in these image processing tutorial articles, some people prefer an alternative approach to such presentations. Although Chapter 4 is devoted to image processing, I'd like to think there is a difference when I discuss a subject.

I describe ideas, but also I try to include a little hands-on experience. Unlike a tutorial that contains little mention of the hardware you might use to duplicate such feats, all of the picture data used in this chapter was digitized on the ImageWise digitizer/transmitter and displayed on its companion display/receiver board (except for the zoom shots, which are displayed on a GT180). You should be able to easily duplicate the process.

I have expanded the original two-part ImageWise hardware project to include two more articles with a little software. Admittedly, I am out of my element, and I ask you to bear with me if I drop a few bits now and then (think of it as poetic license). I couldn't pass up an opportunity to string together such interesting ideas as image processing and colorization—which I can actually demonstrate.

In this chapter, I will focus on image processing. As in the related tutorials, I'll take a digitized image and detect edges, enhance it, filter it, enlarge it, subtract it, and create other more useful images.

First, a quick hardware review of ImageWise will show you what the data is that I am processing (see Chapter 1, parts 1 and 2, for more details).

Picture Format

The ImageWise digitizer/transmitter digitizes a single field of the camera's video signal on-the-fly, converting it into 244 rows of 256 pixels each. The rows are numbered from 0 to 243, and the pixels are numbered from 0 to 255 in each row.

A pixel's brightness is represented by one of 64 gray levels, with a black pixel equal to 0 and a bright white pixel equal to 63. Each pixel requires 1 byte of storage, so there are 62,464 pixel bytes per image. Software adds some additional control-information codes to simplify the display/receiver's job, giving a total of 62,710 bytes in an image.

The digitizer/transmitter compresses the video data using run-length encoding to reduce the time needed to send it over the RS-232 serial link. When the digitizer/transmitter finds a gray-level value repeated more than twice in adjacent pixels (a "run") in the line, it replaces the repetitions with a count. Typical scenes are reduced by 50 percent to 75 percent, with a corresponding speedup in transmission.

The display/receiver accepts RS-232 data, decompresses it into a RAM display buffer, and generates the synchronization signals required to show the images on a standard composite-video TV monitor. The result is a TV picture that looks remarkably like the original scene.

The Personal Computer Connection

Because both the digitizer/transmitter and display/receiver communicate over a standard RS-232 line, you can connect either one to a serial port on a personal computer (the unit can connect to any computer with a serial port, but all my examples use an IBM PC). When the computer is connected to the digitizer/transmitter, it acts as a display/receiver, storing the image data on disk. When it's connected to the display/receiver, it acts as a digitizer/transmitter and sends the stored images out for display.

The computer can accentuate or suppress details in an image by performing simple arithmetic on the numeric values for the pixels. For example, a program can compare two scenes by subtraction, and a count of nonzero values in the result can tell you whether something has moved into (or out of) the picture.

This chapter demonstrates a tool kit of programs that you can use to develop a complete image processing application. The programs are written in Turbo Pascal for an IBM PC, but you can easily convert them for use on other computers. I used an 8-megahertz IBM PC AT with 640K bytes of RAM, a 10-MHz 80287 math coprocessor, and a 1.2-megabyte RAM disk to develop these programs. They will work on any computer that runs Turbo Pascal and has sufficient RAM (about 512K bytes) but might take somewhat longer to run. Because the images are displayed on a TV monitor connected to the display/receiver, you don't need a graphics display on the computer.

Serial Setup

You set the data rate on the serial link using DIP switches on the digitizer/transmitter and display/receiver boards. Although the maximum data rate is 57.6k bits per second, the PC simply can't keep up at that rate with the present software. While I could have used some computer assembly code to tweak the critical loops, I felt it was better to use a more easily

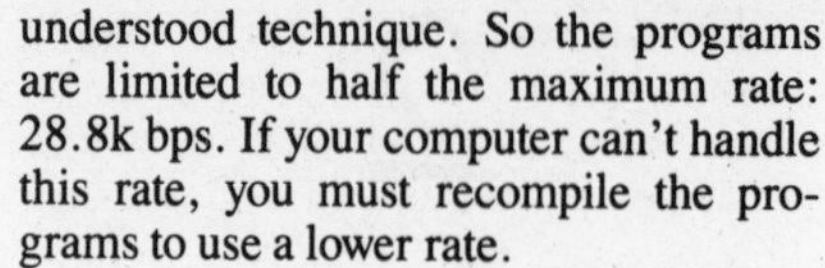

understood technique. So the programs are limited to half the maximum rate: 28.8k bps. If your computer can't handle this rate, you must recompile the programs to use a lower rate.

Only two programs actually communicate with the ImageWise boards. The Grab program prompts the digitizer/transmitter to send an image and stores it on disk. The Show program reads the disk file and sends it to the display/receiver. Both use the COM1 serial port, so you'll have to swap cables when you use each program. (I used a serial-port switch box, but you can easily recompile the programs to grab images from COM1 and show them on COM2.)

Photo 1: *An image captured by the ImageWise digitizing system.*

Photo 2: *I created this image by multiplying the pixels in photo 1 by 2 using the Multiply program.*

Photo 3a: *The digitized image of a circuit board.*

Photo 3b: *I have added something new.*

Photo 3c: *You can use the Subtract program to discover what has changed.*

Photo 4: *You can run the Thresh program on the image in photo* **3c** *to remove background clutter.*

One of the first tasks I have is undoing one of ImageWise's features. Although the compressed data format reduces the transmission time, it's not well-adapted to image processing. The programs must examine every image pixel, something that's not easily done with run-length-encoded data. So Grab decompresses the images before it stores them on disk, creating a 62,720-byte file for each picture. There are 62,710 image and control bytes, with 10 padding bytes added to fill out the file's last 128-byte block.

The Show program and the display/receiver can handle either run-length-encoded or expanded files, so there's no problem sending them to the display/receiver, except for the increased transmission time. The Compress and Expand programs convert between the two formats.

Taking Pictures

In addition to the ImageWise digitizer/transmitter and display/receiver boards, you'll need a TV camera and monitor, a tripod for the camera, and some RS-232 and video cables. A color TV camera will work fine, even though the digitizer/transmitter is designed for monochrome. If you see herringbone patterns on the display/receiver, install the Filter jumper on the digitizer/transmitter to remove the color information from the camera signal.

A zoom lens is a great help because you can adjust the focal length to fill the screen with the scene. If you are taking pictures of small objects, you might also need a macro lens or attachment. Most consumer TV cameras come with a macro-focusing zoom lens, so you're probably in good shape if you have one.

I captured the scenes in this article using a monochrome camera equipped with a 15- to 75-millimeter zoom lens. I used a 75-watt desk lamp for illumination. The camera lens was usually opened wide to f/2.1.

The first rule of photography is to get enough light on the subject. While you can use light meters and judgment, checking the actual results is better. The Histo program analyzes an image and reports on the number of pixels having each of the 64 possible brightness levels. Figure 1 shows the output of Histo for the image in photo 1.

The large peak is created by the desktop and background areas that are all more or less the same shade. There are relatively few black areas (near 0) and relatively few white areas (near 63). The peaks on every other pixel count indicate a little bit of noise in the A/D circuits.

Notice the small number of pixels brighter than about 30. Although it's better to increase the amount of light on the scene, you can achieve a similar result by multiplying each pixel by a constant. Photo 2 is the same as photo 1, with each pixel multiplied by 2 using the Multiply program. This is nearly equivalent to increasing the exposure by one f-stop, thus doubling the brightness.

Figure 2 is Histo's output for the image in photo 2; notice that the pixel values are all even (multiples of two) except for the pixels that "stuck" at 63. The brighiest areas of photo 2 look flat because they are all the same value. Increasing the illumination would have filled in the odd-numbered pixels.

When you're setting up a new picture-taking session, always use Histo to make sure you're getting enough light on the scene. It's all too easy to twist the brightness knob on the monitor, which doesn't do anything for the digitizer/transmitter.

What's New and Different?

One of the more interesting things you can do with two images is to find the differences between them. Photo 3a shows a small circuit board, and photo 3b has something new added. By using the Subtract program to produce the image in photo 3c, you can see exactly what changed.

Often there will be minor, inconsequential differences between the images. You can see some background clutter in photo 3c resulting from small differences in lighting and position. Regardless of how careful you are, these differences will occur. What you need is a program to get rid of the irrelevant details.

The Thresh program sets pixel values below a specified threshold level to 0 (black). Running Thresh to remove all pixels below 40 gives the image in photo 4. In addition to suppressing the clutter, Thresh removed the face inside the helmet. This should serve as a reminder that Thresh is concerned only with the brightness of each pixel: Because the face pixels are less than 40, they are set to 0 just like the background clutter.

Photo 4 contains only the parts of photo 3b that aren't in photo 3a, but there are some shadows and reflections in addition to the figurine. The pixel values represent the brightness of the scene and don't "know" whether they are part of an interesting object or the background. Your image-recognition software must distinguish between the actual objects and their shadows and reflections.

Inspection Applications

I know that many of you are interested in using video for inspection, so the next example shows what's needed to compare two pictures to find differences. Since I occasionally digress, I thought inspecting printed circuit boards for missing components was a suitable example.

A critical ingredient in any inspection task is a reference standard that "looks right." All other items are compared to that standard; anything different is regarded as an error. Of course, the differences have to be visible to be detected.

The image in photo 5a is the reference circuit board (anything less than a perfect image is due to lousy lighting). Photo 5b shows a test board with one IC missing. Notice that the ICs are darker than both the board and the silk-screen print below them, but that the capacitors are lighter than the board.

The Compare function is the same as Subtract, except that it returns the absolute value of the difference. Thus, any change will show up as a bright pixel. Photo 5c shows the results of using Compare to process the images of the reference and test boards. A lot of background clutter is due to minor variations in the boards, the lighting, and positions. A simple threshold won't remove the clutter because some of it is quite bright. The trick is to know where the important

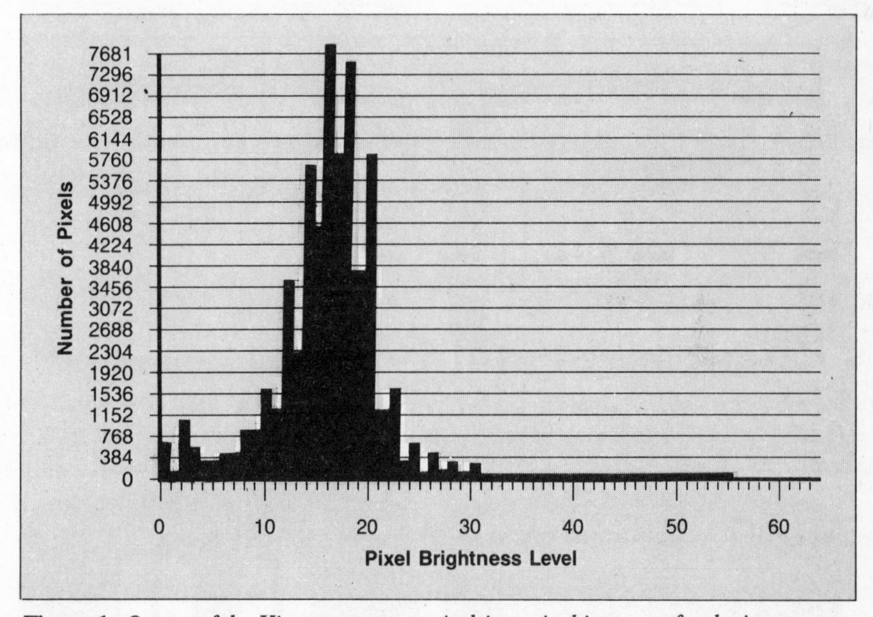

Figure 1: *Output of the Histo program: a pixel-intensity histogram for the image in photo 1.*

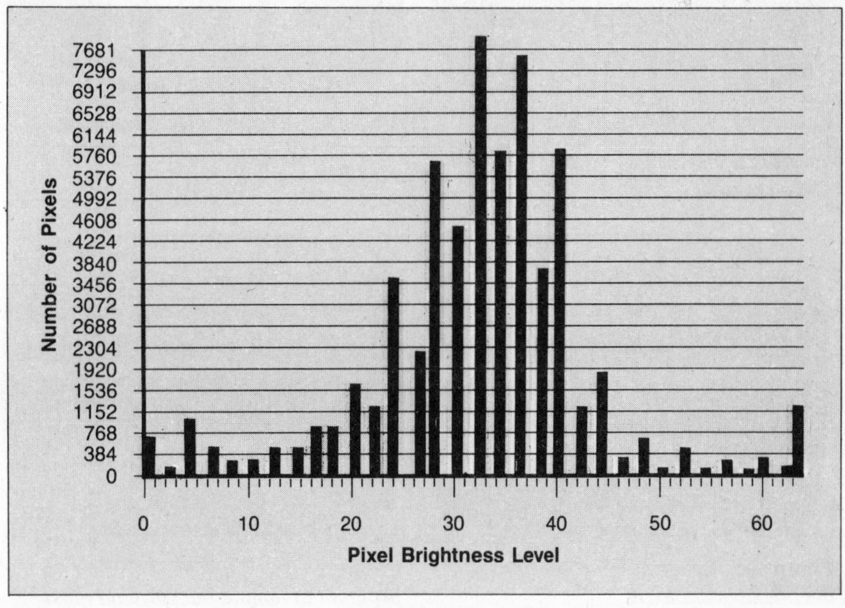

Figure 2: *If you process the image in photo 2 through the Histo program, this is what you get. Notice that all pixels are multiples of 2.*

areas of the picture are and ignore the rest.

Photo 6a shows a mask with bright areas surrounding each component location, prepared by putting white tape on a blank printed circuit board and processing the image to remove the board traces. The Mask program will suppress any pixels in one image lying outside the masked areas in a second image. The image in photo 6b is the result of using the Mask program with the image in photo 6a on the image in photo 5c.

The final step is to apply Thresh to reveal the missing IC in photo 7. What you're seeing is the white silk-screen image printed on the board. It is difficult to see the difference between a dark gray IC and a dark blue circuit board; anything you can do to increase the contrast will help. You will also need to ensure that the two images are accurately aligned and lighted to reduce background clutter. A fixture to hold the boards at an exact location relative to the camera and lights will be essential. You can use Histo, Compare, and Thresh to set up.

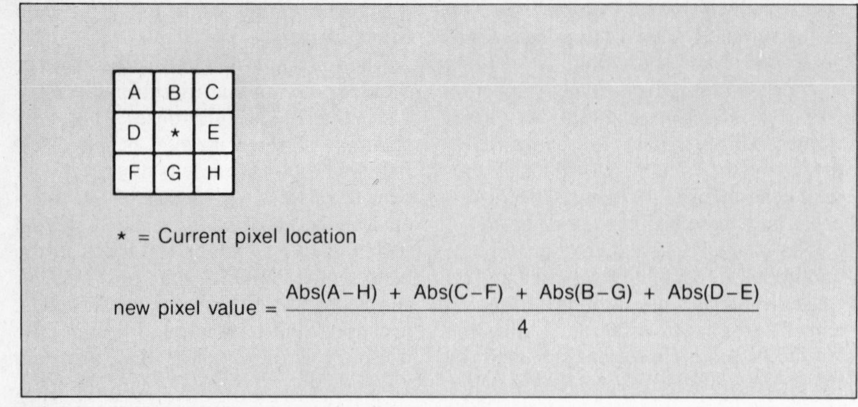

Figure 3: *The formula used by the Edge program.*

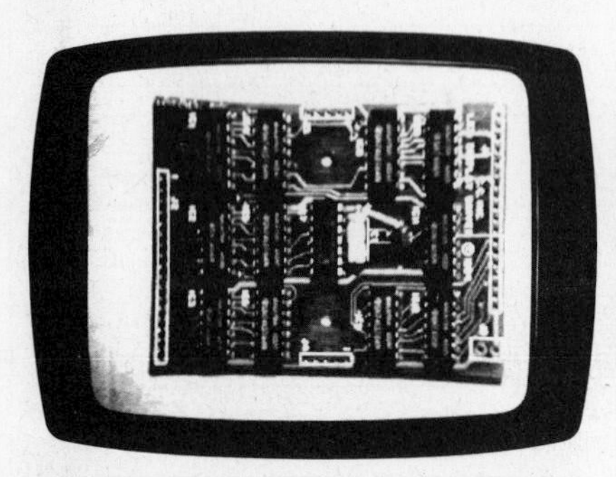

Photo 5a: *Using the Compare program, I have processed the reference image.*

Photo 5b: *This was done with a test image from which I have removed an IC.*

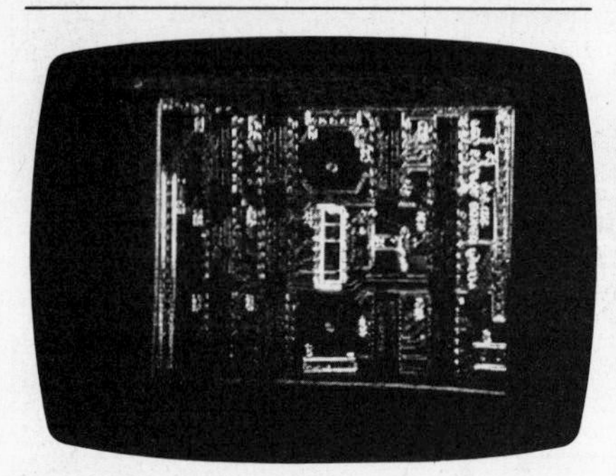

Photo 5c: *The missing IC stands out.*

Photo 6a: *I've prepared a mask image and used it to isolate the important elements in photo **5c**.*

Photo 6b: *The result of processing the image in photo **5c**.*

Photo 7: *Here, I use the Thresh program to remove background noise from the image in photo **6b**. The missing IC in **5b** now stands out clearly.*

Edges and Filters

In some cases, you might be interested in the location of the edges of an object. For example, you might want to know that a pattern is correctly positioned without caring what color (or gray shade) it is. The Edge program produces an image that contains the difference between a pixel's neighbors, calculated as shown in figure 3. A sharp junction between a light and a dark area will result in a bright line, while a uniform area will be reduced to black. The actual shades are not important, only the differences between them.

The Edge routine is a bit more complex than Compare. It finds the absolute value of the differences between the eight pixels surrounding the current pixel in all four directions: vertical, horizontal, and the two diagonals. This is a simple example of a more complex operation called a convolution, which you can use to identify other features in an image.

I restricted Edge to a 3 by 3 set of pixels to reduce the amount of time required to get the answer: It works well enough for these examples. You might want to experiment with a 5 by 5 or larger array, which will let you identify edges more precisely, particularly diagonals at other than 45 degrees. You can also identify the direction of the edges by removing the absolute value function. An edge-detector algorithm that performs this operation (see figure 3) is

$$\text{edge} = \frac{(A-F) + (B-G) + (C-H)}{3}$$

This operation would return a positive value for the horizontal edge between an upper, bright object and a lower, dark object. Reverse the two objects, and the sign becomes negative. Because the results must be returned as pixel values, you will need to add a fixed offset before setting the final pixel value.

Usually, you will have to multiply the result of Edge by 2 or 3 to make all the edges visible on the monitor. Thresh can then suppress all the "soft" edges. Photo 8b is the result of running Edge on the image in photo 8a, multiplying by 3, then using Thresh to remove pixels below 30.

It's also possible to remove edges and textures. Filter averages four pixels using the algorithm shown in figure 4 to produce the output image. Compare the images in photos 9a and 9b to see how Filter reduces "crispness" and fine details. This can be useful if you have an object with fine detail that is not needed by the rest of the processing.

Intruder Alert!

One obvious application for image processing is in a security system that can compare two images, decide when something has changed enough to warrant human inspection, and sound an alert (or fire the laser, or whatever).

You've seen most of the pieces already:

1. Grab a reference image.
2. Grab a test image.
3. Compare the images.
4. Thresh the result to remove clutter.
5. Count the number of changed pixels.
6. If the count is high enough, take action.
7. Replace the reference image with the test image.
8. Go to step 2.

The key program is Count, which examines an image and counts the number of pixels that exceed a threshold level. The preceding image processing steps must create an image with high-intensity pixels identifying the intruder. When the count is high (corresponding to a new shape of a person on the screen), it's time to sound the alarm.

If the images don't differ by too much, you use the test image as the reference image for the next loop. This lets the system cope with small, slow changes in lighting and motion. Obviously, you could defeat this system by easing slowly into the picture, but in practice it's hard to fool.

WATCHDOG.BAT (see listing 1) combines the programs we've used so far to automate that process. The batch file can

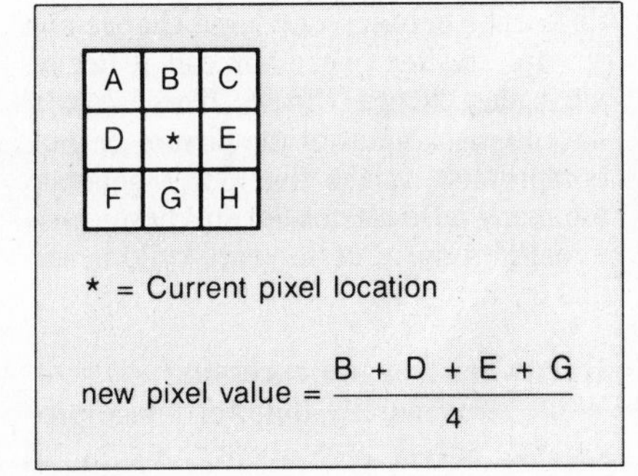

Figure 4: *The formula used by the Filter program.*

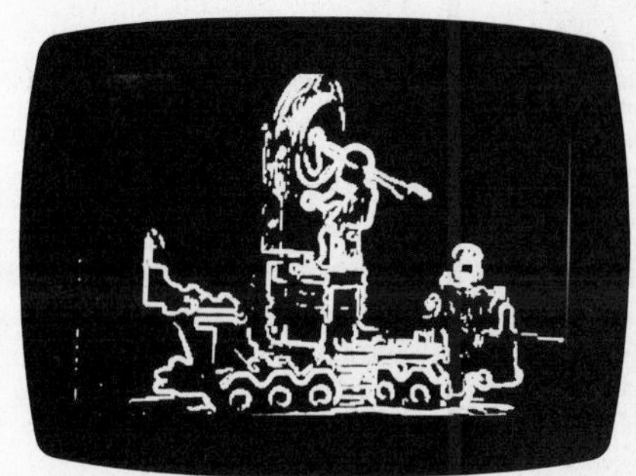

Photo 8: *Edge detection with ImageWise. The image in* **(a)** *is processed through Edge, Multiply, and Thresh to produce* **(b)**.

Photo 9: *Running the Filter program on* **(a)** *produces the image in* **(b)**.

Photo 10: *An image digitized on the ImageWise digitizer/transmitter is displayed on the GT180* **(a)**. *This image is magnified 2 times* **(b)**, *4 times* **(c)**, *and 8 times* **(d)**.

Image Processing Routines

The following is a list of the programs described in this chapter, plus some additional image processing software you might find interesting. File specs between angle brackets are optional. Results will be stored in the first file spec if the target file spec is omitted. An n indicates a numeric value. These programs are available from the Circuit Cellar BBS, BIX, and BYTEnet.

ADD.PAS pic1 pic2 <pic3>
Function: pic3 = pic1 + pic2.

COMPARE.PAS pic1 n <pic2>
Function: pic2 = pic1 if pixel >= n
= 0 otherwise.

COMPRESS.PAS pic1 <pic2>
Function: pic2 is the run-length-encoded version of pic1. Compressed files cannot be used by the other programs.

COUNT.PAS pic1 n
Function: DOS ERRORLEVEL variable = number of pixels >= n.

DUMPER.PAS pic1
Function: Produces formatted print dump of pic1 for hand analysis; use redirection to send output to a disk file.

EDGE.PAS pic1 <pic2>
Function: pic2 contains edge-intensity information from pic1.

EXPAND.PAS pic1 <pic2>
Function: pic2 is the non-RLE version of pic1. Expanded files are required by the other programs.

FASTDOG.PAS $n1$ $n2$
Function: Watches a scene, reports an intruder when $n2$ changed pixels exceed $n1$.

FILTER.PAS pic1 <pic2>
Function: pic2 is a low-pass filtered version of pic1.

GRAB.PAS pic1 /n /c
Function: Accepts picture from transmitter board and stores the expanded data in pic1. Switch /n prevents showing the picture on the receiver; switch /c stores the image without expanding it.

HISTO.PAS pic1
Function: Displays a pixel-intensity histogram for pic1.

INVERT.PAS pic1 <pic2>
Function: pic2 = 63 − pic1.

MASK.PAS pic1 pic2 <pic3>
Function: pic3 = pic1 if pic2 > 0
= 0 otherwise.

MULTIPLY.PAS pic1 n <pic2>
Function: pic2 = pic1 * n.

SHOW.PAS pic1
Function: Sends pic1 to the display board.

SUBTRACT.PAS pic1 pic2 <pic3>
Function: pic3 = pic1 − pic2.

THRESH.PAS pic1 n <pic2>
Function: pic2 = pic1 if pic1 >= n
= 0 otherwise.

Listing 1: *The WATCHDOG.BAT program.*

```
ECHO off
REM Syntax is:
REM   WATCHDOG brightness pixels
REM brightness is COUNT's threshold level
REM pixels is # of pixels >= brightness, in units of 100
REM   WATCHDOG 10 4
REM will alarm when 400 pixels or more are brighter than 10
REM runs best with image files on a RAM disk!
ECHO Make sure serial cable is connected to transmitter
PAUSE
:newref
GRAB ref /n
:newtest
GRAB test /n
COMPARE ref test deltas
COUNT deltas %1
IF errorlevel %2 goto gotcha
ECHO no intruder so far...
ERASE ref
RENAME test ref
GOTO newtest
:gotcha
ECHO --- Intruder alert!!! ---
ECHO Switch serial cable to receiver for display
PAUSE
SHOW test
ECHO Switch serial cable to transmitter
PAUSE
goto newref
```

examine one picture every 30 seconds or so, which might be adequate for most purposes. (If you need more speed, I have a faster program called FASTDOG.BAT in the downloadable software.)

The Count program returns the number of qualifying pixels (divided by 100) in the DOS ERRORLEVEL variable to let the IF statement decide whether an intruder is present. You should replace the ECHO statement with a program that does something useful, like turn on the lights, sound a loud alarm, or whatever you choose.

You'll need to do some experimentation to pick the best values for the threshold and count levels. Count can't tell the difference between one large change and several smaller ones, nor can it decide what the change "looks like." You'll have to mask areas of the picture or pick compromise values that don't generate too many false alarms but still never miss a real intruder. Put on your skulking suit and try to fool it.

Hardware Image Processing

While we generally think of image processing solely as software-dependent tasks, many of the newest graphics-display chips incorporate some of these functions in hardware. Most prominent among such features is the hardware zoom or image-expansion function. The GT180 color graphics board I presented in the November 1986 Circuit Cellar has a hardware zoom that can expand an image up to 16 times.

Photo 10a shows a standard-resolution 256- by 244-pixel picture (no, it's not me this time) digitized on the digitizer/transmitter board and displayed in 16-level gray scale on a GT180 high-resolution graphics-display board. Because the GT180 has a resolution of 640 by 480, the lower-resolution digitized picture fills only the top left corner, but it expands to fill and then overflow the screen as it is zoomed. Photo 10b is 2 times magnification, photo 10c is 4 times, and photo 10d is 8 times the original image.

Conclusions

As anyone who owns a TV camera can attest, video is fascinating. Until now, small computer users haven't been able to work with pictures of the real world because the video hardware was frightfully expensive. With the hardware and software I've provided, you can take digital pictures, enhance them to pick out interesting objects, and save them for later. I'm sure you'll find many more ways of tweaking the video.

The complete source code for all the programs described in the text box on page 28 is available from the Circuit Cellar BBS, BIX, BYTEnet.

Special thanks to Ed Nisley for his expert collaboration on this project.

Editor's Note: Steve often refers to previous Circuit Cellar articles. Most of these past articles are available in book form from BYTE Books, McGraw-Hill Publishing Company, P.O. Box 400, Hightstown, NJ 08250, (1-800-2-MCGRAW).

Ciarcia's Circuit Cellar, Volume I covers articles in BYTE from September 1977 through November 1978. *Volume II* covers December 1978 through June 1980. *Volume III* covers July 1980 through December 1981. *Volume IV* covers January 1982 through June 1983. *Volume V* covers July 1983 through December 1984.

The following items are available from

Circuit Cellar, Inc.
4 Park St., Suite 12
Vernon, CT 06066
(203) 875-2751

1. ImageWise digitizer/transmitter board experimenter's kit. Contains digitizer/transmitter printed circuit board, 11.05-MHz crystal, programmed 2764 EPROM with transmitter software, and CA3306 flash A/D converter and manual with complete parts list.
DT01-EXP $99

2. ImageWise display/receiver board experimenter's kit. Contains gray-scale display/receiver printed circuit board, 11.05-MHz crystal, programmed 2764 EPROM with receiver software, Telmos 1852 video D/A converter, manual with complete parts list, and an IBM PC 2.0 disk containing sample digitized images and test patterns.
DR01-EXP $99
DT01-EXP and DR01-EXP together $179

3. ImageWise digitizer/transmitter full kit. Contains all digitizer/transmitter components, including printed circuit board, 64K bytes of static RAM, IC sockets, crystals, programmed 2764, CA3306 flash A/D converter, manual, and IBM PC 2.0 disk containing utility routines for storing and displaying and downloading image files using an IBM PC. Does not include power supply or case.
DT01-KIT $249

4. ImageWise display/receiver full kit. Contains all gray-scale display/receiver components, including printed circuit board, 64K bytes of static RAM, IC sockets, crystals, programmed 2764, Telmos 1852 video D/A converter, manual, and an IBM PC 2.0 disk containing sample digitized images and test patterns. Does not include case or power supply.
DR01-KIT $249
DT01-KIT and DR01-KIT together $489

ImageWise is also available assembled. Call CCI for source and availability of assembled boards and complete systems, black-and-white TV cameras, 32K-byte static RAM chips, and power supplies.

All payments should be made in U.S. dollars by check, money order, MasterCard, or Visa. Surface delivery (U.S. and Canada only): add $3 for U.S., $6 for Canada. For delivery to Europe via U.S. airmail, add $10. Three-day air freight delivery: add $8 for U.S. (UPS Blue), $25 for Canada (Purolator overnight), $45 for Europe (Federal Express), or $60 for Asia and elsewhere in the world (Federal Express). Shipping costs are the same for one or two units.

There is an on-line Circuit Cellar bulletin board system that supports past and present projects. You are invited to call and exchange ideas and comments with other Circuit Cellar supporters. The 300/1200/2400-bps BBS is on-line 24 hours a day at (203) 871-1988.

3

BUILD AN INFRARED REMOTE CONTROLLER

A custom hand-held infrared transmitter and receiver

In the last few years, I have installed a lot of automatic features in my home and the Circuit Cellar. Besides the sophisticated alarm system, computer-controlled wood stove, and perimeter lighting system (see "Living in a Sensible Environment" in the July 1985 BYTE), the Home Run Control System (HCS) has significantly increased the convenience of living in my house (see "Build the Home Run Control System" in the April through June 1985 BYTEs). I have the usual remote-controlled stereos, TVs, FM radios, etc., but it is the small things like lights that automatically go on and off as I walk through the house and a voice-synthesized central monitoring system that I can call and hear a verbal status report when I'm away that have forever erased any pleasure derived from a manually oriented existence.

Just when I thought I had come to grips with electronic living, I ran into two new problems: automated living user intervention anarchy (ALUIA) and hand-held infrared controller overpopulation (HIROP). Generally speaking, these are high-tech diseases common among gadget-happy technocrats and overzealous inventor-authors who are insane enough to try to install all their designs into one house.

ALUIA is like gridlock. So many things are controlled within an environment that the only way to activate something is through the control system (if you can remember which system is controlling what device at a particular time). Unless you physically intervene, the control sequence will remain as set, and you must live with the consequences. Perhaps you have a timer controlling a series of outside lights, but one night you want them on at different times. Do you reset the automatic timer to a new cycle, override the automatic system manually at each setting, or disconnect the automatic system? The problem with many automatic systems is that they make little provision for unpredicted and unprogrammed user behavior.

In the case of the HCS, I tried to provide for such possibilities. Rather than just a timer that turns lights on and off, for example, the HCS can turn a light on in a room as long as it senses your presence via a motion detector connected to one of its hard-wired inputs. Such a control system adapts itself to your behavior—up to a certain point! It cannot control what it cannot sense, nor can it be expected to accommodate all your wishes through passive sensors. If you want the stereo to go on when you walk into the room, it could be coordinated with the automatic lighting, but would you want it to come on even when you merely pass through the room?

Some automatic control functions are more applicable to lighting than to stereo equipment. Is the only alternative manual control if the system is not fully automatic? Not if it is designed to allow independent user input in addition to automatic control.

This control gridlock comes about when you and the system have different ideas as to what should be happening. Sometimes the only alternative is to design a system with shared control—one that executes a preprogrammed sequence but accommodates itself to selective manual in-

Photo 1: *Enclosed IRCOMM transmitter and receiver units.*

tervention. By allowing you a means to override or direct the activities of the control system, the system retains the flexibility of independent control in the absence of user directives and lessens its potential for becoming a frustrating obstacle as your needs change.

Perhaps you want the stereo to automatically come on when you walk in the room any time between 5 and 6 p.m., but not at any other time unless you physically turn it on. You could easily program the HCS to use the motion sensor in the room to activate the stereo and lock out execution of the order except between 5 and 6 p.m. But how do you manually turn the stereo on yet still advise the control system of your action? The system needs to know that the stereo should be ON, even if it has been turned on manually. Any well-designed control system like the HCS

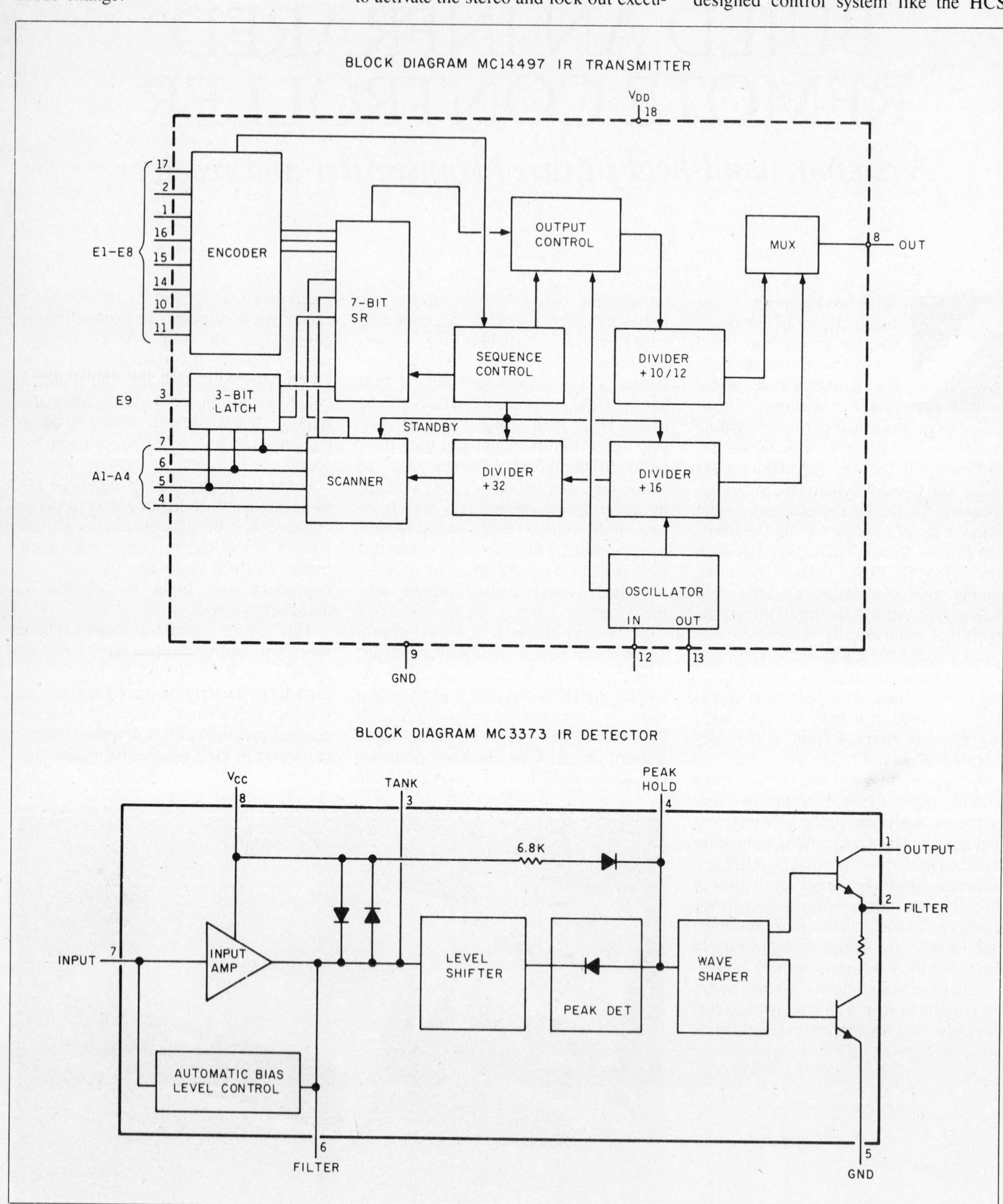

Figure 1: *Block diagrams of the chips used in the IRCOMM.*

periodically "refreshes" or retransmits the current status (ON or OFF) to all controlled devices, reducing concern about false activation/deactivation of the BSR X-10 remote-control modules by transients. The best way to coordinate such automatic/manual control is to do everything through the control system.

I don't expect you to run down to your Circuit Cellar and type a manual override every time you want to turn the stereo on. Since we are intending a direct and specific control action, "turn the stereo on," it only makes sense to use a direct input to the control system similar to a motion detector. A simple push button connected to one of the HCS inputs can turn on the stereo. When you press it, the control system is actually turning the stereo on, and it will coordinate this with its other functions.

While ultimately successful, if you follow this tack, you will soon find that it falls short as an easily implementable procedure. The interest in using a BSR X-10 wireless remote-control device with the HCS in the first place was to eliminate the need for wires. The more convenient alternative would be a wireless remote-control device that communicates specific commands from you to the HCS or other computerized control system. A typical example of such a device is the infrared remote control used with most TV sets.

This chapter's project is the design and construction of a custom hand-held infrared transmitter and receiver, called the IRCOMM (see photo 1). The transmitter circuit can be constructed as a small inexpensive hand-held controller or expanded to implement a 62-key wireless keyboard. The receiver is equally uncomplicated and intended to provide a convenient link between the user and the home control system.

An unfortunate side effect of creating the IRCOMM is that it adds one more IR remote-control unit to the pile you probably have and contributes to HIROP, as I stated earlier. Therefore, in Chapter 4 I will make amends for contributing to IR remotes on every table and chair with my own form of population control.

Generally Speaking

My primary consideration in the design of IRCOMM was to use it with the HCS. Therefore, hundreds of remote function keys and a 20-mile effective range were of little importance as design criteria. Much like remote controls for TVs or VCRs, the IRCOMM controller needed only to be short-range and command basic functions like "stereo system power ON," "surround sound system power ON," "projection TV ON," "mood lights ON," "entertainment system all power OFF," "room-to-room sound tracking GO," etc.

When we speak of the functions that a remote controller performs, we are actually describing what the device being controlled (a TV, for example) does as a result of your pressing a key on the remote. The remote control is nothing other than a wireless keyboard. When you press a key, a stream of data is transmitted either as an ultrasonic, radio-frequency, or infrared signal. Present-day consumer electronic devices primarily use infrared signaling because of its low cost and limited interference with other remote-controlled appliances (like accidentally turning on the TV in the next apartment).

Consumer electronic devices use infrared signaling because of its low cost and limited interference with other remote-controlled appliances.

Infrared controllers generally use pulse position modulation (PPM) or pulse code modulation (PCM). The actual technique used is significant only to a person designing a receiver/decoder.

In designing the IRCOMM, I tried to keep both my needs and the intelligence

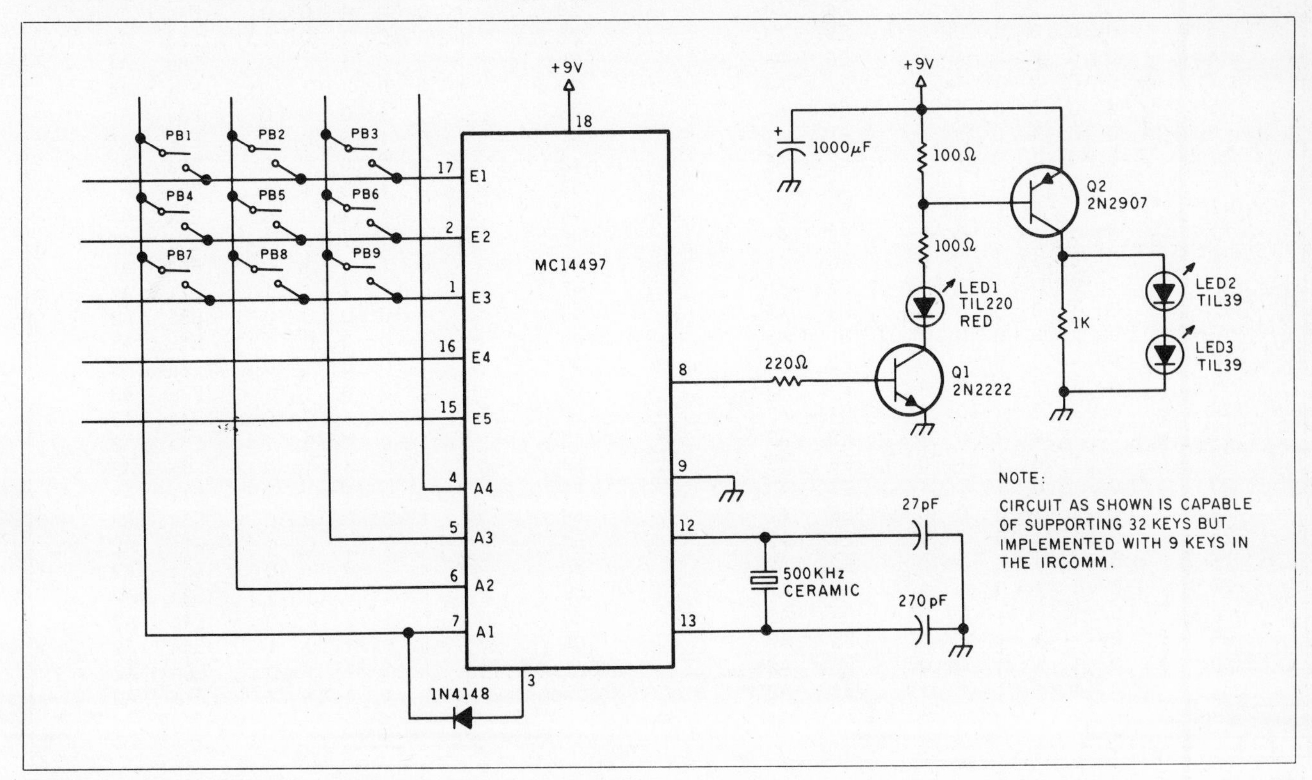

Figure 2: *Schematic of the IRCOMM hand-held transmitter.*

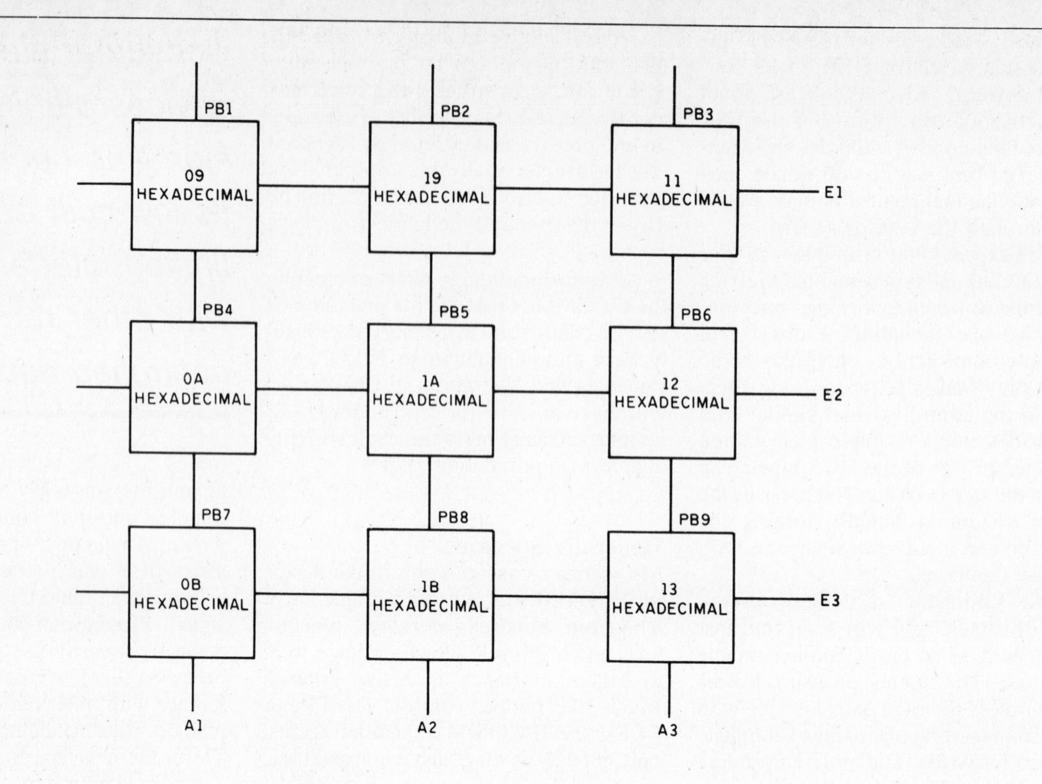

Channel	Code Word						Keyboard		Channel	Code Word						Keyboard	
	F	E	D	C	B	A	In	Out		F	E	D	C	B	A	In	Out
0	0	0	0	0	0	0	E8	A4	32	1	0	0	0	0	0	E8a	A4
1	0	0	0	0	0	1	E1	A4	33	1	0	0	0	0	1	E1a	A4
2	0	0	0	0	1	0	E2	A4	34	1	0	0	0	1	0	E2a	A4
3	0	0	0	0	1	1	E3	A4	35	1	0	0	0	1	1	E3a	A4
4	0	0	0	1	0	0	E4	A4	36	1	0	0	1	0	0	E4a	A4
5	0	0	0	1	0	1	E5	A4	37	1	0	0	1	0	1	E5a	A4
6	0	0	0	1	1	0	E6	A4	38	1	0	0	1	1	0	E6a	A4
7	0	0	0	1	1	1	E7	A4	39	1	0	0	1	1	1	E7a	A4
8	0	0	1	0	0	0	E8	A1	40	1	0	1	0	0	0	E8a	A1
9	0	0	1	0	0	1	E1	A1	41	1	0	1	0	0	1	E1a	A1
10	0	0	1	0	1	0	E2	A1	42	1	0	1	0	1	0	E2a	A1
11	0	0	1	0	1	1	E3	A1	43	1	0	1	0	1	1	E3a	A1
12	0	0	1	1	0	0	E4	A1	44	1	0	1	1	0	0	E4a	A1
13	0	0	1	1	0	1	E5	A1	45	1	0	1	1	0	1	E5a	A1
14	0	0	1	1	1	0	E6	A1	46	1	0	1	1	1	0	E6a	A1
15	0	0	1	1	1	1	E7	A1	47	1	0	1	1	1	1	E7a	A1
16	0	1	0	0	0	0	E8	A3	48	1	1	0	0	0	0	E8a	A3
17	0	1	0	0	0	1	E1	A3	49	1	1	0	0	0	1	E1a	A3
18	0	1	0	0	1	0	E2	A3	50	1	1	0	0	1	0	E2a	A3
19	0	1	0	0	1	1	E3	A3	51	1	1	0	0	1	1	E3a	A3
20	0	1	0	1	0	0	E4	A3	52	1	1	0	1	0	0	E4a	A3
21	0	1	0	1	0	1	E5	A3	53	1	1	0	1	0	1	E5a	A3
22	0	1	0	1	1	0	E6	A3	54	1	1	0	1	1	0	E6a	A3
23	0	1	0	1	1	1	E7	A3	55	1	1	0	1	1	1	E7a	A3
24	0	1	1	0	0	0	E8	A2	56	1	1	1	0	0	0	E8a	A2
25	0	1	1	0	0	1	E1	A2	57	1	1	1	0	0	1	E1a	A2
26	0	1	1	0	1	0	E2	A2	58	1	1	1	0	1	0	E2a	A2
27	0	1	1	0	1	1	E3	A2	59	1	1	1	0	1	1	E3a	A2
28	0	1	1	1	0	0	E4	A2	60	1	1	1	1	0	0	E4a	A2
29	0	1	1	1	0	1	E5	A2	61	1	1	1	1	0	1	E5a	A2
30	0	1	1	1	1	0	E6	A2	62(EOT)	1	1	1	1	1	0	E6a	A2
31	0	1	1	1	1	1	E7	A2	Not transmitted	1	1	1	1	1	1	E7a	A2

Figure 3: *Codes generated by the MC14497 and the IRCOMM keypad.*

of the BYTE readership in mind. What I am presenting should be considered as a model and a sample application of infrared remote control and not as the only way to implement it. The usual article approach to this subject is to buy off-the-shelf remote-control chip sets designed for the TV industry. Such an approach is valid, but it better serves the author than the user. TV remote-control chips are designed for a specific application, and their receivers are often bus or multiplexed output devices. Additional glue logic is frequently added to provide 1-of-24 signal lines or 4- to 16-bit decoded outputs. Whatever the decoding technique employed, the resulting receiver outputs must still be read through a parallel input port. Ever try to find a parallel input port on your IBM PC?

The Circuit Cellar IRCOMM

We aren't intending to use the IRCOMM to mimic a TV remote control, so why bother to spend the money or carry the overhead of decoding circuitry intended for TVs? A better alternative is to merely condition the incoming signal and allow the control computer to decode the signal. Motorola manufactures a pair of general-purpose CMOS IR remote-control chips, the MC14497 and the MC3373, that fit the bill exactly (see figure 1).

Figure 2 is the schematic of the handheld IRCOMM transmitter. As I have it shown, the MC14497 is hard-wired for use with up to 32 keys, AM modulation, and a logic 1 start bit. The MC14497 is a CMOS biphase PCM remote-control transmitter chip in an 18-pin package. In standby mode it draws a mere 10 microamperes and operates anywhere within a range of 4 to 10 volts. Transmission and internal timing are controlled with a 500-kilohertz ceramic resonator.

The basic configuration of the MC14497 will support 32 keys (described as channel 0 through channel 31). With two additional diodes and switches between pins 3 to 6 and 3 to 5, the capacity can be increased to a maximum of 62 keys. Either option is far more than I needed. Perhaps more as a result of the plastic box I had on hand than any calculated requirement, I ended up with 9 keys (PB1–PB9) connected as shown. If you want more keys, simply add more push buttons at the cross points of the Ex and Ax lines.

The IRCOMM has both infrared and visible outputs when it transmits. A two-transistor driver circuit simultaneously pulses visible LED1 and infrared LEDs, LED2 and LED3. The MC14497 transmits either in FSK (frequency-shift keying) or AM mode. We are restricted to AM transmission because of the receiver I used, so that is all I will address.

Figure 3 shows the layout and hexadecimal codes for the 9-key IRCOMM. Since I arbitrarily selected the matrix lines they do not correspond to channels 0, 1, 2, etc. (Since the receiver cares only *which* nine channels it has to identify—not that they be sequentially ordered—keys can be placed anywhere in the matrix.)

Figure 4 illustrates what the PCM coding looks like as it is transmitted. Biphase PCM is relatively easy to read once you get the hang of it. The most important part in reading PCM is keeping track of the bit times and noticing at what point the 0-to-1 or 1-to-0 logic transition occurs during the bit time. If the pulse burst is sent/received during the first half bit time, the bit is a logic 1. Conversely, if the pulse occurs during the second half bit time, the bit is a logic 0.

When a key is pressed, the transmitter sends an AGC (automatic gain control) burst lasting a half bit time, a start bit (logic 1), and a 6-bit PCM data word. The purpose of the AGC pulse preceding the PCM data is to set up the AGC loop in the receiver in time for the start bit. The 6 data bits are designated as A (least significant bit) through F (most significant bit) and are shifted in LSB first. While each bit is represented as a logic 0 or 1 level pulse having a duration of 0.5 or 1 millisecond, the actual output of the transmitter is a 41.66-kHz pulse burst for the duration of any logic 1 level. Only after the data is conditioned by the receiver will it appear as discrete logic levels.

The IRCOMM repeatedly transmits the same code as long as the key is pressed. When the key is released, a channel 62 EOT (end of transmission) code is automatically sent. Channel 63 is not used.

The fact that the coding is simplified allows some license to be taken with the receiver circuitry.

The IRCOMM Receiver

As you can see, the IRCOMM chip transmits an easily recognizable and repeatable code. Many TV-style IR remote chips send the pulse burst once or send it once and repeatedly transmit the EOT signal. It is easy to read the code from the IRCOMM with an oscilloscope. Try reading your TV remote manually.

The fact that the coding is simplified allows some license to be taken with the receiver circuitry. Rather than the expensive LSI hardware often required in pulse position coding, only the bit timing is relevant. To acquire this data, we merely convert the 41.66-kHz pulse bursts to TTL logic levels through an envelope detector and apply this signal to a computer that monitors the bit timing. I could decode the signal all in the hardware, but if the end result is still to connect it to a computer,

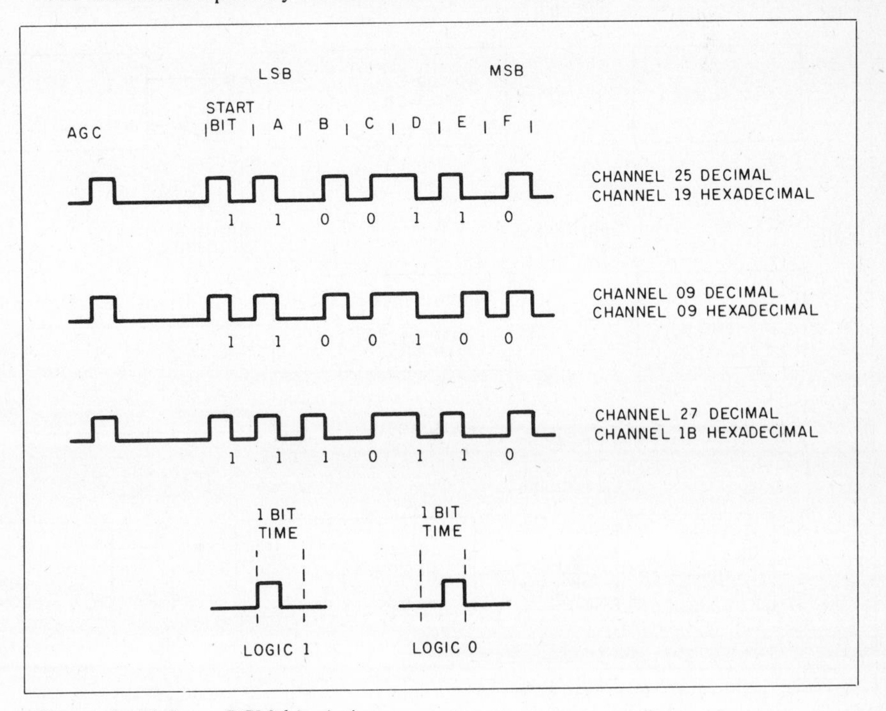

Figure 4: *Biphase PCM bit timing.*

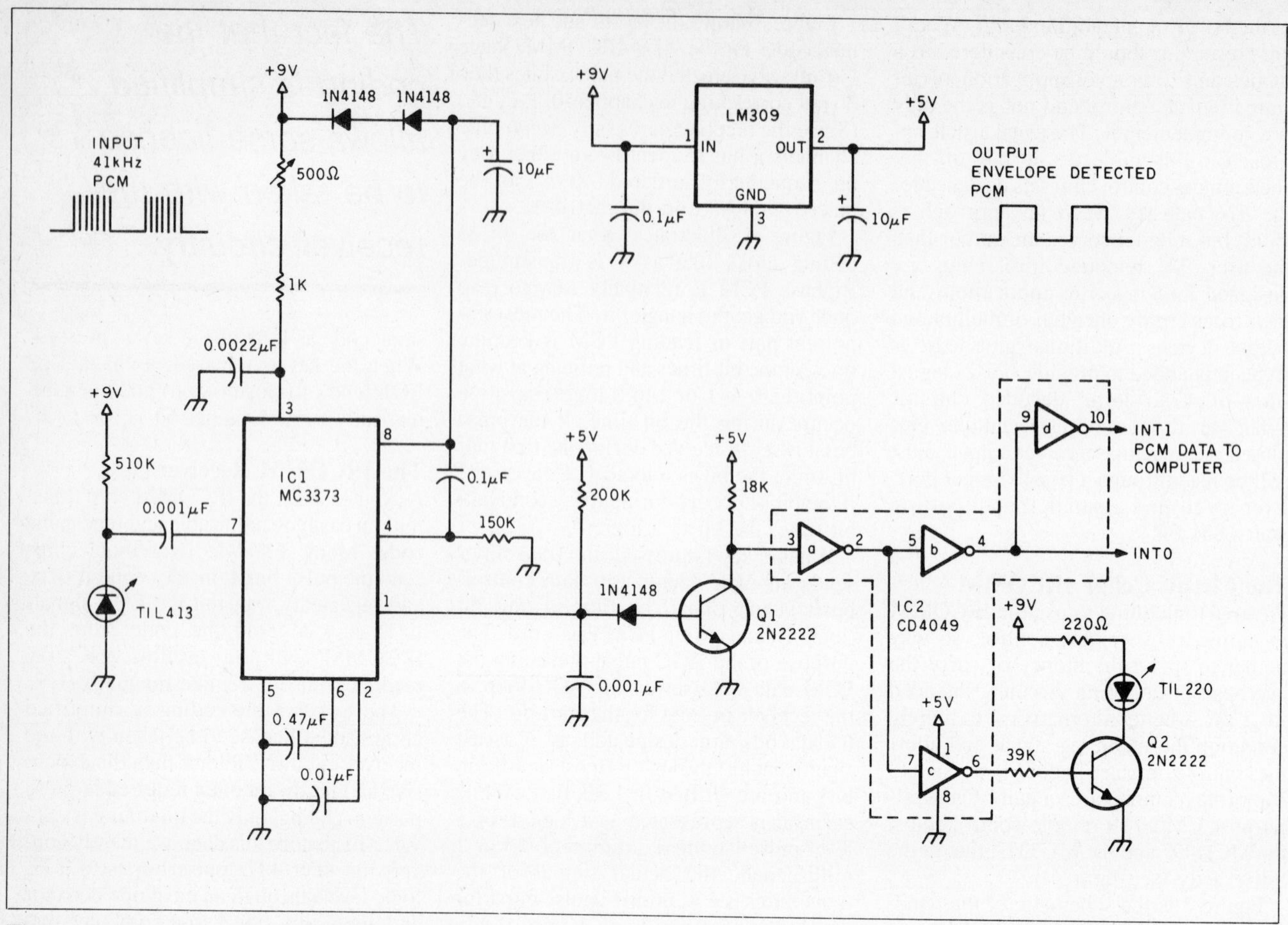

Figure 5: *Schematic of the IRCOMM receiver.*

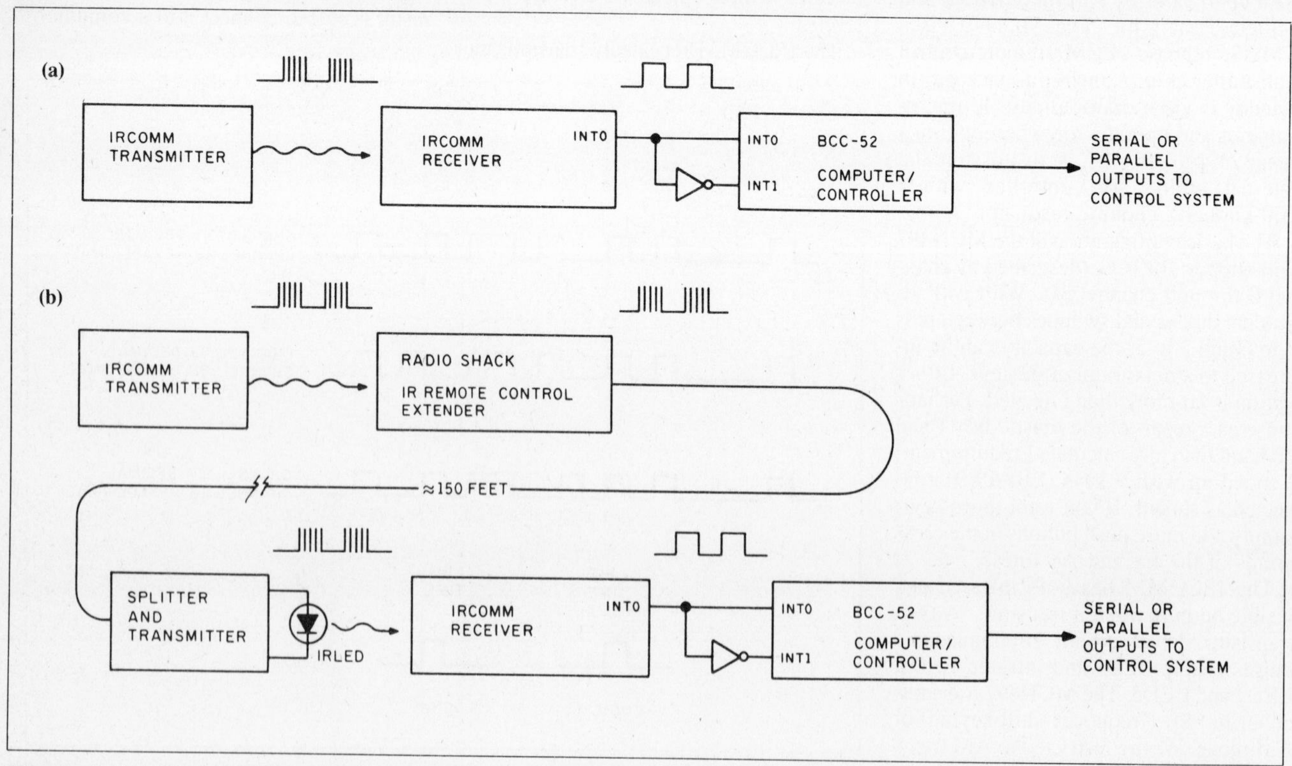

Figure 6: *(a) Direct hookup of the IRCOMM to the BCC-52 computer; (b) hookup using the IR remote-control extender.*

why not use the computer to decode the raw transmission in the first place? I never thought I'd be advocating a software solution, but I support reality.

The Motorola MC3373 wideband amplifier-detector chip is designed for use with infrared pulse-burst transmissions. The entire receiver circuit, shown in figure 5, is a two-chip envelope detector and TTL level shifter. An AC-coupled photodiode receives the infrared pulses from the IRCOMM transmitter and amplifies them. When an infrared signal of approximately 40 kHz is perceived, the output goes low. Q1 inverts this signal and applies it to a series of 4049 CMOS inverters that are capable of driving the LSTTL (low-power Schottky transistor-transistor logic) input load of the computer and lighting an additional visible LED so that you can see that data is being received.

Decoding the PCM in Software

As I mentioned earlier, my intended application of the IRCOMM was to add remote-control features to my home control system by using it to trigger direct inputs to the HCS. Of course, the HCS was not designed with facility for IR remote control, but it does have 16 parallel input lines that can be used to trigger events. To use the IRCOMM with the parallel input of the HCS, however, another computer must be interposed between them. This special-function computer translates the IRCOMM receiver's PCM output into 9 (or 32 if you used that many keys) parallel signal lines that are attached to the HCS. I chose the BCC-52 computer/controller for this task (see "Build the BASIC-52 Computer/Controller" in the August 1985 BYTE). Figure 6a illustrates a block diagram of the connection.

The BCC-52 is programmable in BASIC or 8051 assembly language. To achieve the processing speed necessary to analyze PCM bit times, we must use an assembly language routine. The flowchart of the conversion process is given in figure 7, and a complete source listing of the necessary routine is provided in listing 1.

I used the BCC-52 because it was convenient and cost-effective. Since it is bus-compatible with a variety of A/D, display, and power-control peripherals that have also been presented as Circuit Cellar projects, this suggests a far more powerful future application for the IRCOMM remote control. The IRCOMM and the computer that decodes the PCM signal are

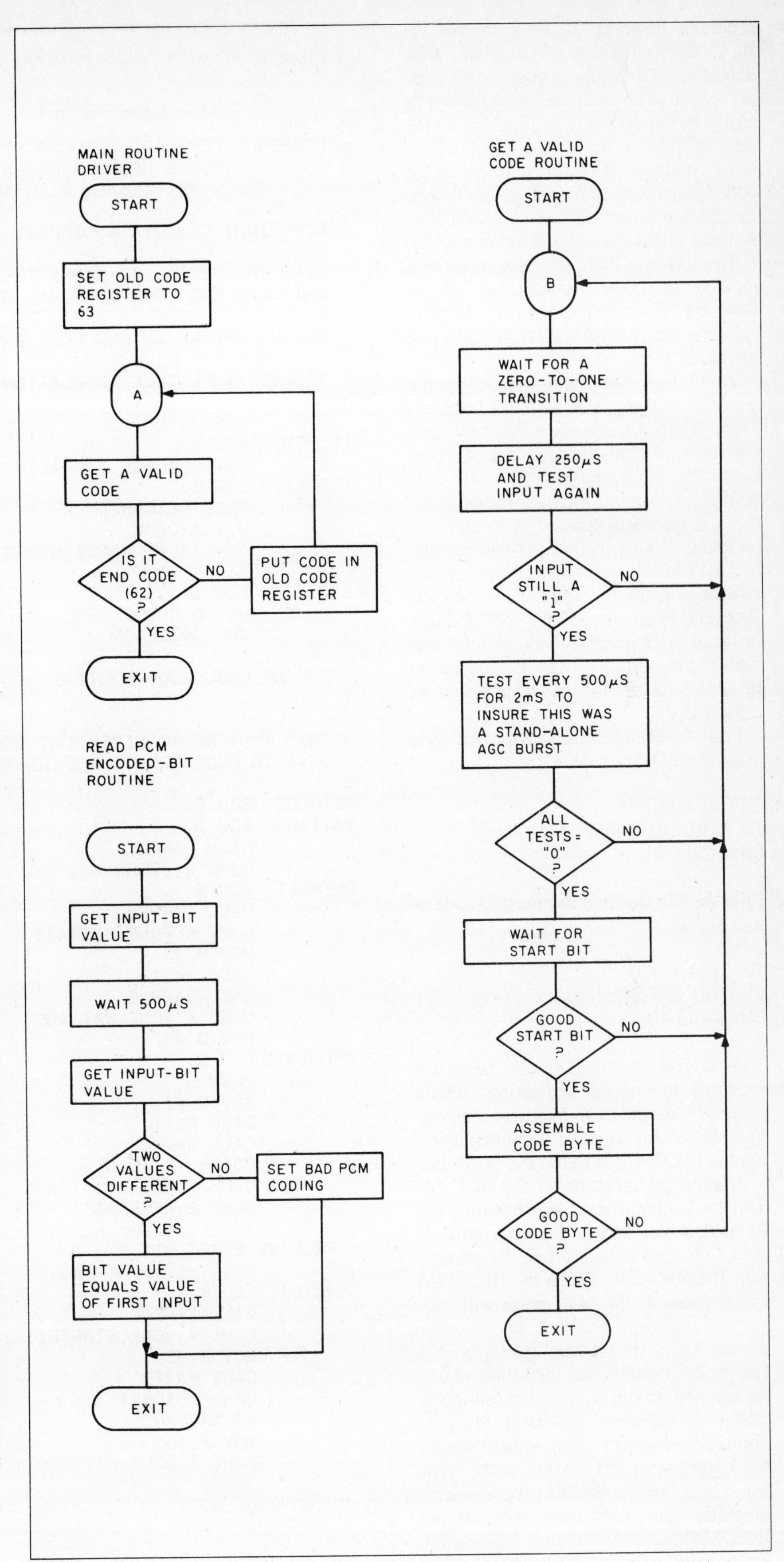

Figure 7: *Flowchart of the PCM data-receiver routine.*

separate, however. I chose to use the BCC-52 so that the complete flow diagram and source code could be presented as an example. It could just as easily be connected to an input bit on the parallel printer port of an IBM PC or other computer. If anyone implements the code for another computer, please upload the routine to my BBS at (203) 871-1988 so that I might share it with other readers.

The 100-byte PCM decoding routine is located at 4000 hexadecimal in an EPROM. The output of the IRCOMM receiver is attached to the INT0 input line of the BCC-52.

There are two entry points to the decoder routine: 4000 and 4011 hexadecimal. When you execute a CALL4000H, the software reads the incoming IR data but waits for a channel 62 EOT code before returning to BASIC. The code for the key that was pressed is in memory location 1F hexadecimal. If you execute a CALL4011H instead, the software returns immediately after it receives the first valid PCM signal. This code is stored in location 19 hexadecimal. A three-line BASIC program is all that is required to print the received code using either of these calls.

Print the hexadecimal key pressed code after the EOT is received:

```
10 CALL4000H
20 PH0. DBY(1FH)
30 GOTO 10
```

Print the hexadecimal key pressed code as it is received:

```
10 CALL4011H
20 PH0. DBY(19H)
30 GOTO 10
```

Of course, it would hardly be worth your time to consider dedicating a whole computer like the BCC-52 just to scan one input bit. A far superior way of connecting the IRCOMM is to use the PCM data to trigger an interrupt on the BCC-52. This is accomplished by inverting the PCM data and connecting this signal to the INT1 input in addition to the connection already made. When the IRCOMM key is pressed, the AGC pulse will trigger an interrupt and call the decoder routine. Using the interrupt goto (ONEX1) and the GET command, the keyboard can be scanned on the fly without waiting at an INPUT statement, and fast BASIC programs with input provided either through the keyboard or IRCOMM can be written. The following BASIC program demonstrates the combined use of the IRCOMM and the keyboard. It prints any key pressed, either on the keyboard or the terminal, as it's entered:

Listing 1: *Assembly language listing for the PCM data-decoder routine.*

```
;
; LEDPCM - PCM DATA DECODER ROUTINE FOR READING IR LED
;          CODES USING A BCC-52 COMPUTER/CONTROLLER
;          - WRITTEN BY WILLIAM D. CURLEW
;
; COPYRIGHT CIARCIA'S CIRCUIT CELLAR 1986
;
; READ THE SERIAL IR RECEIVER INPUT. REMEMBER THE LAST CODE
; UNTIL AN END CODE (DECIMAL 62) IS RECEIVED. RETURN THE
; REMEMBERED CODE IN REGISTER 1FH. A RETURNED CODE OF 63
; DECIMAL MEANS AN END CODE WAS THE FIRST THING DETECTED.
;
; ASSUMES AMPLITUDE MODULATION MODE IS BEING USED.
;
          ORG 4000H
LEDPCM    EQU $
          MOV OLDCODE,#INVALID    ; SET UP OLD CODE AS 63
CODELOOP  EQU $
          CALL GETCODE            ; GET A CODE
          MOV A,CODE              ; PUT CODE IN A REG
          CJNE A,#ENDCODE,STORE   ; IF NOT END CODE, STORE & WAIT
          RET                     ; RETURN TO CALLER
STORE     EQU $
          MOV OLDCODE,CODE        ; SAVE CODE AS OLDCODE
          JMP CODELOOP            ; AND WAIT FOR THE NEXT CODE
;
; END OF CODE LOOP ROUTINE
;
;
; THIS ROUTINE RECEIVES THE NEXT VALID CODE.
; INVALID CODE GROUPS ARE IGNORED
;
GETCODE   EQU $
WAITAGC   EQU $                   ; WAIT FOR AGC BURST
          CALL SAMPLE             ; GET THE VALUE AT THE INPUT PORT
          CJNE A,#ZERO,WAITAGC    ; WAIT UNTIL IT IS A 0
ONEWAIT   EQU $
          CALL SAMPLE             ; GET ANOTHER VALUE
          CJNE A,#ONE,ONEWAIT     ; IF NOT ONE, WAIT FOR A 1
          MOV B,#1                ; SET DELAY TO 250 US
          CALL WAIT               ; AND WAIT FOR THAT TIME
          CALL SAMPLE             ; RE-TEST INPUT BIT
          CJNE A,#ONE,WAITAGC     ; IF NOT 1, FALSE AGC BURST.
          MOV B,#3                ; SET UP FOR 3 * 500 US
AFTERAGC  EQU $
          PUSH B                  ; SAVE ON STACK
          MOV B,#2                ; SET DELAY TO 500 US
          CALL WAIT               ; AND WAIT FOR THAT TIME
          CALL SAMPLE             ; CHECK INPUT VALUE
          POP B                   ; RECOVER COUNT VALUE
          CJNE A,#ZERO,WAITAGC    ; IF 1, FALSE WAIT AFTER AGC
          DJNZ B,AFTERAGC         ; CHECK IF DONE WITH WAIT
;
; RECEIVE START BIT
;
STARTBIT  EQU $
          CALL SAMPLE             ; GET INPUT BIT
          CJNE A,#ONE,STARTBIT    ; IF NOT 1, TRY AGAIN
          MOV B,#1                ; SET DELAY TO 250 US
          CALL WAIT               ; AND WAIT FOR THAT TIME
          CALL READBIT            ; DO PCM BIT INPUT
          JC WAITAGC              ; IF INVALID PCM, START OVER
          MOV A,IN1               ; GET BIT VALUE
          CJNE A,#ONE,WAITAGC     ; IF NOT 1, BAD START BIT
;
; READ CODE BITS
```

```
;
READCODE EQU $
         MOV CODE,#00          ; RESET CODE BYTE
         MOV B,#6              ; DO 6 BIT INPUTS
READLOOP EQU $
         PUSH B                ; SAVE B VALUE
         MOV B,#2              ; WAIT 500 US
         CALL WAIT             ; AND WAIT FOR THAT TIME
         CALL READBIT          ; GET PCM BIT VALUE
         JC BADCODE            ; IF INVALID PCM, THROW AWAY CODE
         CLR C                 ; CLEAR CARRY FLAG
         MOV A,IN1             ; GET PCM BIT VALUE
         CJNE A,#ZERO,LOAD1    ; IF BIT=1 THEN DO 1
         JMP ROTATE            ; ELSE JUST ROTATE
LOAD1    EQU $
         SETB C                ; SET CARRY FLAG
ROTATE   EQU $
         MOV A,CODE            ; GET CODE BYTE
         RRC A                 ; ROTATE XTER RIGHT THROUGH CARRY
         MOV CODE,A            ; STORE IN CODE REG
         POP B                 ; RECOVER BIT COUNT
         DJNZ B,READLOOP       ; IF NOT ALL BITS, DO AGAIN
         RR A                  ; ROTATE TWO MORE
         RR A                  ; BITS RIGHT
         MOV CODE,A            ; STORE IN CODE REG
         CLR C                 ; CLEAR CARRY FLAG
         JMP CODEEND           ; AND EXIT
BADCODE  EQU $
         POP B                 ; RECOVER B REG FROM STACK
         SETB C                ; BAD CODE INDICATOR
CODEEND  EQU $
         RET                   ; RETURN TO CALLER
;
; END OF CODEEND ROUTINE
;
;
; THIS ROUTINE RECEIVES A VALID PCM BIT
; CARRY IS SET IF THE PCM ENCODING IS NOT VALID
;
READBIT  EQU $
         CALL SAMPLE           ; GET INPUT VALUE
         MOV IN1,A             ; STORE IN IN1 REG
         MOV B,#2              ; WAIT ANOTHER 500 US
         CALL WAIT             ; AND WAIT FOR THAT TIME
         CALL SAMPLE           ; GET ANOTHER SAMPLE
         CJNE A,IN1,GOODPCM    ; IF 2 INPUTS <>, GOOD PCM CODE
BADPCM   EQU $
         SETB C                ; SET CARRY FLAG (BAD PCM)
         JMP READEND           ; AND EXIT
GOODPCM  EQU $
         CLR C                 ; CLEAR CARRY FLAG
READEND  EQU $
         RET                   ; RETURN TO CALLER
;
; END OF READBIT ROUTINE
;
;
; THIS ROUTINE SAMPLES THE INPUT BIT
;
SAMPLE   EQU $
         MOV A,INPORT          ; READ BYTE AT PORT
         ANL A,#ONE            ; MASK OFF OUR BIT
         RET                   ; AND RETURN TO CALLER
;
; END OF SAMPLE ROUTINE
;
```

Continued on page 40

```
10 ONEX1 40
20 A=GET : IF A<>0 THEN
   PRINT A
30 GOTO 10
40 PRINT"IRCOMM INTERRUPT
   RECEIVED" : GOSUB 100
50 PRINT"INTERRUPT
   PROCESSED"
60 RETI
100 CALL4000H
110 PH0. DBY(1FH)
120 RETURN
```

Extending the Capabilities

With most infrared remotes, the range is limited to about 25 to 30 feet. Extending the range beyond that involves more powerful transmitters and more sensitive receivers. While I was considering doing just that, I came across a Radio Shack product called the Video Remote Control Extender that seems to adequately solve the problem.

The Extender (catalog number 15-1289) is an IR repeater. As shown in figure 6b, it consists of an infrared receiver, amplifier, and transmitter. One end, located in a room where you might also use the IRCOMM, is the receiver and the amplifier. At the other end is a splitter box with an attached infrared LED. The splitter is connected to the receiver with antenna wire. When a pulse burst is received, it is amplified and conveyed through the antenna wire to the splitter box where it is retransmitted via that IRLED. It is easy to visualize the HCS mounted in the Circuit Cellar with the IRCOMM and BCC-52 next to it. The Extender would be upstairs in the entertainment room.

More than one Extender can be used (I have three on the IRCOMM). Since their outputs are IRLEDs aimed at the IRCOMM receiver, Extenders from other rooms can also be used with their IRLEDs mounted next to each other all aimed at the IRCOMM.

In Conclusion

I'm becoming as dependent now on the IRCOMM as I am on my automatic lighting. With it, the entertainment room comes alive in a programmed and orderly manner. Without it and the HCS, I bump into cold, dark walls of a house bathed in utter silence.

Indeed, I've solved the problem of coordinating the control of the multitude of electronic boxes in the entertainment room, but I've created an overabundance of IR remotes. The result is that I've created a terminal condition of HIROP.

In Chapter 4 we'll throw away the IRCOMM and the rest of your remotes and replace them with a single Circuit Cellar IR Master. The IR Master is a trainable remote controller that has the capacity to retain the

```
; THIS ROUTINE DELAYS 250 MICROSECONDS FOR EACH
; COUNT IN THE B REGISTER
;
WAIT       EQU $
           PUSH B                ; SAVE B REG COUNT
DELAY250   EQU $
           MOV B,#DELAYCNT       ; LOAD WITH DELAY COUNT VALUE
LOOP250    EQU $
           DJNZ B,LOOP250        ;  BURN UP CYCLES
           POP B                 ; RECOVER COUNTS
           DJNZ B,WAIT           ; IF NOT DONE, WAIT MORE
           RET                   ; RETURN TO CALLER
;
; END OF WAIT ROUTINE
;
; SYSTEM EQUATES
;
; REGISTER ALIASES
;
IN1        EQU 18H               ; FIRST HALF PCM INPUT
CODE       EQU 19H               ; CURRENT CODE
OLDCODE    EQU 1FH               ; LAST CODE READ
;
; CONSTANTS
;
ZERO       EQU  0                ; ZERO BIT VALUE
ONE        EQU  4                ; PORT 3, BIT 2 MASK VALUE
DELAYCNT   EQU  111              ; 250 US CONSTANT
ENDCODE    EQU  62               ; END CODE FROM TRANSMITTER
INVALID    EQU  63               ; INVALID CODE SETTING
;
; HARDWARE PORTS
;
INPORT     EQU P3                ; MEMORY MAPPED INPUT PORT
;
; END OF GLOBAL EQUATES
;
; END OF LEDPCM PROGRAM
;
ZZZZ       EQU $
           END
```

command functions of up to 16 (yes, 16!) independent IR remotes. It uses a scrolling LCD to indicate command function and control device. A single execute ("DO IT") key is the only command button.

Special thanks to Bill Curlew for his software expertise.

The BCC-52 Computer/Controller is available from Micromint Inc., 4 Park St., Vernon, CT 06066, for $189 plus shipping. To order, call (800) 635-3355; for information, call (203) 871-6170.

There is an on-line Circuit Cellar bulletin board system that supports past and present projects. You are invited to call and exchange ideas and comments with other Circuit Cellar supporters. The 300/1200/2400-bps BBS is on-line 24 hours a day at (203) 871-1988.

Editor's Note: Steve often refers to previous Circuit Cellar articles. Most of these past articles are available in book form from BYTE Books, McGraw-Hill Publishing Company, P.O. Box 400, Hightstown, NJ 08250, (1–800–2–MCGRAW).

Ciarcia's Circuit Cellar, Volume I covers articles in BYTE from September 1977 through November 1978. *Volume II* covers December 1978 through June 1980. *Volume III* covers July 1980 through December 1981. *Volume IV* covers January 1982 through June 1983. *Volume V* covers July 1983 through December 1984.

4

BUILD A TRAINABLE INFRARED MASTER CONTROLLER

This device can control all your home entertainment equipment

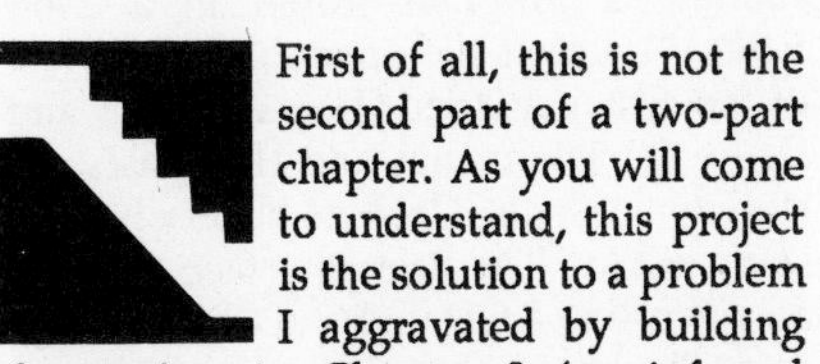

First of all, this is not the second part of a two-part chapter. As you will come to understand, this project is the solution to a problem I aggravated by building the project in Chapter 3 (an infrared remote control for my home control system). Confused? Let me explain.

While people residing in warm climates tend toward Jacuzzis and hot tubs, some of us who live in colder climates prefer not to tempt fate and brave the elements for about six months of the year. Of course, I could succumb to the winter sports thing. You know, skiing, skating, snow this, and snow that, but it would be much too great a chore at this stage to reorient my sedentary lifestyle to enjoy northeast winters. I hibernate like most indigenous mammals and wait for the color outside the window to metamorphose from white to green.

About a year ago, I decided that holing up in the cellar for six months a year was antisocial. While the isolation proved beneficial in coming up with great projects for the summer and fall issues, I did find that by the time March rolled around, I looked very much like a bear that was leaving his cave, and I communicated just about as well.

In an attempt to improve the quality of winter life and break the cycle of hibernation, this last year I decided to spend some of the time aboveground (upstairs) in an environment that allowed me to observe the realities of my existence (through the windows) and absorb the cumulative knowledge of our culture (watch TV).

In layman's terms, I built a media room. Not just a TV den, mind you, but a room where I could be immersed in a synthesized environment so far from the ice and snow that six months seemed like overnight. Of course, this audiovisual experience was tastefully produced by massive amounts of electronic equipment.

The beautiful scene of the tropical island was accurately reproduced on a Kloss 2000 projection TV. You'd think you were sitting next to that tinkling waterfall as the music moves above and around you in complete surround sound. And when the warm breeze of island spring (actually the heat wafting from the seven amplifiers) is tumultuously interrupted by a hurricane faithfully reproduced with 2400 watts of Nakamichi audio power through a pair of B&W 808s (180 pounds each), two Speakerlab subwoofers, and 11 Canton surround speakers, you feel like the walls are about to explode. Sometimes it is good not to have neighbors.

Enough of warm breezes. I now had a new problem. In addition to all the audiovisual stuff, there were a couple of VCRs, an FM tuner, and a CD player. All this equipment required the 14 remote handheld controls shown in photo 1. Media rooms are a great idea, but you can't expect people to glue a dozen remotes on a long board. There had to be a better way.

IR Master Controller to the Rescue

This chapter's project, an infrared Master Controller that takes charge of all your gadgets, can prevent "controller clutter." It "learns" the infrared signals for each function and plays them back on command. It uses a six-button keypad to select the device and functions, shown on a two-line LCD, and a single button, Do It, to execute what's selected.

I am not the first person to design a trainable remote control. More than a year ago, I bought a similar device made by General Electric, called Control Central. This device could be trained to simulate the functions of four remotes.

Control Central and similar commercial units have two major shortcomings. First, all the acquisition, data-reduction, processing, and memory circuitry is contained in the single hand-held unit. Given the finite physical size of today's integrated circuitry, there is a limit to the capacity of such a device that allows it to still be cost-effective. Second, it is designed for use by a mass audience assumed to have a finite set of electronic devices. The buttons have predesignated nomenclature, so it is not user-programmable.

You can still train your GE controller to simulate the remote control for your CD player. The Mute button on the GE unit could be trained to be the Auto Repeat on your CD player remote, for example. Unfortunately, every time you want to repeat a CD, you'll have to remember to press Mute since there is no Repeat button on the GE.

I am not criticizing the GE Control Central. I am merely making a case for designing something different for a very vertical, gadget-happy, affluent audience: BYTE readers. Why tie a design to the lowest common denominator. Instead, yell "let them eat cake" and demand the remote to end all remotes: the Circuit Cellar Master Controller!

The shortcomings of the GE and other trainable remotes are the strengths of the Master Controller. Rather than attempt to contain all the necessary intelligence and processing circuitry, the Master Controller temporarily utilizes an external computer

Infrared controllers are not compatible because each manufacturer speaks in a different language.

as a user-programmable interface. Also, rather than having buttons with fixed-function nomenclature, the Master Controller incorporates a scrolling LCD to identify unit designations (devices) and functions (commands). Device designations like "Bedroom VCR" or "Nakamichi preamp" and commands like "CD repeat/all" or "slow motion" are used instead of remembering what the Mute button was supposed to do.

The Master Controller uses an IBM PC for training. After that, it is battery-operated and completely independent. The IBM PC is connected to the Master Controller via an RS-232 interface and is used to set up menus of devices (receivers, CD players, tape decks) and functions for each device (turn on, play forward, etc.). After a menu is downloaded to the Master Controller, each function is "taught" and tested. Next, the completed menu and synthesis data are then uploaded to the IBM PC and stored on disk (in case you want to load it into another Master Controller or add another device later without retraining all of them).

The Master Controller's IBM PC program can also combine sets of infrared signals once they are trained for their respective devices. I can now use a single Master Controller button to turn on the audio system, route the output to the living room, select the CD player, move to the third selection, and repeat it forever. Compared to other commercial controllers, the Master Controller solves the IR remote plague hands down. Because it uses an external computer for functional modifications by the user, more room is available for its ultimate task. Instead of four remotes, the Master Controller can be trained to simulate the functions of 16 individual remote controls complete with descriptive command designations.

An Infrared Introduction

Most infrared remote controls are functionally similar. The microprocessor in the remote controller creates a stream of bits that is turned into on/off pulses of IR light from an IR LED. An IR-sensitive photodiode in the receiver turns the light pulses back into an electrical signal from which the original bits can be extracted.

The IR LED's fast on-and-off action creates a carrier signal. The carrier is then turned on and off to form the individual bits of the message. Each controller uses a different carrier frequency, sets different bit timings, and assigns different meanings to the bits in the message. The reason that controllers are not compatible is that there is no standard for the format of the bits in the message. Each manufacturer speaks in a different language.

The Master Controller sidesteps this problem by simply recording and playing back the infrared signals without attempting to decode the messages. It's just like a tape recorder. You can record English, Russian, and Spanish on the same tape because they all occupy the same frequency bands, and you don't have to understand the languages to play them back.

This scheme works because of a limited range of differences in the IR signals. The controllers I've tested had carrier frequencies ranging from 32 to 48 kilohertz. Each message bit has between 10 and 30 carrier cycles, and there are two different carrier modulation systems: pulse-width modulation and pulse-position modulation. In Chapter 3 I explained pulse-width modulation. Pulse-position modulation works by determining the time when a bit occurs relative to a fixed starting point.

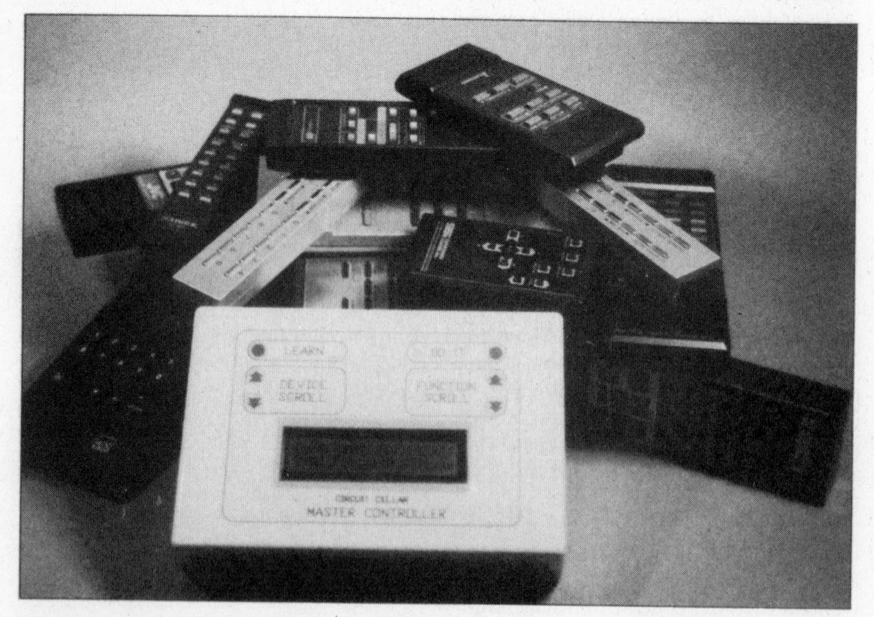

Photo 1: *The six-button Master Controller can duplicate the functions of the 14 controllers shown in the background.*

The Heart of the Master

An Intel 8031 single-chip microprocessor running a program stored in a 2764 EPROM directs the operation of the rest of the circuitry. (See photos 2 and 3 and figure 1.) The menus and IR signals are stored in a single 32K-byte battery-backed static RAM. The user interface consists of a two-line LCD and a six-button keypad. The keypad is either a simple membrane matrix or individual keys arranged in a matrix that is scanned by the 8031. This eliminates the need for a keyboard encoder. Two keys each are used for device and function up/down scrolling on the LCD. A fifth button, Do It, executes the device/function command appearing on the display. A sixth button, Learn, is used for training.

The LCD has 20 characters on each of two lines. The interface to it requires only six wires: four data bits, one timing strobe, and an address line. The display's internal character generator converts ASCII data into character dots, so the 8031 can communicate directly in ASCII.

A TIL413 photodiode converts the IR signals from other remote controllers into a discernible logic signal. Because the Master Controller and the remote are placed close together during training, there is no need for the sophisticated signal processing that's required to detect weak IR signals across a room. The Master Controller photodiode circuitry was designed to accept strong IR signals only. You should position the remote-control unit within a few inches of the Master Controller's photodiode. If it's too far away, the Master Controller will "see" nothing. If it's too close, the Master Controller will receive a distorted signal. A little experimenting with each controller will locate the correct position.

An LM311 comparator converts the photodiode input signal to a TTL-level signal. A 74LS164 shift register samples the output from the LM311 at a 1-megahertz rate and converts the data into parallel format. The 8031 reads the shift register every 8 microseconds while it is learning a new IR signal. This data is stored in RAM for later analysis.

At the transmitting end, the process is reversed. Although the 8031 is a fast microprocessor, it cannot generate both the carrier and bit timing of the IR signals instantaneously. To lighten the processing overhead, an 8254 programmable interval timer controls the IR carrier frequency and duty cycle as well as the duration of each message bit. The 8031 sets up the 8254's registers for each bit of the IR message.

A pair of TIL39 infrared LEDs produce the IR signal. Because the human eye cannot see IR light, a visible LED is connected in parallel as an indicator. The LEDs are switched by a field-effect transistor driven by a standard logic gate. The FET is an efficient way to interface logic levels with real-world devices because it directly translates an input voltage into an output current.

Power

Power is an important consideration in any battery-operated device. The Master Controller was designed to use either 74LS or 74HC devices at 5 volts. The 5 V is derived from a 6-V battery (four AA cells) using a special low-dropout voltage regulator. While LS takes considerably more power than HC, the duty cycle is low. The Master Controller need only be powered up long enough to set the device and function and press Do It. It can be shut off afterward. Admittedly, I could have spent more time developing automatic power up/down circuitry, but it would have complicated the design and added more software. Feature-specific circuit tailoring will have to wait.

Turning the power on and off is not a problem. The 8031's system software is contained in a 2764 EPROM, and the LCD and IR data are contained in battery-backed RAM. The memory is a 32K by 8-bit static low-power CMOS RAM chip. The backup circuit consists of two 3-V lithium batteries and a Dallas Semiconductor DS1210 battery-backup controller chip. The DS1210 senses loss of the +5-V supply voltage and automatically write-protects the RAM as it switches power to the battery. The second battery is necessary only if the first one fails.

Signal Processing

As you can see from the schematic in figure 1, most of the Master Controller's functions are done in software. It's worthwhile to look more closely at the processing required for the learning and reproduction of the IR signals. The IR carrier frequency is about 40 kHz, giving a period of about 25 μs. The particular frequency used by a controller must be measured precisely because each microsecond of error changes the reproduced frequency by about 4 percent. While this doesn't sound like much, when the Master Controller reproduces the IR signal, the receiver could completely ignore it. The reason for this is that the IR receivers in consumer electronic gear must detect faint IR signals.

Generally, the receivers use a phase-locked loop, tuned to the remote unit's carrier frequency. The PLL can handle a 10 to 20 percent frequency error, but the design margins include errors due to temperature, voltage, and other effects.

Photo 2: *Component side of the Master Controller, showing the 2764 EPROM and 8031 CPU (center left and right). The backup batteries and Dallas Semiconductor DS1210 battery controller are in the bottom center and right.*

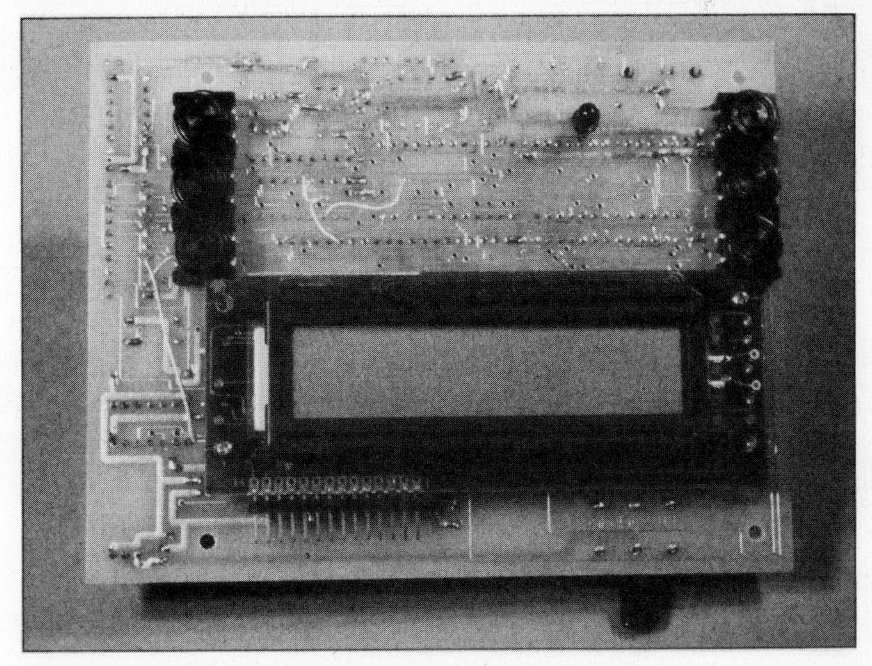

Photo 3: *The etch side of the Master Controller showing the 2-line by 20-character LCD and six control buttons.*

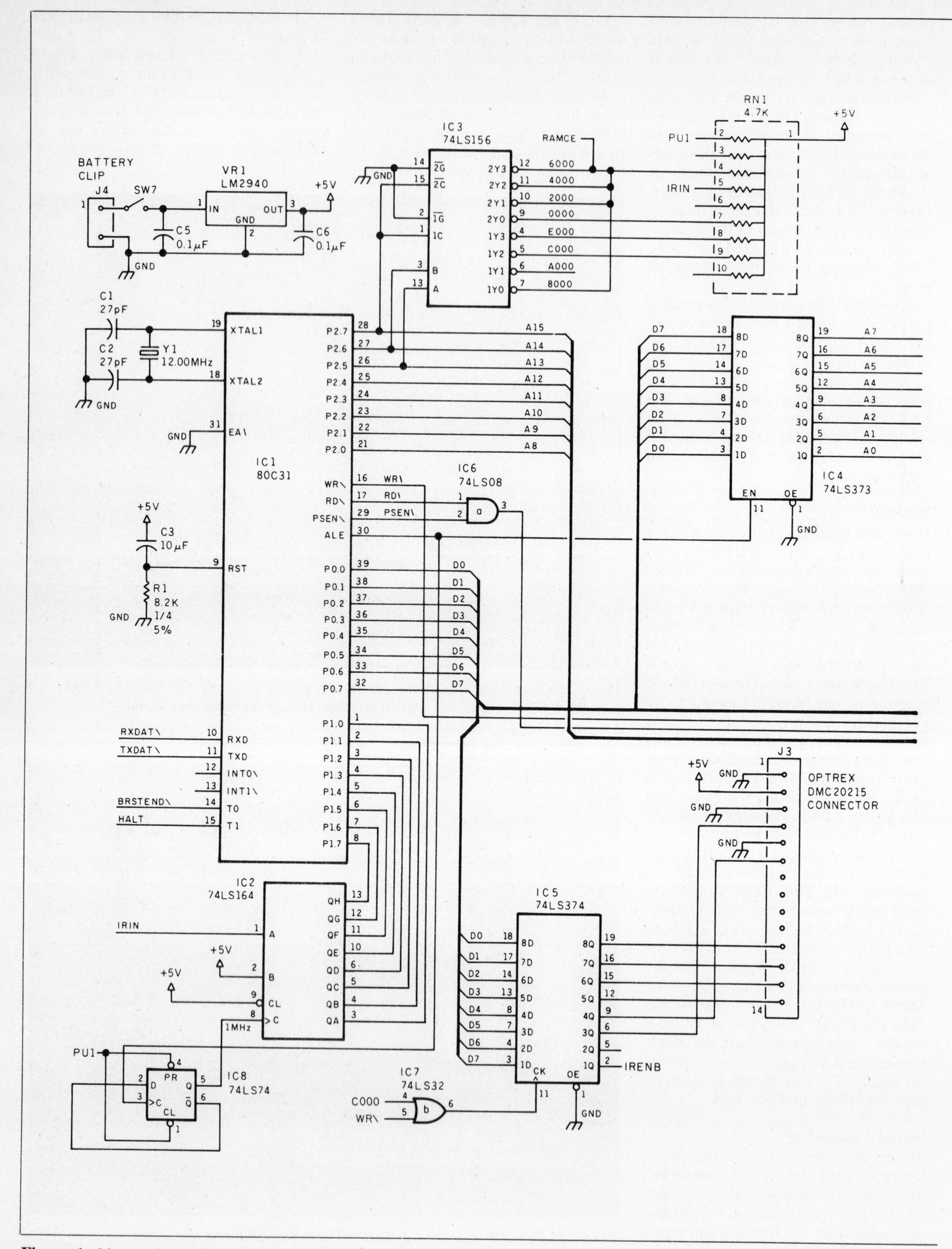

Figure 1: *Master Controller schematic diagram.*

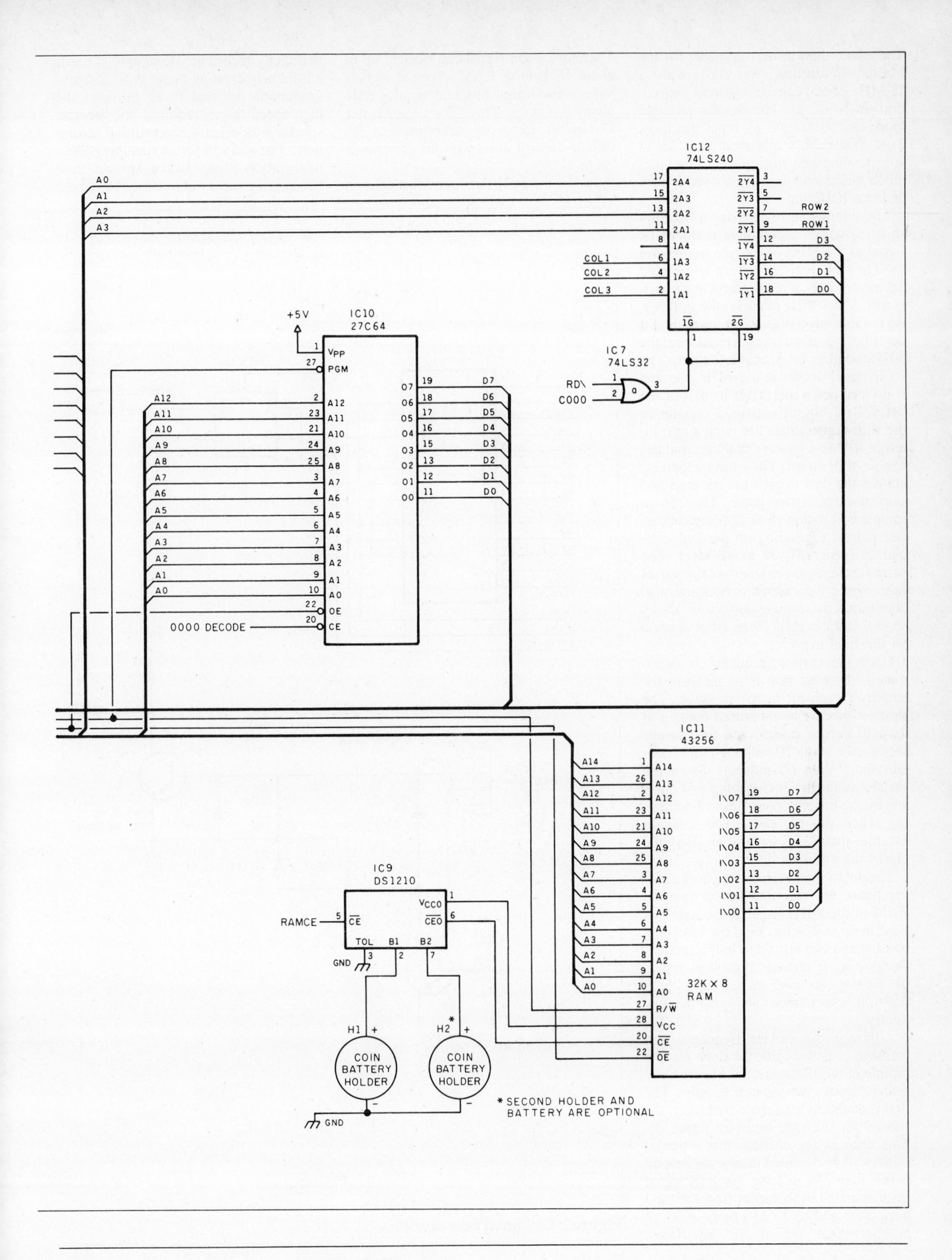

IC12
74LS240
A0
A1
A2
A3
2A4
2A3
2A2
2A1
1A4
1A3
1A2
1A1
1G
2G
2Y4
2Y3
2Y2
2Y1
1Y4
1Y3
1Y2
1Y1
ROW2
ROW1
D3
D2
D1
D0
COL1
COL2
COL3
IC7
74LS32
RD\
C000
+5V
IC10
27C64
VPP
PGM
A12
A11
A10
A9
A8
A7
A6
A5
A4
A3
A2
A1
A0
OE
CE
0000 DECODE
O7
O6
O5
O4
O3
O2
O1
O0
D7
D6
D5
D4
D3
D2
D1
D0
IC11
43256
A14
A13
I\O7
I\O6
I\O5
I\O4
I\O3
I\O2
I\O1
I\O0
32K × 8
RAM
R/W
VCC
CE
OE
IC9
DS1210
RAMCE
VCCO
CEO
TOL
B1
B2
GND
H1
H2
COIN BATTERY HOLDER
COIN BATTERY HOLDER
* SECOND HOLDER AND BATTERY ARE OPTIONAL

The whole margin isn't available for the Master Controller. An 8031 with a 12-MHz crystal can execute most instructions in 1 or 2 μs. The shortest possible loop used to sample an input pin takes 2 μs. While this might seem very fast to you, reading and storing the value takes much longer and necessitates some form of input buffering.

The solution involves using a 74LS164 shift register to accumulate 8 bits of IR signal at a 1-MHz clock rate (see figure 2). The shift register in turn is sampled once every 8 μs, a requirement that's easily met. The 1-MHz clock for the shift register (and also the 8254, which I'll describe in a moment) is derived from the 8031's address latch enable output. The ALE signal occurs at a 2-MHz rate and is divided down to 1 MHz by half of the 74LS74 flip-flop. Exactly 32 samples of the shift register data are copied into internal RAM, a process that accumulates 256 μs of IR signal. The software then examines the data to pick out the start and stop of each carrier pulse. The 256-μs sample will include 8 to 12 complete carrier pulses, depending on the exact frequency (generally, 32 to 48 kHz). The software averages the length of the pulses to compute the carrier period and also determines the average duty cycle. Using several samples reduces the effect of noise on the final average.

Once the carrier frequency is determined, the next step is to measure the length of each bit in the message. The shortest bits we've measured contain at least 10 carrier pulses, and the average seems to be about 20 (although some contain more than 60 pulses). Given the variability in carrier frequency and pulse length, the main problem lies in determining when the bit ends. The software assumes that 32 μs without an IR signal marks the end of a bit.

The duration of each bit and the following pause are recorded in the external RAM of the 8031. Each IR message can contain up to 256 bits (and the following pauses) and can last up to half a second. Because most remote controllers repeat the message as long as the key is held down, it is very important to tap that key lightly.

Although the carrier frequency and message analysis could be done on one sample of the IR signal, the Master Controller requires two separate samples. The first is analyzed for carrier frequency, the second for message bits (see figure 3). This reduces the chances that a partial signal will be recorded in case the first bit is less than 256 μs long. Each IR signal is summarized by its carrier frequency and duty cycle and up to 256 pairs of 16-bit numbers that record the bit times. Therefore, each signal can occupy up to about 1K byte of RAM. Typical signals have a few dozen bits and require only about 100 bytes. This allows the Master Controller to easily accommodate 16 remote-control units with 16 commands each.

Just as the 8031 isn't quite fast enough to directly record the IR signals, it needs a little help creating them. An 8254 programmable interval timer provides the high-speed logic required to generate signals with microsecond timing resolution. The 8254 PIT contains three identical timers that can be set up in a bewildering variety of modes. The Master

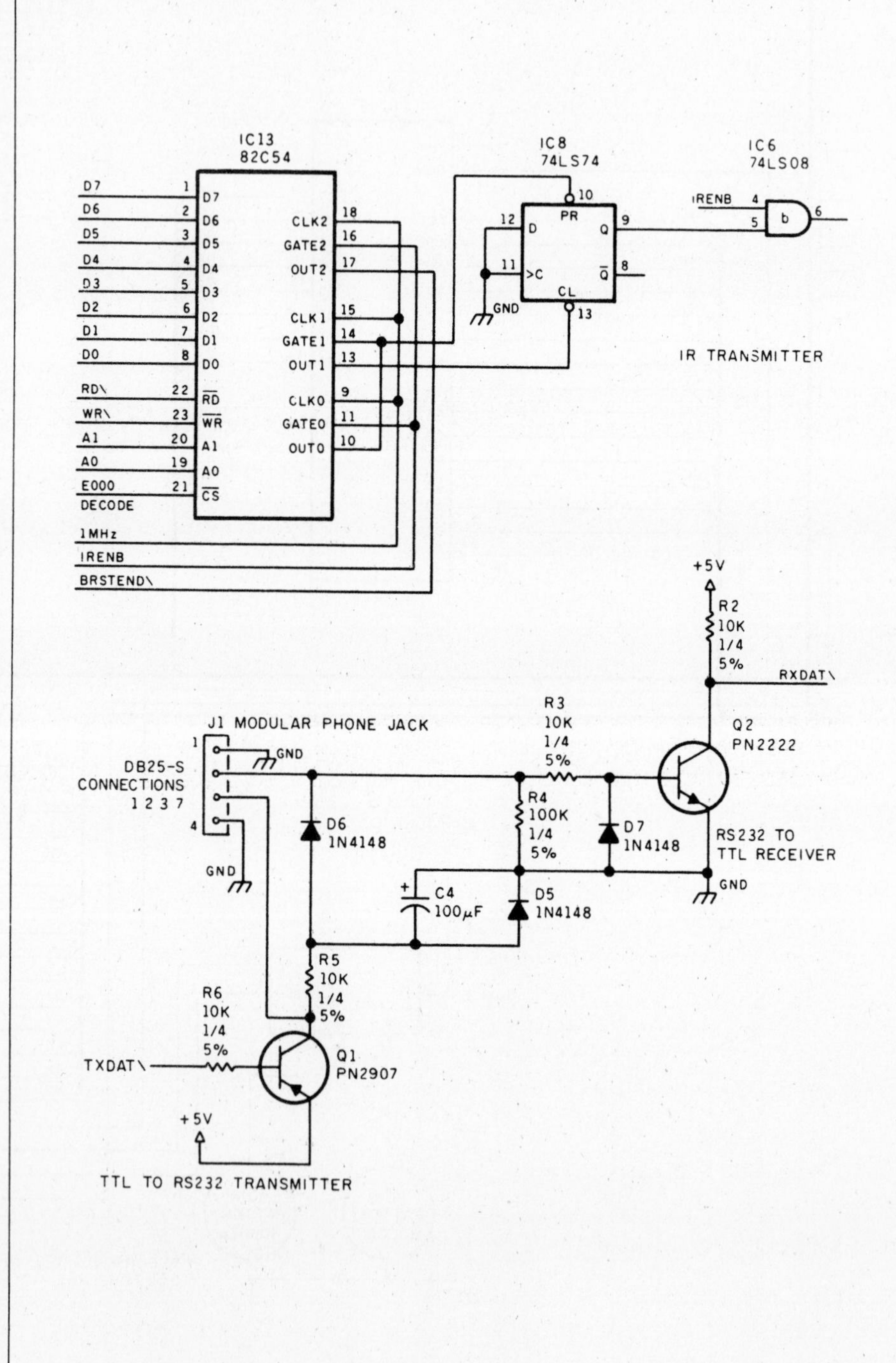

Figure 1: *Continued from page 45.*

Controller software uses all three of them, as well as a timer inside the 8031, during the IR playback.

Timers 0 and 1 in the 8254 (IC13) set the IR carrier frequency and duty cycle, respectively. Timer 2 determines the duration of each message bit, and the 8031 timer controls the pause following each bit. The first two timers are set once at the beginning of the message, while the last two are set for each bit. The times are stored in external data RAM accessed only when the 8031 is running.

Because the 8254 produces a pulse only at the end of each timer's count, a 74LS74 (IC8) is used to create the actual IR pulses. The Timer 0 pulse (pin 10 of IC13) sets the LS74's output at the start of the carrier cycle, with Timer 1 (pin 13 of IC13) resetting the output at the end of the carrier pulse. The LS74's output is combined with a gating signal (IRENB) and sent to the IR LED drivers. As you can see, a great deal of code is required to handle the IR signal analysis. The code is written in 8031 assembly language.

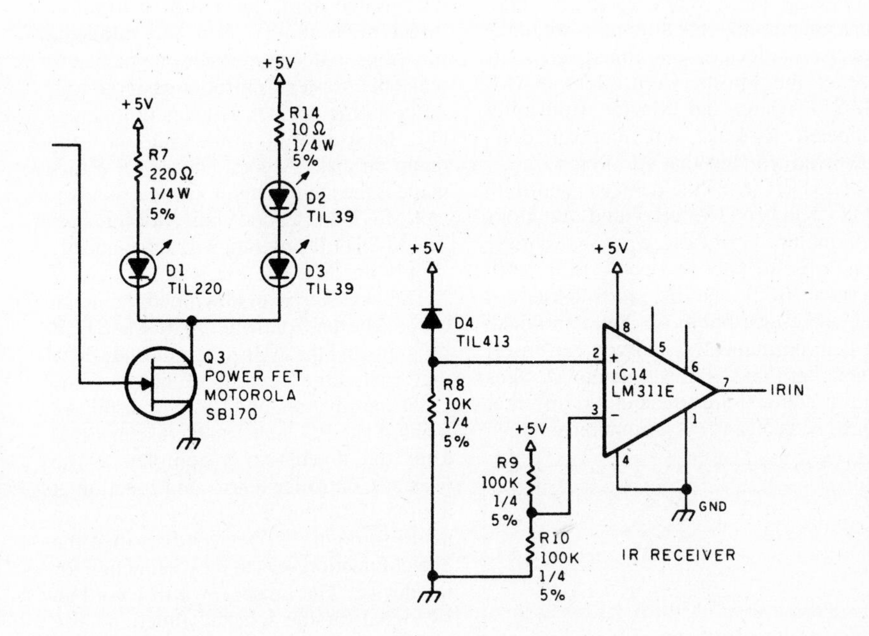

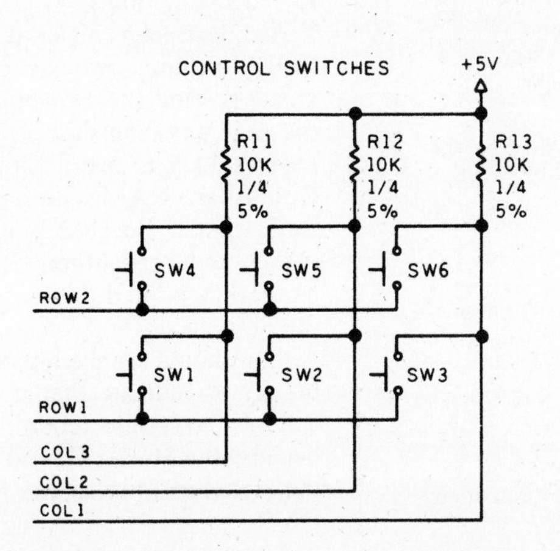

The PC Connection

Using an IBM PC to create menus for the Master Controller may seem like overkill, but it really simplified the logic. The PC has a full keyboard and display, disk storage, and, best of all, high-level programming languages. Writing a PC program is much easier than writing an 8031 program, so I decided to put as little code in the Master Controller as possible.

The PC program (called MASTER) provides three main operations: creating and editing menus of devices and functions, saving and loading these menus in disk files, and transferring them to and from the Master Controller. (See photos 4, 5, and 6.) The Master Controller is connected to the PC only when uploading and downloading menus. Under normal use, the Master Controller doesn't have any wires trailing out of it.

I'll have to admit to taking a little poetic license in the design of the RS-232 circuit. Rather than include a separate negative-voltage power supply for the RS-232 signal levels, I used a diode and capacitor to "borrow" the negative voltage from the PC's transmitted data line. A pair of transistors are simpler than the power-hungry level converters normally used to translate between logic levels and RS-232 levels. This is an important consideration in battery-operated devices. Because I know the communication will always be with an IBM PC, a worst-case, tolerant RS-232 circuit is not a necessity.

The serial connector is an RJ-11 telephone jack instead of the usual 25-pin DB-25 connector. Only three wires are required: data from the PC, data to the PC, and signal ground. Because the MASTER program and the Master Controller were designed together, they use an efficient method of passing data that doesn't require the normal RS-232 RTS/CTS and DSR/DTR status lines.

The 8031 serial interface includes a bit-rate generator. The exact bit rate depends on a number programmed into a register as well as the frequency of the 8031's clock crystal. I used a 12-MHz crystal to get the highest resolution possible for the IR signal-processing circuitry, but that's not the optimum crystal for the serial interface. As a result, the 8031 transmits data

The MASTER program, which is written in Turbo Pascal, can create menus for up to 16 devices.

to the PC at 10,417 bits per second. If you're familiar with normal RS-232 data rates, you'll recognize that 10,417 isn't one of the choices. Fortunately, the PC's serial port also determines the bit rate from a number in a register. The closest match is 10,473 bps, but everything works just fine.

Here I should point out that, although the link between the Master Controller and the PC is called RS-232, it is surely not standard. For example, using mismatched bit rates and voltage levels is acceptable only if you've carefully checked out the consequences and assured yourself that both ends of the connection are still compatible. If you elect to build a Master Controller, you shouldn't try to stretch the limits of the connection too much. A 50-foot cable probably won't work at all!

The Master in Action

Perhaps an example of how to use the Master Controller is in order. I'll show how to set up the first menu, then how to combine IR signals to produce customized effects.

The MASTER program, written in Turbo Pascal, can create menus for up to 16 devices: receivers, CD players, tape decks, and so on. Each device can have up to 16 functions (on/off, play, rewind, volume up, etc.). While up to 256 functions are possible, the ultimate limit to the number of devices and functions is the size of the Master Controller's RAM. MASTER and the Master Controller cooperate to make sure that you don't download a menu that's too big.

MASTER treats the devices (Bedroom VCR, Kitchen TV, etc.) and functions (volume up, power on, etc.) as a collection of lists. Function keys let you "cut" an item from one list and "paste" it elsewhere (see photo 4). You can delete an item permanently, and you can insert a new item and give it a name. Devices and functions are treated as different items, so you can't cut a function and then paste it into the device list.

Table 1 shows the complete list of MASTER function keys on the IBM PC. As an example, you might use the MASTER screen for the Sony RM-S750 controller menu. The Tape Deck line in the Devices list is highlighted, and the Functions list details all the tape deck's functions. The word "new" at the end of each function indicates that the IR signals have not been learned yet (see photo 6).

The Remote Keyboard item in the Devices list contains the general functions like power on/off and volume up/down as well as the digits from 0 to 9 to allow direct radio tuning. You can duplicate functions under more than one device to make the Master Controller easier to use. After all the devices and function names have been entered, you should save the menu on disk. A good choice for a filename is the manufacturer's model number, so RM-S750 is a good choice for this one. The MASTER program will automatically supply an .MC file extension.

The next step is to download the menu to the Master Controller. The MASTER program and the 8031 program first verify that each other exists, exchange some status information, and finally transmit the menu. (The PC cable can be disconnected after the download is complete.) The Master Controller learns one function at a time. Use the Select Device keys to scroll through the Devices list, then the Select Function keys to pick a function for the device. The Learn key will record an IR signal for the selected function. You can test the signal and relearn it until it's correct, but you can learn only one signal for each function.

As I described earlier, the Master Controller requires two samples of the IR signal to find the carrier frequency and message bits. You should tap the remote's keys quickly to avoid filling the Master Controller's RAM with repetitions of the same signal. Because most remote controllers will repeat their IR signal as long as the key is held down, you should tap the Remote key and release it immediately. You should see the remote's LED blink briefly to indicate that it sent a signal. There's no point recording repetitions because the Master Controller will repeat the signal as long as you hold down the Do It key.

After the second tap, you can test the Master Controller's stored signal by aiming its IR LEDs at your VCR or TV system and pressing the Do It key. If the function works correctly, the Master Controller has a valid IR signal in RAM. You can repeat the learning process by tapping the Learn key again. Once you have a good signal stored in RAM, tap any key other than Learn or Do It to return to the

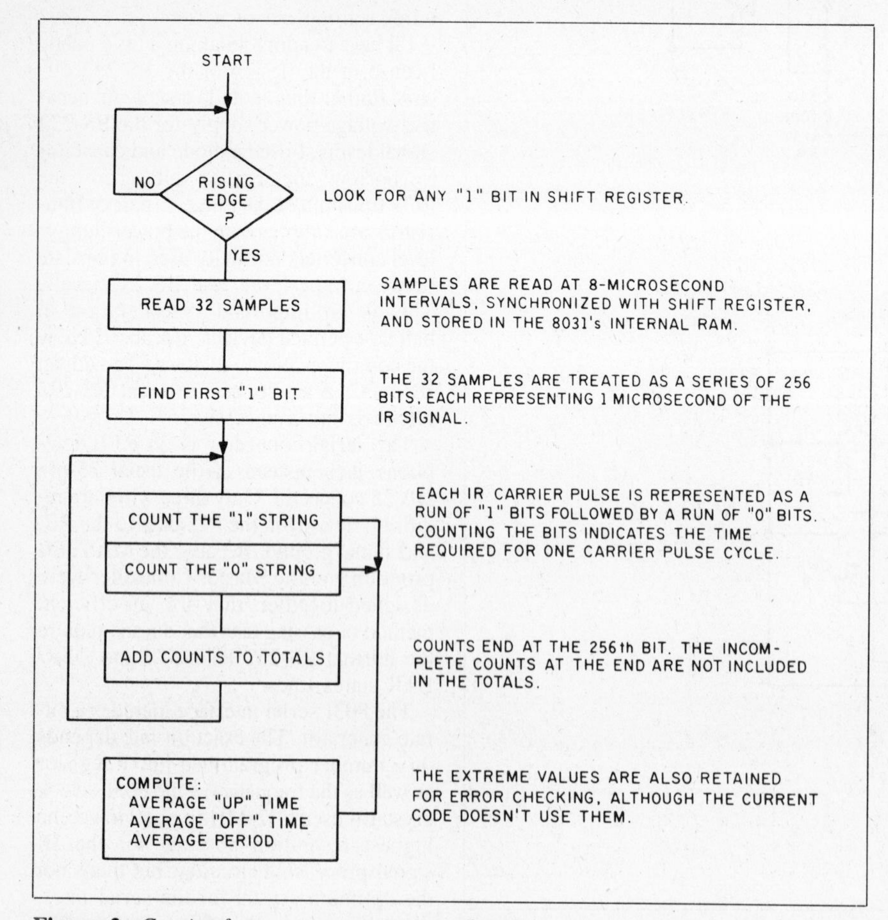

Figure 2: *Carrier-frequency measurement flowchart.*

normal display. Select the next function and repeat the learning process again. The cycle takes only a few seconds once you get the hang of it.

When all the functions are learned, reconnect the RS-232 connection to the PC and upload the menu using the MASTER program. The word "new" after each function is replaced with the length of the IR signal in bits. The Sony RM-S750 produces three repeats of 13 bits for each key press, so each function shows a length of 39. Other remotes will, of course, have different signal lengths.

Menu Modifications

At this point, you have a Master Controller menu that duplicates the functions of the RM-S750 as its first device. The MASTER program can combine the IR signals for two functions to produce the same effect as pressing two keys on the remote in sequence. This comes in handy for operations that you normally do in sequence, like turn on the power to the FM tuner and select your favorite station, for example.

To combine two IR signals, first "cut" one signal from the function menu by pressing F7. Then position the cursor over the other signal and press F10. Notice that the signal length is now the sum of the two old signals. You might want to use F9 to change the function name to reflect the new signal.

You can combine any number of function signals, with the only restriction being that the total length of the combined signals cannot exceed 1024 bits. The "cut" signal (or signals) is put at the end of the combined sequence. (Because the Master Controller uses a single carrier frequency for an entire IR signal, a combined signal may not work correctly if the signals came from different controllers.) MASTER will warn you if the carrier frequencies differ by more than about 10 percent but will allow you to shoot yourself in the foot. The Master Controller will use the first signal's carrier frequency for all the combined signals.

Some tape decks require pressing two keys (usually Record and Play) simultaneously to start recording. Generally, you can't get the same effect by pressing the Record key followed by the Play key. The reason is that the remote sends out a different message when the two keys are pressed simultaneously than it does for either of them separately. If you combine the Record and Play signals using Master, it won't work any better than the two separate keys will. You must "learn" the correct signal by pressing the two keys simultaneously. You've got to be quick on the keys to avoid filling the Master Con-

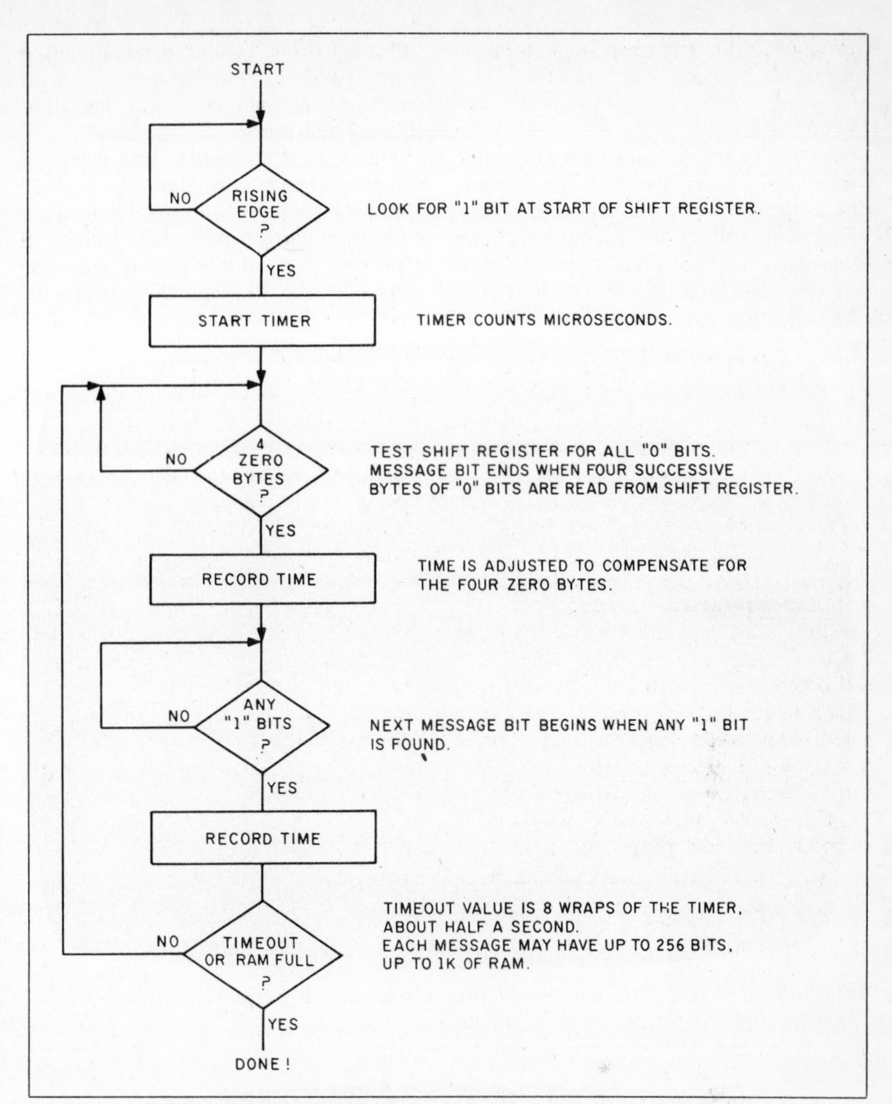

Figure 3: *Signal-capture flowchart.*

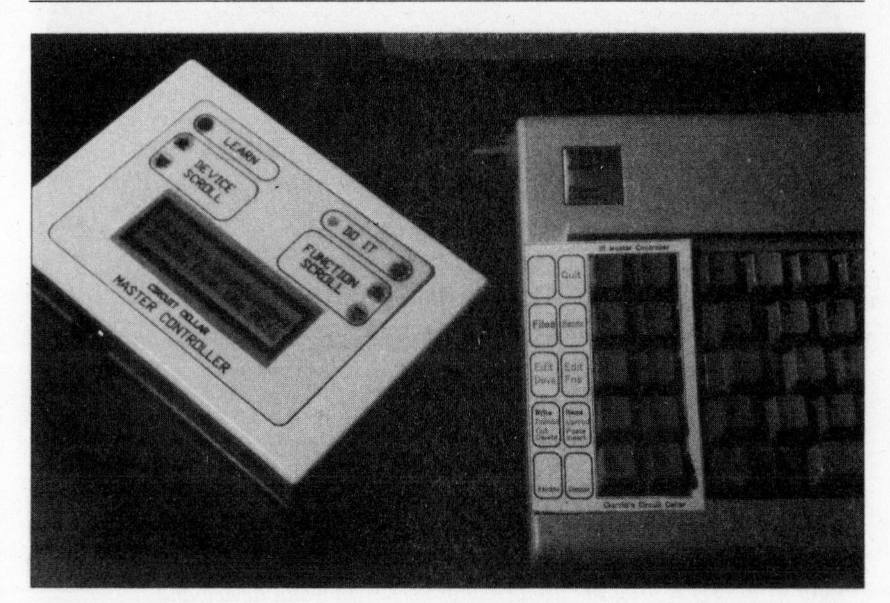

Photo 4: *The Master Controller and the IBM PC keyboard showing the MASTER program function-key template.*

troller's RAM with repetitions, though. As usual, practice makes perfect.

Conclusion

Once you've recorded the basic controller functions, you can use MASTER to combine them in wonderful ways. Although I've been calling the menu selections "devices" and "functions," you don't have to. You might wind up with a device called "Coming Home" with functions ranging from "Tired" to "Exhilarated" to turn on your system and select just the right lighting and music. Get the idea?

The Master Controller was designed and prototyped as a Circuit Cellar project. While it has some obvious and immediate consumer market potential, without a clear goal in mind it is hard to convince someone to go through the expense of manufacturing it (especially producing a custom enclosure). I have only a short time between projects, and I don't have the time to speculate on the eventual market niche or the specific configuration the Master Controller will take (Sharper Image, are you listening?). However, unless there is some way to evaluate the present device, another generation of the Master Controller will never be built.

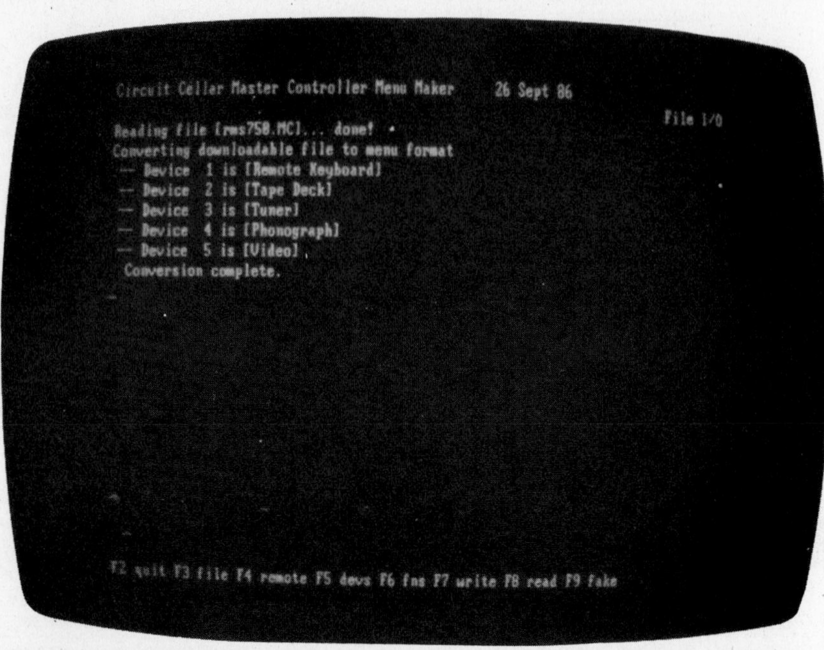

Photo 5: *MASTER program menu-maker screen with a five-device menu. Line 25 displays the function-key menu.*

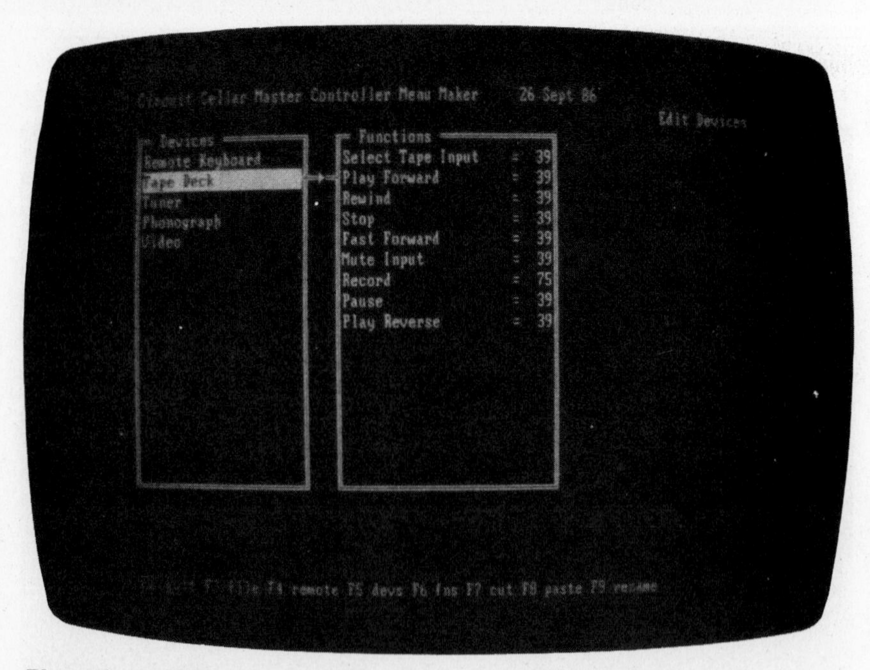

Photo 6: *MASTER program device-editor screen showing device and function menus. The Master Controller has just been trained to emulate the Sony RM-S750 controller.*

To facilitate these evaluations, I've made a printed circuit board for the Master Controller. My intention is to populate a few more and circulate them in the proper consumer channels. While the Master Controller is not available as a kit per se, these PC boards are available if you want to build your own plague antidote.

If you don't mind a little hand-wiring, I encourage you to build the Master Controller from scratch, and I will support your efforts as usual. A hexadecimal file of the executable code for the 8031's system EPROM and the Turbo Pascal source code for the IBM PC are available for downloading from my bulletin board at (203) 871-1988. Alternatively, you can send me a preformatted PC disk with return postage, and I'll put all the files on it for you (the hexadecimal file could be used with my serial EPROM programmer, for example). Of course, this free software is limited to noncommercial personal use.

Finally, I apologize if Master Controller is someone's trademark. It seemed an obvious descriptive name for the project, but there is no way for me to know whether it has been taken without a costly trademark search. That might be necessary eventually, but it is premature at present. Perhaps I should just call it ROVER (Response to Obnoxious, Valueless, and Extraneous Remotes).

Special thanks to Ed Nisley for his talented contributions to this project.

The 2-line by 20-character LCD is available from Jordan Technology Inc., P.O. Box 362, Lexington, MA 02173, (617) 863-8898.

The DS1210 is available from Dallas Semiconductor, 4350 Beltwood Pkwy., Dallas, TX 75244, (214) 450-0400.

Many of the individual components are available from JDR Microdevices, 1224 South Bascom Ave., San Jose, CA 95128, (800) 538-5000; Jameco, 1355 Shoreway

Table 1: *IBM PC function keys and their MASTER program functions.*

IBM PC function keys and their MASTER program functions

Rule of thumb: F2 through F6 select what you want to do
F7 through F10 actually do something
F1 does nothing
To get out of MASTER, press F2

F3 selects file I/O, then:

F7	writes the current menu to disk
F8	reads a menu file from disk
F9	creates a menu with simple names and random IR data (may be too big to fit into the remote RAM, though; this is of no use to anyone other than me)

F4 selects remote I/O, then:

F7	downloads the current menu to the remote
F8	uploads the remote's menu

F5 selects device editing, then:

F7	cuts a device and puts it on the junk list; if there was a device on the junk list, it's gone now
Ctrl-F7	deletes a device forever

F8 pastes a device from the junk list. If there's no device on the junk list, creates a new one:

F9	lets you rename a device
F6	selects function (and message) editing
F7	cuts a function and puts it on the junk list; if there was a function on the junk list, it's gone now
Ctrl-F7	deletes a function forever
Alt-F7	wipes out the IR message data for the current function
F8	pastes a function from the junk list; if there's no function on the junk list, creates a new one
Ctrl-F8	inserts a new function, leaving junk list alone
F9	lets you rename a function
F10	tacks the IR message from the junk list onto the end of the current function, leaving the junk list unchanged, so you can combine it with any number of functions

Rd., Belmont, CA 94002, (415) 592-8097; DigiKey, P.O. Box 677, Thief River Falls, MN 56701; and Advanced Computer Products Inc., 1310 East Edinger, Santa Ana, CA 92705, (714) 558-8813.

Editor's Note: Steve often refers to previous Circuit Cellar articles. Most of these past articles are available in book form from BYTE Books, McGraw-Hill Publishing Company, P.O. Box 400, Hightstown, NJ 08250, (1-800-2-MCGRAW).

Ciarcia's Circuit Cellar, Volume I covers articles in BYTE from September 1977 through November 1978. *Volume II* covers December 1978 through June 1980. *Volume III* covers July 1980 through December 1981. *Volume IV* covers January 1982 through June 1983. *Volume V* covers July 1983 through December 1984.

The following is available from

Circuit Cellar, Inc.
4 Park St., Suite 12
Vernon, CT 06066
(203) 875-2751

Master Controller experimenter's kit that includes printed circuit board with ROM-resident software on a 2764 EPROM, detailed parts list, and IBM PC programs on disk.
Order MC Exp. Kit $54

All payments should be made in U.S. dollars by check, money order, MasterCard, or Visa. Surface delivery (U.S. and Canada only): add $3 for U.S., $6 for Canada. For delivery to Europe via U.S. airmail, add $10. Three-day air freight delivery: add $8 for U.S. (UPS Blue), $25 for Canada (Purolator overnight), $45 for Europe (Federal Express), or $60 (Federal Express) for Asia and elsewhere in the world. Shipping costs are the same for one or two units.

There is an on-line Circuit Cellar bulletin board system that supports past and present projects. You are invited to call and exchange ideas and comments with other Circuit Cellar supporters. The 300/1200/2400-bps BBS is on-line 24 hours a day at (203) 871-1988.

5

THE BCC180 MULTITASKING CONTROLLER

PART 1: THE HARDWARE

Using a Hitachi CPU, Steve comes up with this multitasking single-board computer

Necessity is indeed the mother of invention. I have been known to stretch that adage on occasion, but my track record is pretty consistent. Many people think that I study trends in computer technology, intensively investigate reader interest, and carefully formulate a writing strategy that results in the projects you see. While I do consider all those factors, the actual selection process is considerably less complex. If I need it, I build it.

Again, I am at the point where I need to configure a new piece of controller hardware or resort to less popular alternatives. I am presently installing and testing a video motion and tracking system that I may document as a future project. (McGraw-Hill's lawyers will probably hyperventilate when I start discussing the "laser targeting" section, but that's a story for another time.)

Using eight video cameras, the system senses motion and triggers specific control actions depending upon what it "sees." While real video recognition is still a bit in the future, coordinating all the control decisions presently generated—even from the uncompleted video unit and a multitude of hard-wired sensors—is becoming a monumental task.

Generally, I would code these kinds of control applications in interpreted BASIC on a board like my BCC52 (see the August 1985 Circuit Cellar). I could then use all its bus-compatible peripherals for the control and sensor I/O.

However, given the magnitude of the task, I thought a BASIC interpreter would be too slow unless it was liberally salted with assembly language calls. Either I had to write more assembly language code (I'm not enamored with programming as it is), dedicate a large computer to the task (an expensive alternative), or design a small controller that was both fast and powerful enough to accomplish the task (sure, why not).

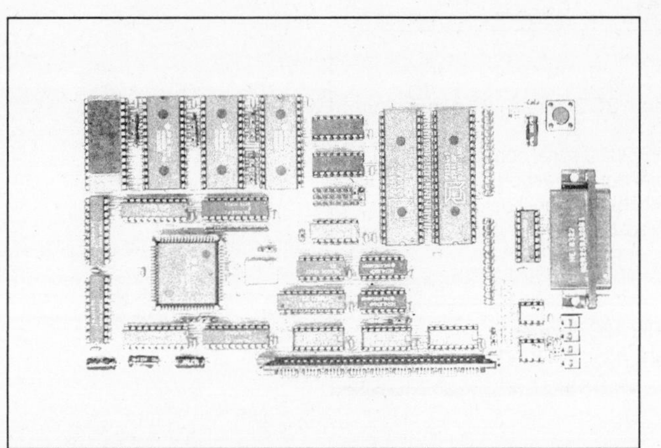

The BCC180 Computer/Controller

This new controller is called the BCC180 (table 1 lists its specifications). Designed from the ground up for efficiency and performance, the BCC180 uses the same 64180 CMOS Z80 instruction-compatible processor as my SB180 and SB180FX computers (see the September 1985 Circuit Cellar). Configured primarily for process control, the BCC180 uses the same 44-pin I/O expansion bus as the BCC52. All the BCC bus peripherals that I've described over the years will work nicely.

The BCC180 also contains a substantial amount of on-board I/O. It has six parallel ports and three serial I/O ports, and it communicates command and control decisions serially via RS-232C, RS-422, or RS-485. It can accommodate up to 384K bytes of on-board memory, which can be pure application code, monitor and application code, or a resident high-level language and application code.

BASIC-180

The most significant aspect of the BCC180 is its new approach to high-speed, high-level-language programming. Like the BCC52, I dictated that the BCC180 would have a ROM-resident BASIC. Unlike the BASIC-52 interpreter (albeit fast by most standards), the BCC180 has a *compiled multitasking* BASIC—BASIC-180—written by Softaid Inc. (8930 Route 108, Columbia, MO 21045). BASIC-180 was configured and adapted specifically for the 64180, and for the BCC180 in particular.

Unlike many generic BASICs that have 64K-byte ceilings, BASIC-180 uses the BCC180's hardware in the most efficient manner to optimize performance, and it can address and utilize a full megabyte of program space. (I've installed BASIC-180 as an EPROM rather than mask-programming it onto the processor. You can remove or switch the BASIC-180 EPROM at any time to allow the BCC180 to function completely in Z80 or 64180 assembly language code or another high-level language like C or Pascal.)

By using a multitasking BASIC compiler, I will have enough performance to continue my video-control project, plus the added benefit of a user-friendly software development environment. BASIC-180 can run up to 32 independent program tasks of up to 32K bytes each *concurrently* and, while task complexity does affect execu-

Table 1: *Specifications for the BCC180.*

Processor

Hitachi HD64180, an 8-bit CPU in a 68-pin PLCC package
Superset of Z80 instruction set, including hardware multiply
Integrated memory-management unit
Dynamic RAM refresh
Wait-state generator
Clocked serial I/O port
Two-channel direct-memory-access controller
Two-channel asynchronous serial-communication interface
Two-channel 16-bit programmable reload timer
12 interrupts
6.144-MHz and 9.216-MHz system operation

Memory

Up to 384K bytes of total memory on-board
128K bytes of either static RAM (62256) or EPROM (27256)
Optional 256K-byte dynamic RAM SIMM
Full-function 8K-byte ROM monitor included

I/O

Console RS-232C serial port with automatic data transfer rate selectable to 38,400 bps
Peripheral serial port, 150 through 38,400 bps, selectable RS-232C, RS-422, or RS-485
48 bits of bidirectional parallel I/O
64K-byte I/O space available through the BCC bus edge connector

Power Supply Requirements

+5 V +/− 5 percent @ 700 mA (fully populated with LSTTL)
+12 V +/− 20 percent @ 30 mA
−12 V +/− 20 percent @ 30 mA
12-V supplies are required only for RS-232C operation

Dimensions and Connections

4.5- by 8.5-inch board
Dual 22-pin (0.156-inch) edge connector
Compatible with all Micromint BCC-series I/O expansion boards
25-pin DB-25S connector for RS-232C serial console I/O
20-pin header for RS-232C serial peripheral port
Four screw terminals for RS-422/RS-485 serial peripheral port
Two 26-pin headers for six bidirectional parallel ports

Operating Conditions

Temperature: 0-50 degrees C (32-122 degrees F)
Relative humidity: 10-90 percent, noncondensing

tion speed, is a real screamer.

I'll go into benchmarks later, but if you are familiar with BASIC-52 and anxious for some comparisons, I'll give you a quick one: At 6.144 MHz, BASIC-180 executes an integer variable `FOR...NEXT` loop benchmark approximately 100 times faster than BASIC-52 does!

BASIC-180 comes in two flavors: disk-based, for development on an SB180/SB180FX, and ROM-based, for development on the BCC180 board. Using the disk-based version, you can create and—to a certain degree—test programs on the SB180/SB180FX. This lets you use a full-screen editor for writing source code and a disk drive for saving the code. You obviously can't test a program that requires any BCC180-specific I/O operations on the SB180, but you can test fundamental operations without change.

Once you've written the code on the SB180 and you've verified that it's syntactically correct, you can compile it into one or more binary files and burn them into an EPROM (with the Circuit Cellar serial EPROM programmer, perhaps) or send the files directly to the BCC180 (with the monitor ROM installed) for testing in RAM or programming into an EPROM there.

If you don't have an SB180 for development or prefer to do all the development on the BCC180, you can use the ROM-based version of BASIC-180. It supports all the features of the disk-based version with a few modifications: Instead of saving program source code to disk, the ROM-based compiler saves it to EPROM. Only as much of the EPROM is programmed as is necessary to store the source code, so multiple programs (or versions of the same program) can be saved to the same EPROM. This is often referred to as write once, read many (WORM) storage. When you fill the EPROM up, you can simply erase it and use it again.

Additionally, the compiler can program the object code directly into an EPROM. You can then use this EPROM to replace the BASIC-180 ROM for auto-start applications. You might also want to compile the object code into RAM, where you can execute the program immediately.

The BCC180's Hardware

The BCC180 uses the same Hitachi HD64180 (or Zilog Z180) microprocessor used on my SB180 and SB180FX computers (see figure 1 for the BCC180's schematic). Briefly, this chip executes the complete Z80 instruction set, plus a few new instructions (including an 8-bit multiply).

The chip contains an on-board memory management unit (MMU), a built-in direct-memory-access (DMA) controller with two DMA channels, two asynchronous serial ports, one synchronous serial port, two 16-bit programmable reload timers, and eight internal and four external interrupt sources with a built-in interrupt controller.

The HD64180 can address up to 1 megabyte of memory and 64K I/O ports. Since the BCC bus has only 16 address bits, I decided that all memory would be resident on the main board and that all transactions the BCC180 carried out through the bus would be I/O-based. As a result, I tried to squeeze as much memory as possible onto the board.

The BCC180 contains four 28-pin sockets (IC10 through IC13) addressed in 32K-byte increments, starting at physical address 00000 and going through 1FFFF hexadecimal. Each socket will accommodate either a 27256 EPROM or a 62256 static RAM chip. This lets you burn the control program into one or more EPROMs and place it at low memory for execution upon reset.

You can use zero-power RAM (static RAM that contains its own battery) or SmartSockets (sockets that contain a battery) with static RAM chips plugged into them in the remaining sockets to provide inexpensive, nonvolatile storage. A 74LS138 (IC14) decodes each socket's address.

The four sockets just described allow up to 128K bytes of static RAM storage, but what about applications that need a lot of temporary storage? Dynamic RAM is ideally suited in cases where large amounts of storage are needed in a small area but that doesn't have to be battery-backed.

Dynamic memory arranged on a single in-line memory module (SIMM) is becoming more popular these days, so I added a SIMM socket to the BCC180, decoded to start at physical address 40000 hexadecimal. A 256K-byte SIMM brings the BCC180's total on-board system memory up to 384K bytes. An address multiplexing circuit made up of three 74LS158s (IC15 through IC17) switches the SIMM address lines between row addresses and column addresses. ME\ generates RAS\, and the flip-flop circuit made up of IC18 through IC20 generates CAS\.

Unfortunately, 384K bytes must be the upper limit for the time being. While the current HD64180s can address 1 megabyte of memory, they provide only 8-bit refresh (1 megabyte needs 9-bit refresh). Keeping possible future developments in mind, however, I've wired the SIMM socket to accommodate a 1-megabyte SIMM. If a new version of the HD64180 becomes available that provides 9-bit refresh, you'll be able to plug a 1-megabyte SIMM into the socket and bring the total system memory up to 896K bytes (since the SIMM's addressing starts at 40000 hexadecimal, we must throw away 256K bytes of the 1-megabyte SIMM).

Next, if a process-control computer is going to be useful, it must be able to deal with real-world inputs and outputs. For that

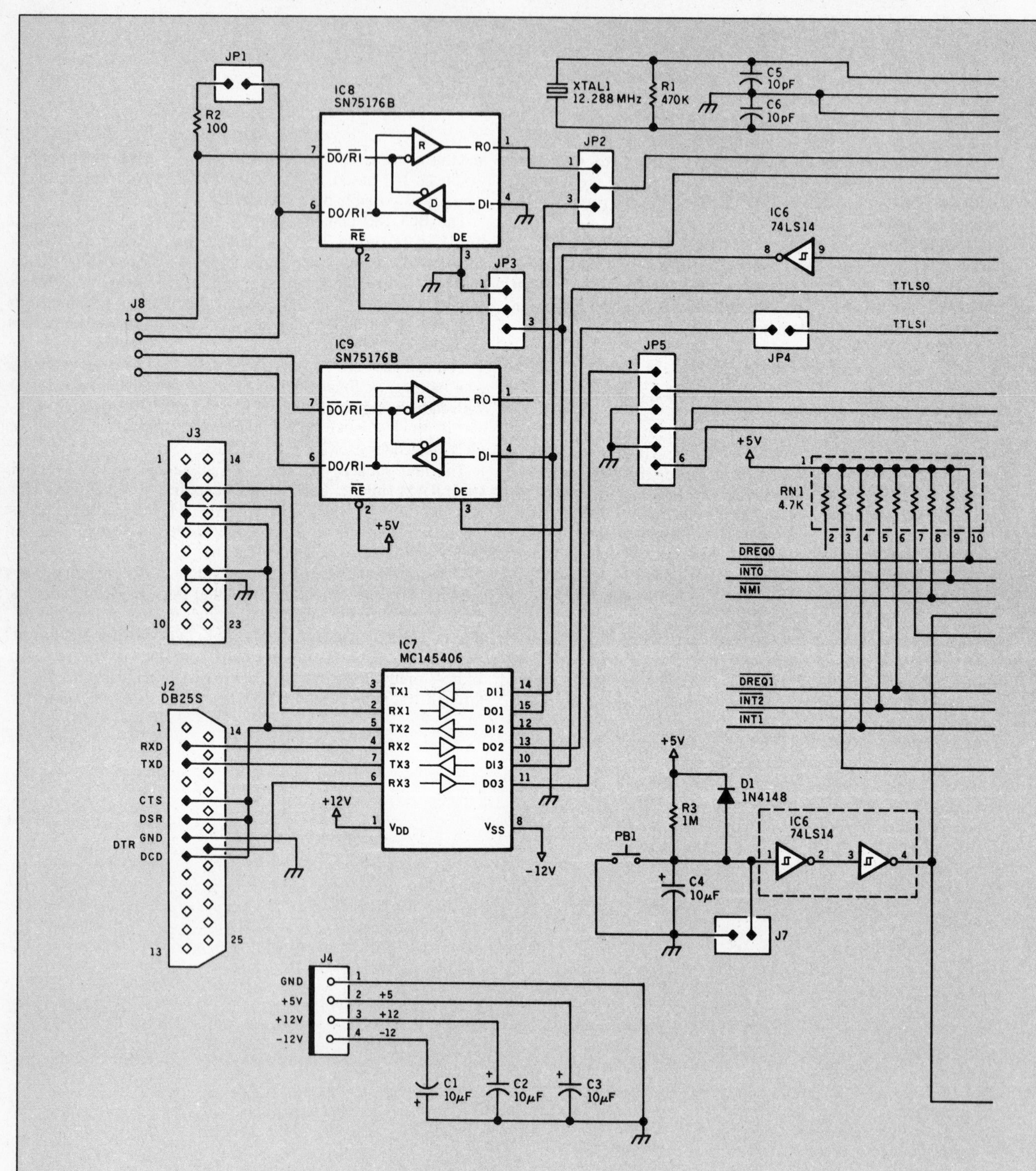

Figure 1: *Schematic for the BCC180 computer/controller.*

reason, two 8255 peripheral interface adapters (PIAs) are on the BCC180. Each 8255 has three 8-bit parallel I/O ports that can be individually configured for input or output, for a total of 48 bits of parallel I/O on the board (available on two 26-pin Berg-type connectors, J5 and J6).

In figure 1, IC23 and IC24 are the 8255s, and IC19, IC21, and IC22 decode an I/O address for each chip. You can select the addresses for the 8255s using jumpers JP10 and JP11.

Besides having parallel I/O, the BCC180 also has serial I/O. The serial ports let you communicate, via terminal, with the BCC180 and let it access external data-collection devices. Two asynchronous serial ports are built into the HD64180.

Serial port 1 uses an MC145406 (IC7) to convert TTL-level signals to RS-232C levels and is connected to a standard DB-25 connector (J2). Normally, you would connect an external termi-

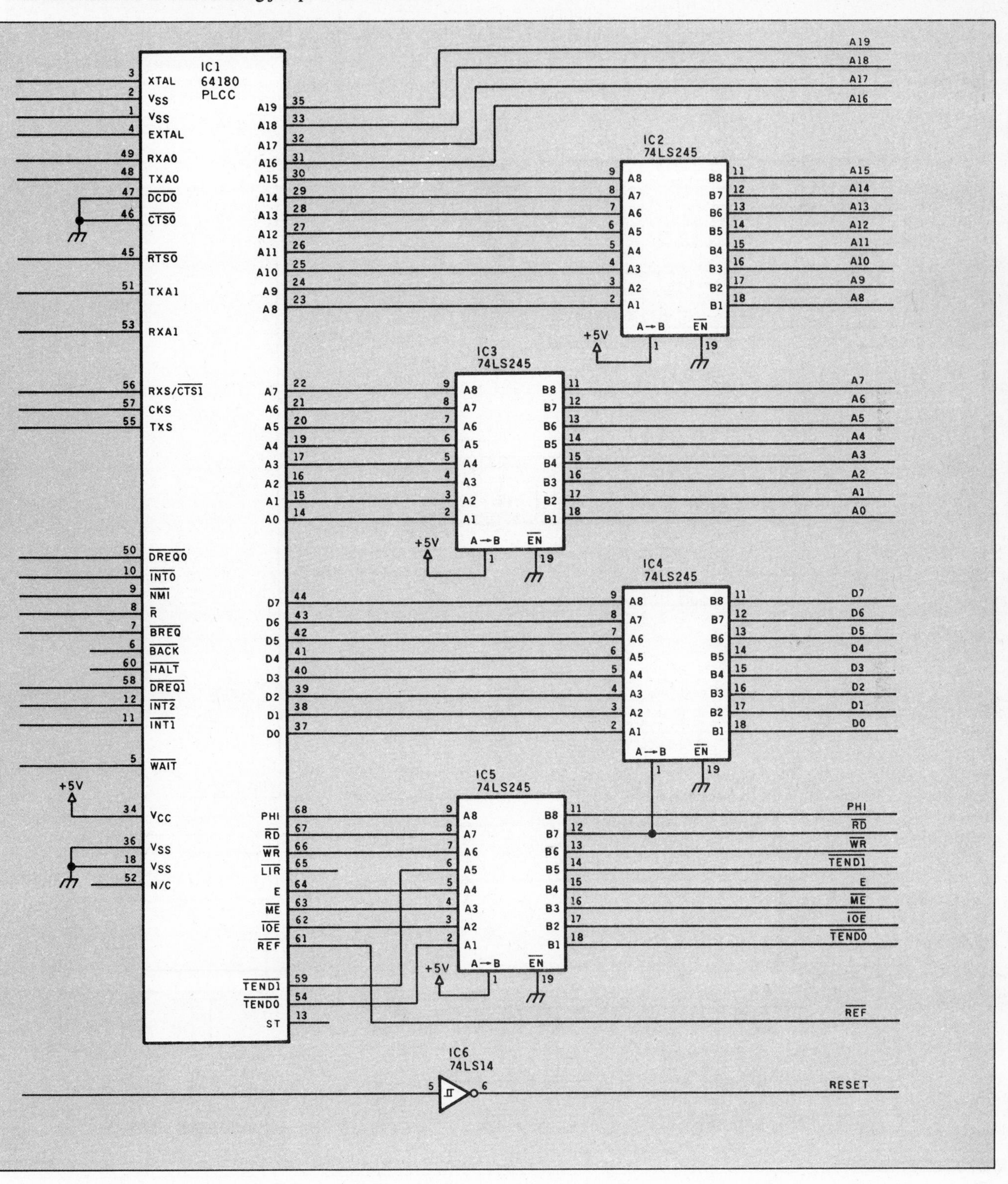

nal to J2. I've also connected serial port 1 to the BCC bus to allow TTL-level communication directly.

You can use the second asynchronous serial port (port 0) with one of three interfaces: RS-232C, RS-422, or RS-485. If you connect a jumper between pins 2 and 3 of JP2, port 0 passes its signals through the MC145406 for use as an RS-232C port.

J3 connects port 0 to the outside world in this configuration. When JP2 has a jumper between pins 1 and 2, port 0 communicates through the two SN75176B chips (IC8 and IC9) for use in either an RS-422 or an RS-485 application.

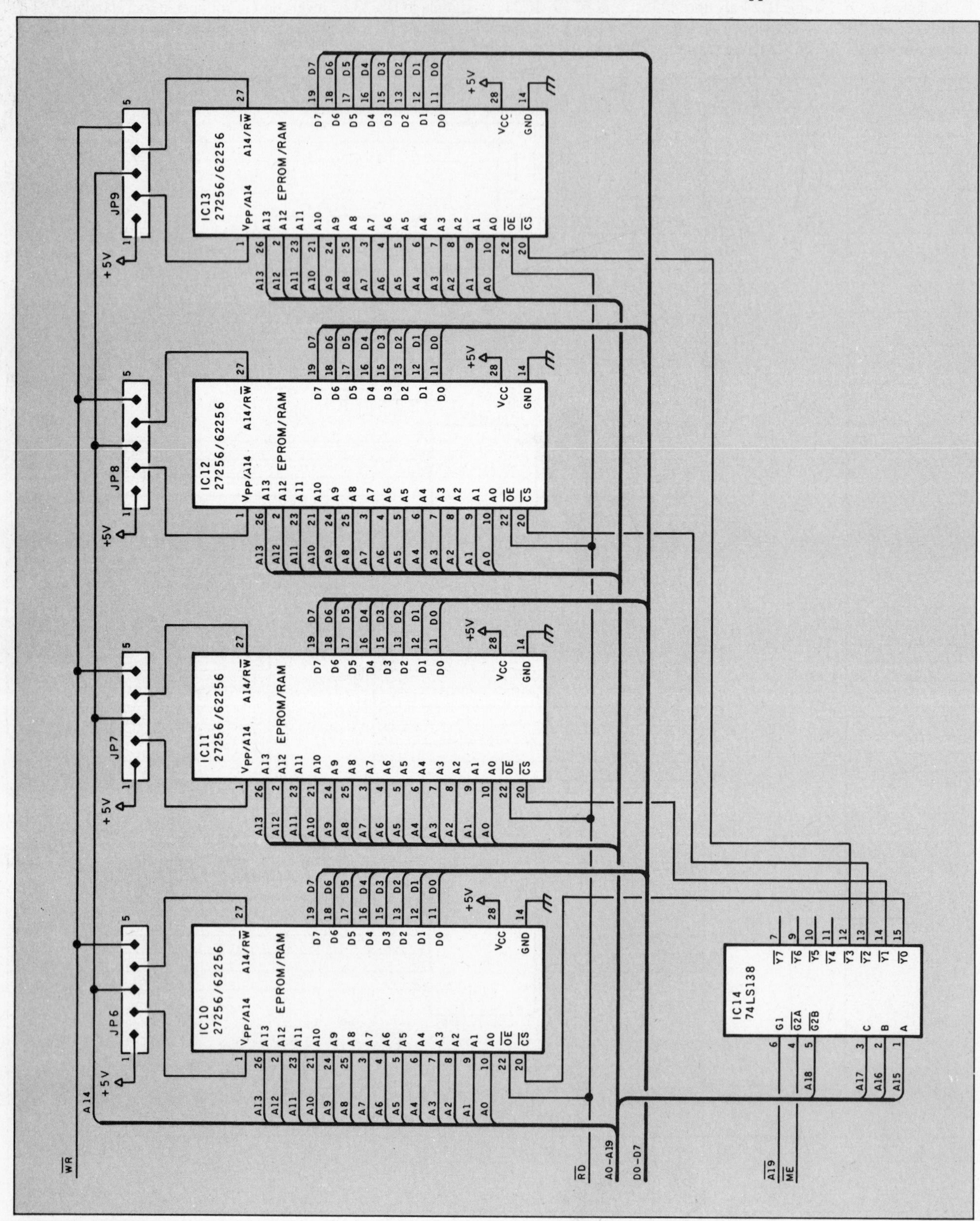

Figure 1: *Continued.*

Due to their relatively high noise immunity over long distances, RS-422 and RS-485 are becoming popular for use in communicating between remote data-collection sites and a central controller. Unlike RS-232C, which is single-ended (one wire is tied to ground, and a voltage varies on the other), RS-422 and RS-485 use balanced lines for data transmission.

In a balanced line, the voltage differential between the two wires is what's important, rather than the absolute voltage referenced to ground. The absolute voltage of the pair of wires referenced to ground can be anywhere from −7 volts to +12 V, and it won't affect the operation of the connection. The twisted-pair telephone line running into your home is an example of a balanced line. In RS-422 uses, separate transmit and receive pairs allow full-duplex operation, and each line has just one driver and one receiver. Its setup is similar to RS-232C, in that it's used mostly for point-to-point connections.

RS-485, on the other hand, is usually used in a party-line configuration. A single twisted pair connects numerous devices, and each device has a driver and a receiver connected to the same pair of wires. Only one driver can be active at a time, and all the receivers can be active at once. It's up to the software

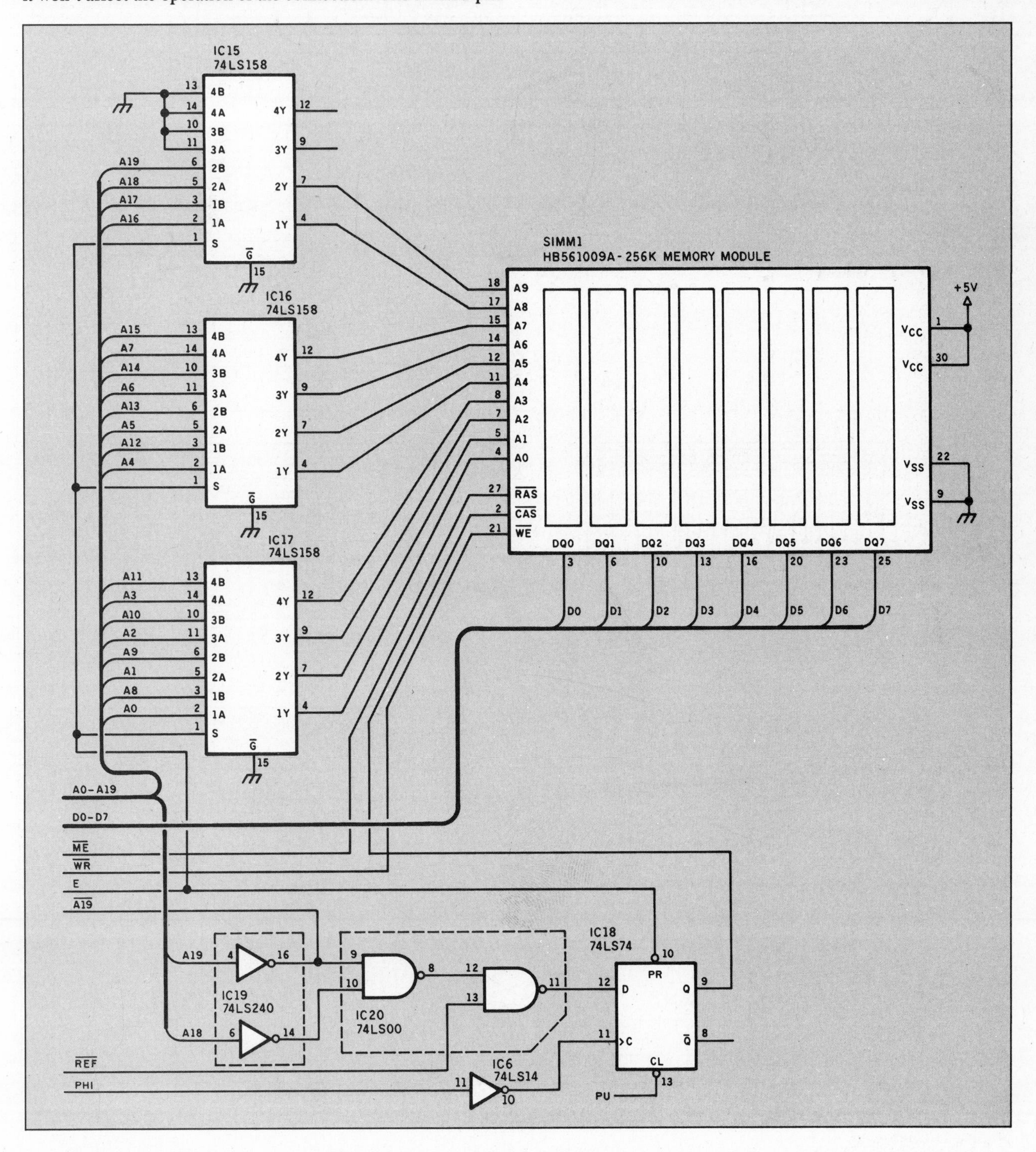

designer to implement a protocol. Although it can operate only in half-duplex, it is a simple and inexpensive way to implement a local-area network (LAN).

Although the BCC180's SN75176B is intended primarily for use in RS-485 applications, since RS-485 is really just a specialized use of RS-422, this driver IC will work well in most RS-422 applications. If you place a jumper between pins 1 and 2 on JP3, port 0 is set up for double-pair, full-duplex RS-422 operation. Placing a jumper on JP3 between pins 2 and 3, and tying together pins 1 and 3 and pins 2 and 4 on J8, configures the board for single-pair, half-duplex RS-485 operation.

The BCC Bus

In the early days of microcomputers, Intel wanted to increase the capability of its microprocessors without increasing the number of pins needed on the chip. The company started using a method

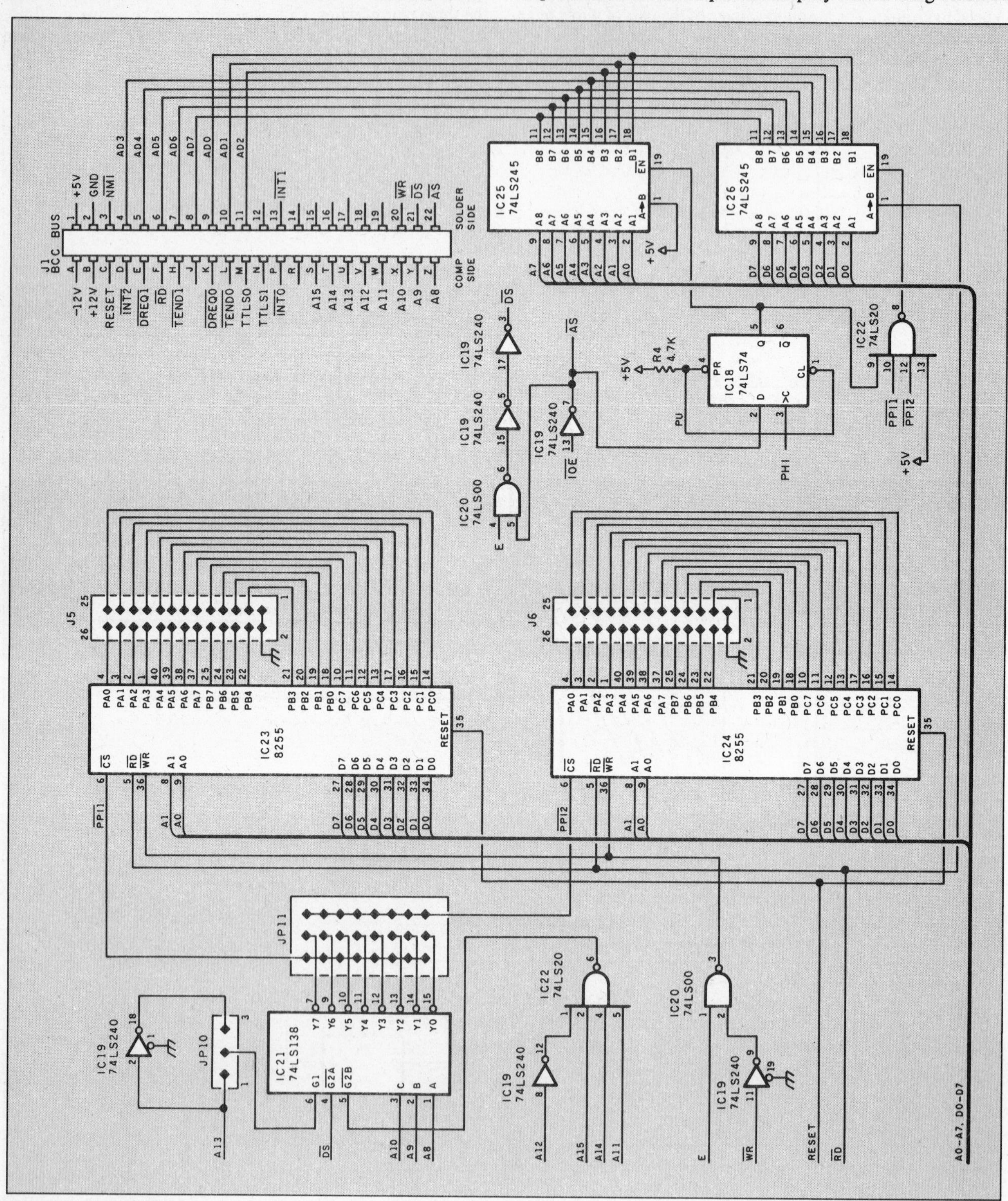

Figure 1: *Continued.*

known as multiplexing to place the eight low-order address lines on the same pins as the eight data lines. During the first clock cycle of a machine cycle, the high-order address is placed on the high-order address lines, and the low-order address is placed on the combined address/data lines.

When the address is stable, the microprocessor provides a strobe signal so that the low-order address bits can be latched into an external buffer. During the rest of the machine cycle, the system can use the same address/data lines for data since the low-order address bits have been latched.

When Zilog was started by several former Intel employees, some of Intel's design philosophies must have followed. While address/data multiplexing wasn't used on the Z80, it was used on the Z8. When I designed the BCC11 computer/controller (see the July 1981 Circuit Cellar) using the Z8 as the processor, I included the same multiplexed address/data lines in my definition of the BCC bus. Any peripheral card that you plug into the BCC bus must include the external latch mentioned above to latch the low-order address bits so the same lines can be used for data.

The BCC52 used the Intel 8052. Since the chip was from Intel, it had a multiplexed address/data bus like the Zilog Z8, and it was an easy task to attach it to the BCC bus. Consequently, all the peripheral boards that had been designed since the introduction of the BCC11 could function with the BCC52.

When I decided to make a BCC bus–compatible board using the 64180, it presented a bit of a problem. Since the HD64180 does not have a multiplexed address/data bus, I had to create a multiplexed bus interface.

The 8052 generates an address-strobe signal (AS\) and a data-strobe signal (DS\). When the address is stable and ready to be latched, the processor generates a low-to-high transition on AS\. During a write cycle, when data is stable on the bus, the processor generates a low-to-high transition on DS\ to tell the peripheral that it can read the data.

Likewise, during a read cycle, when the processor reads the data bus, it generates a low-to-high transition on DS\ to indicate that it's done with the data being presented to it.

A look at the HD64180's timing diagrams shows that the I/O enable line (IOE\) goes from high to low at the end of T1 (the first clock cycle) to indicate the start of an I/O cycle. (Remember, we want all bus transactions to be I/O-based.) Since the address is stable at the beginning of T1, IOE\ was a perfect candidate for use in generating AS\. Indeed, all it takes is an inverter to create the needed AS\ signal.

Generating DS\ is a little trickier, but not much. For that, I employed the ever-popular and ever-mystifying E signal. Most data sheets won't give you precise information on E's function. The HD64180 data book says nothing more than, "E is a synchronous clock for connection to HD63xx series and other 6800/6500 series compatible peripheral LSI."

Turning to the timing diagrams again, I discovered that, for an I/O read, E goes from high to low at the end of T3 (the last clock cycle) to signify that the processor has read the data bus. For an I/O write, E goes from high to low in the middle of T3 to signify that data is stable. It turns out that this is exactly what we need to generate DS\. Combining E and IOE\ through an AND gate and inverting the result yields the desired active-low DS\ signal.

I used two 74LS245s (IC25 and IC26) with their "B" sides tied together to perform the multiplexing of the address and data lines. RD\ controls the direction line of the data buffer (IC26) so that it can operate bidirectionally, while the address buffer (IC25) is hard-wired for output-only operation.

To control the buffer-enable lines, we delay AS\, using a 74LS74 flip-flop (IC18). When a machine cycle starts, AS\ is low and passes through the flip-flop, enabling the address buffer and disabling the data buffer. After AS\ goes high, whatever other devices are on the BCC bus have latched the address bits, so the BCC180 disables the address buffer and enables the data buffer. To avoid race conditions and to allow for a small hold time, we don't disable the address buffer until the next rising edge of PHI after AS\ goes high.

This corresponds to the start of T2 (the second clock cycle) and provides plenty of time for the data to propagate through the data buffer before it is needed. At the completion of the machine cycle, when AS\ goes low again, the address buffer is reenabled by the rising edge of the first clock cycle of the next machine cycle. Since the address bus isn't stable until after the next machine cycle has started, we've preserved the address setup time.

The rest of the signals on the bus are straightforward. Most of them are connected to the HD64180, with the outputs going through buffers. Along with the two used for multiplexing the address and data bus (IC25 and IC26), a total of six 74LS245s are used. Needless to say, this is a well-buffered board. All the inputs are pulled high using 4.7-kilohm resistors.

The BCC180 Monitor ROM

Now that we have some hardware, we need something to make it go. I've already alluded to the special multitasking BCC180 BASIC that I'll begin describing in Chapter 5, part 2, but the system needs something at a lower level so that we can exercise all parts of the machine without writing driver programs or purchasing the BASIC. For that purpose, part of the software for the BCC180 includes a monitor ROM.

The monitor provides functions that let you inspect memory, change memory, access I/O devices, and read and program EPROMs (see table 2). Veteran SB180 users will note the similarity between this monitor and the one on the SB180.

When you've installed the monitor ROM in the BCC180 and applied power to the controller, the system sends `BCC180` to the terminal at 9600 bits per second. This message will display clearly on a terminal properly set for 9600 bps.

However, if you've set the terminal for some rate other than 9600 bps, pressing Return tells the BCC180 the terminal's actual data transfer rate. The system will then display an opening banner (at the proper data transfer rate) showing the amount of RAM and ROM in the system and give you a command prompt. Once in the monitor, you can obtain a full help screen by typing ?.

From the monitor, you can fill memory with a byte value, copy blocks of memory from one location to another and verify that the copy was performed properly, display sections of mem-

Table 2: *The ROM monitor provided with the BCC180 is a complete set of utilities and debugging aids. You invoke commands using a single character.*

BCC180 ROM Monitor

A — ASCII table	N — New command
B — Bank select	O — Output port
C — Copy EPROM	P — Printer select
D — Download hexadecimal file	Q — Query memory
E — Emulate terminal	R — Read EPROM
F — Fill memory	S — Set memory
G — Goto program	T — Test system
H — Hexmath	U — Upload hexadecimal file
I — Input port	V — Verify memory
J — Jump to ROM language	W — Write EPROM
L — List memory	X — Examine CPU registers
M — Move memory	Y — Yank I/O registers

ory on the terminal, and modify individual memory locations. You can also search memory for a particular series of bytes.

The `Bank` command lets you set the 64K-byte bank of memory on which the above commands operate. (The system requires this command since the software is aware of only 64K bytes of memory, but the external address bus can access up to 1 megabyte of memory.)

The monitor lets you directly access I/O devices, both on the BCC180 board and on the BCC bus. Using a series of `Input` and `Output` commands, you can check a board that's just been plugged into the bus without having to write and debug a program.

Another useful function of the monitor is its EPROM programming support. You can transfer into memory the contents of an EPROM that has been plugged into the programming board, examine and possibly modify the contents, then program the block onto a blank EPROM. You can also send a file in Intel hexadecimal format to the BCC180 and have the computer program an EPROM.

This is the basis of the SB180-based development system I described earlier. On the SB180, you create a hexadecimal file containing the object code, then transfer that file to the BCC180 monitor. You use the monitor to program the final EPROM.

On the miscellaneous side, the monitor has commands that let you examine and modify the HD64180's general-purpose registers and display, with labels, the processor's 64 internal I/O registers. As a help to programmers, the `A` command displays an ASCII table, and `H` can perform simple hexadecimal mathematics.

Experimenters

While the BCC180 is available commercially, I encourage you to build your own. If you don't mind doing a little work, I will support your efforts as usual. A hexadecimal file of the executable code for the BCC180's ROM monitor is available for downloading from my bulletin board at (203) 871-1988. Alternatively, you can send me a preformatted IBM PC or SB180 disk with return postage, and I'll put the file on it for you. Add $5 for a printed copy of the BCC180 manual.

I also have a number of copies of the BASIC-180 development software that, for the price of the manuals and distribution media, I will gladly give to experimenters who build the BCC180. Of course, this free software is limited to noncommercial personal use.

In Conclusion

In Chapter 5, Part 2, I'll finish the hardware program with a description of the BCC180's auxiliary EPROM programmer board and introduce BASIC-180. As I begin talking about BASIC-180, I'll include a tutorial on multitasking.

I'd like to acknowledge and personally thank Ken Davidson and Jack Ganssle for their efforts on the BCC180 project. Ken Davidson's extensive knowledge of the HD64180 helped us avoid the omnipresent hardware design pitfalls, and Jack Ganssle's superb software talents helped explain multitasking in a way that can really be understood.

Editor's Note: Steve often refers to previous Circuit Cellar articles. Most of these past articles are available in book form from BYTE Books, McGraw-Hill Publishing Company, P.O. Box 400, Hightstown, NJ 08250, (1–800–2–MCGRAW).

It's virtually impossible to provide all the pertinent details of a project or cover all the designs I'd like to in the pages of BYTE. For that reason, I have started a bimonthly supplemental publication called Circuit Cellar Ink, which presents additional information on projects published in BYTE, new projects, and supplemental applications-oriented materials. For a one-year subscription (6 issues), send $14.95 to Circuit Cellar Ink. Credit card orders can call (203) 875-2199.

Ciarcia's Circuit Cellar, Volume I covers articles in BYTE from September 1977 through November 1978. *Volume II* covers December 1978 through June 1980. *Volume III* covers July 1980 through December 1981. *Volume IV* covers January 1982 through June 1983. *Volume V* covers July 1983 through December 1984. *Volume VI* covers January 1985 through June 1986.

The following items are available from

Micromint Inc.
4 Park St.
Vernon, CT 06066
For orders: (800) 635-3355
For information: (203) 871-6170
Telex: 643331
FAX: (203) 872-2204

1. A 9-MHz assembled and fully socketed BCC180 computer/controller board with 32K bytes of static RAM, ROM monitor, BASIC-180 development software (same as item 3), and user's manuals. BCC180-1-20 ....$395; for additional 256K DRAM, add $100
2. BCC180 PAK evaluation system. Contains a 9-MHz BCC180 board with 32K-byte static RAM, ROM monitor, BASIC-180 development software, MB08 eight-slot backplane, CC01 10-inch card cage, UPS10 35-watt switching power supply, and user's manuals. BCC180-PAK ....$595; for additional 256K DRAM, add $100
3. BASIC-180 multitasking BASIC compiler for ROM- or disk-based development. Contains both BASIC-180 EPROM for direct use on BCC180 board and BASIC-180 disk for direct use or software development on SB180. Includes 100-page user's manual. Compiled code may be freely used without further license. BASIC-180 DEV $250
4. An auxiliary 27256 EPROM programmer board for the BCC180. BCC180PROG5 ....$89

The following items are available from

Circuit Cellar, Inc.
4 Park St., Suite 12
Vernon, CT 06066
(203) 875-2751

1. A 9-MHz BCC180 computer/controller complete kit with 32K bytes of static RAM, ROM monitor, BASIC-180 development software, and user's manual. BCC180-KIT-20 .... $295
2. BCC180 auxiliary 27256 EPROM programmer board full kit. BCC180PROG5K...$74

For either source above, all payments should be made in U.S. dollars by check, money order, MasterCard, Visa, or American Express. Surface delivery (U.S. and Canada only): add $5 for U.S., $8 for Canada. For delivery to Europe via U.S. airmail, add $14. Three-day air freight delivery: add $10 for U.S. (UPS Blue), $25 for Canada (Purolator overnight), $45 for Europe (Federal Express), or $60 for Asia and elsewhere in the world (Federal Express). Shipping costs are the same for one or two units.

There is an on-line Circuit Cellar bulletin board system that supports past and present projects. You are invited to call and exchange ideas and comments with other Circuit Cellar supporters. The 300/1200/2400-bps BBS is on-line 24 hours a day at (203) 871-1988.

5

THE BCC180 MULTITASKING CONTROLLER

PART 2: EPROMs AND COMPILERS

Using the Hitachi HD64180 CPU, Steve's project is a multitasking single-board computer/controller

In Chapter 5, part 1, I introduced the BCC180 multitasking controller by describing the basic hardware and alluding to the power of its ROM-resident multitasking software. In part 2, I'll finish the discussion of the BCC180 hardware and introduce the BASIC-180 multitasking compiler in more detail.

The BCC180 Auxiliary EPROM Programmer Board

An attractive feature of the BCC52 (see the August 1985 Circuit Cellar) is its on-board EPROM programming capability. Using the BCC52 and a terminal, you can write a program, debug it, and burn it into the final EPROM.

Since the 8052 chip used on the BCC52 is really a full microcomputer (as opposed to a general-purpose microprocessor), it is capable of programming EPROMs connected to the processor's address and data lines. Unfortunately, the HD64180 used on the BCC180 needs additional circuitry to tailor its function to something as specialized as programming EPROMs.

While I could have added this circuitry and exactly duplicated the stand-alone utility of the BCC52, I felt that such an infrequently used option would make the final board overly large and more expensive. Instead, as a compromise of essential flexibility and price/performance, I decided to put the EPROM programming circuitry on a small daughterboard that you use only as needed (see figure 1).

The daughterboard plugs into the J5 and J6 8255 peripheral interface adapter (PIA) parallel-port connectors on the BCC180 (refer to the circuit diagram in Chapter 5, part 1). There are six parallel ports: One is connected to the EPROM's data bus and is used to read data from and write data to the EPROM; two more ports provide the EPROM with an address; 2 bits from a fourth port control the EPROM's CE\ and OE\ lines, and two more bits from that port control power to the EPROM and control the programming voltages; a fifth port reads the daughterboard's identity code.

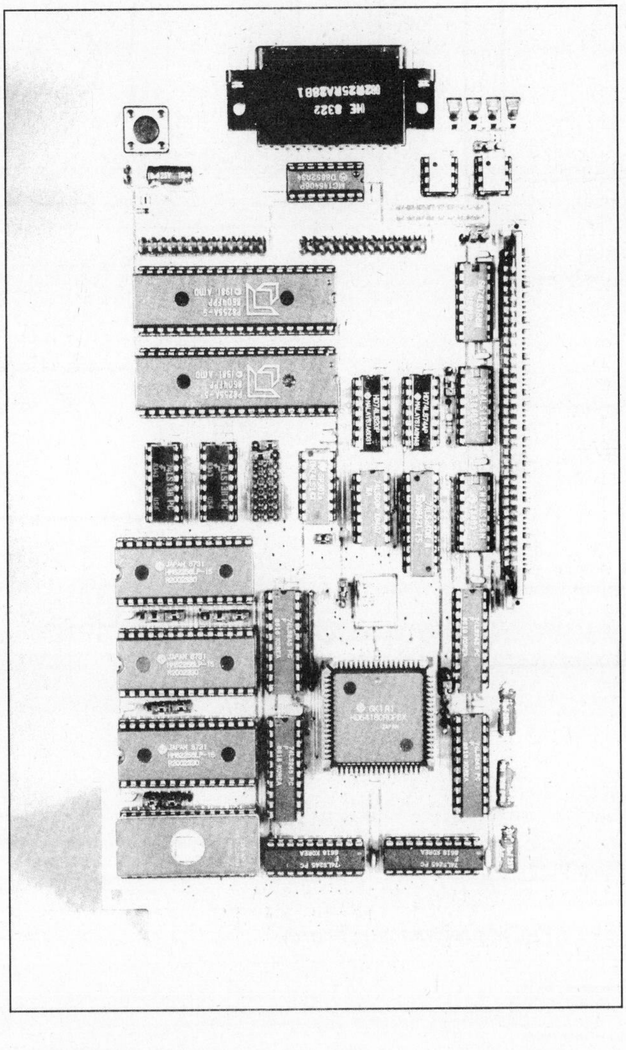

Photo 1: *The BCC180 computer/controller.*

The first problem to overcome when designing an EPROM programmer is where to get the necessary programming voltages. The BCC180 needs +6 volts and +12.5 V to perform fast programming of 27256 EPROMs (see my October 1986 article for information on "fast" and regular EPROM programming techniques).

Instead of using an external +21-V power supply, as I did on the BCC52, this time I had the real estate and incentive (performance usually takes priority over cost on optional boards) to do the job more completely.

The BCC180 EPROM programmer requires only +5 V for its operation (see figure 1). It uses a 78S40 switching regulator to step +5 V up to

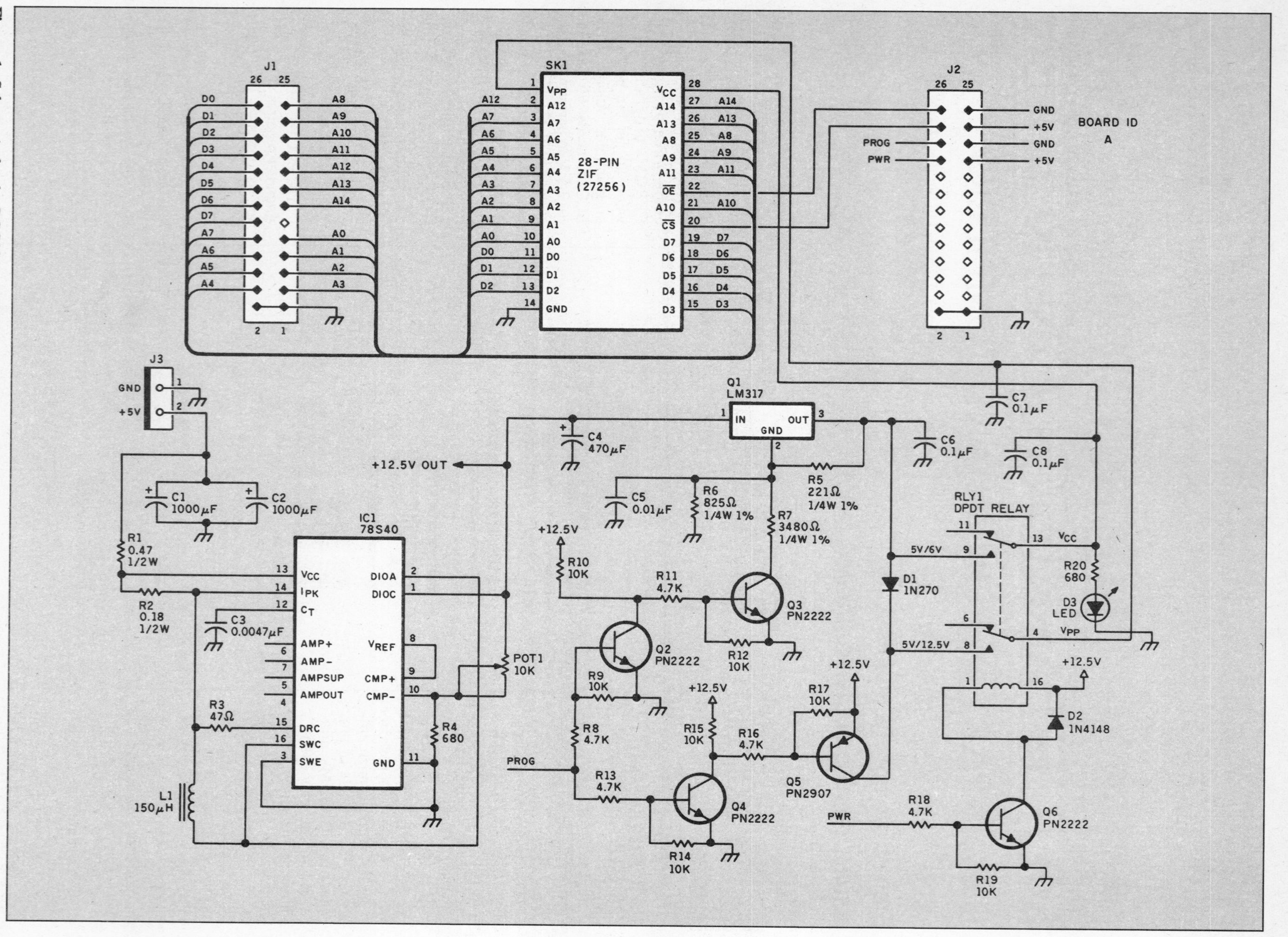

Figure 1: *Schematic for the BCC180's EPROM programming board.*

+12.5 V and regulates +6 V from it.

The 78S40 is a general-purpose switching regulator. It is capable of stepping a voltage up to a higher voltage (+5 V to +12 V), stepping a voltage down to a lower one (+12 V to +5 V), or inverting a voltage to a negative voltage (+12 V to −5 V). I wired it up to convert from +5 V to +12.5 V. A variable resistor on its output fine-tunes the final output voltage.

Once we have the +12.5-V V_{pp} supply, how do we switch the voltages on V_{pp} and V_{cc} between 0 V, +5 V for normal operation, and +12.5 V or +6 V for programming? (The 0 V is necessary for safe loading and unloading of the EPROM in the zero-insertion-force [ZIF] socket.) The easiest way to force V_{pp} and V_{cc} to 0 V is simply to disconnect both pins from the power supply. I used a double-pole, double-throw (DPDT) relay controlled by a bit on the BCC180's parallel ports (PWR) to accomplish this. When the BCC180 is reset, PWR floats low (due to the pull-down resistor, R19) and power to the EPROM stays disconnected. When PWR is set high, transistor Q6 turns on and energizes the relay coil, pulling in the relay and applying power to the EPROM.

The method I used to generate 6 V and to switch the V_{cc} line between +5 V and +6 V is reminiscent of my serial EPROM programmer. The LM317 is an adjustable voltage regulator whose output voltage depends on the value of the feedback resistor (R5) connected between the output and the control input and a second resistor from the control input to ground.

When transistor Q3 turns off, the total resistance from the control input to ground is 825 ohms. This sets the LM317 output to +6 V. When Q3 turns on, the total resistance becomes 667 ohms, causing the output to be regulated to +5 V.

Upon reset, the program voltage control line coming from the BCC180 (PROG) floats low, due to the pull-down resistors R9 and R14. Transistor Q2 stays turned off, so the base of Q3 gets pulled high to +12.5 V. This, in turn, turns Q3 on, connecting R7 to ground. With R6 and R7 in parallel, the LM317 will output +5 V to V_{cc}. When PROG is pulled high, indicating that programming is to begin, Q2 is turned on and forces Q3 to turn off. With Q3 turned off, R7 is removed from the circuit, and the LM317's output voltage (and V_{cc}) becomes +6 V.

To control the switch between +5 V and +12.5 V on the V_{pp} line, I used a slightly different technique. When PROG is low, Q4 is off, allowing the base of Q5 to be pulled to +12.5 V. Q5 is a positive-negative-positive (PNP) transistor, so the high voltage on its base keeps it turned off and its collector floats.

Since we want both V_{cc} and V_{pp} to be +5 V when PROG is low, we can steal +5 V from the circuit described above using a low-drop germanium diode. In normal operation, V_{pp} draws less than 5 milliamperes, so the drop across the diode is only 200 millivolts or so.

When PROG goes high, indicating that programming is to begin, the system turns on Q4, pulling Q5's base to ground. Q5 turns on and allows +12.5 V to get through to the V_{pp} line. On the other side of the diode, V_{cc} goes only to +6 V, so the diode is reverse-biased and stops conducting, effectively isolating V_{cc} from the high voltage on V_{pp}.

The final feature on the programming daughterboard is a board ID number. Right now, since the BCC180 uses only 27256s, that's all this programmer board is designed to handle. But because we'll no doubt have other daughterboards, there is a unique ID number (read via a few bits on port 6) assigned to each daughterboard. This will allow intelligent software to check what board is installed.

In Search of Appropriate Software

Of course, no project these days is just hardware. Like any computer, the BCC180 is not very useful without software. If it were a disk-based system like the SB180, we would need only a BIOS. However, since it is a stand-alone computer, it requires either a cross-development environment or an embedded language with its own operating system.

Before defining what software is required for the BCC180, we should look at the board's typical applications. It is not designed as another generic computer for word processing, spreadsheets, or games. The BCC180 is for embedded applications where it may not be particularly obvious that a computer is part of the system.

For example, a factory-control system will typically use a number of computers distributed around the building. A single BCC180 might control one local process and then be linked by a serial line to a master control computer. Each remote processor independently runs a ROM-resident program directing that processor's activities.

The system's software must fulfill certain common requirements. First, it must start automatically on power-up and execute out of ROM without operator intervention day after day for years. Ideally, you should be able to develop code directly on the BCC180 and then burn it into the EPROM. You could develop larger applications on another computer with disks—like the SB180—then burn the programs into the EPROM and place them on the BCC180.

Second, the software should take advantage of the extended memory of the 64180. Part of the attraction of 16-bit microprocessors is their large memory space, but, in real-time process control applications, an 8-bit computer will often run much faster than a 16-bit machine. The 64180 overcomes the 64K-byte memory barrier by incorporating a memory management unit (MMU) on the chip. The BCC180 has 384K bytes of RAM and EPROM on-board; the software must be able to use this.

Third, the software must be *fast*. Many real-time applications must respond to interrupts or other external events in milliseconds. The BCC180's supporting software must generate code that executes quickly, so that the system won't miss these events.

Finally, while multitasking is only now becoming common in the personal computer world, it has long been an important part of real-time systems. I dictated from the very beginning that any language for the BCC180 must implement multitasking.

Fortunately, I didn't have to start from scratch in finding this "perfect" software. Softaid created a custom-tailored operating system and language for the BCC180 by modifying its MT-BASIC compiler. The result was BASIC-180, which is a comprehensive BASIC specifically designed to meet the needs of the process control industry. It has all the features engineers and programmers have come to expect, like multitasking, floating-point math, multiline user-defined functions, and windowing (see the text box on page 67).

Why Multitasking?

Multitasking is the process of running two or more activities on a single computer at (apparently) the same time. It is important to distinguish it from multiprocessing (or multiprogramming), which is the process of running several activities on several processors at the same time. Multitasking is also not the same thing as "multiuser." A multitasking system is often single-user, although all multiuser systems are multitasking.

We say "apparently" because a single CPU can execute only one instruction at a time. The system performs multitasking by switching the processor between two or more activities at a high rate of speed. If two activities are sharing one CPU, the computer might execute one for 0.01 second, then the other for 0.01 second, and then switch back to the first. Over the course of time, each activity gets 50 percent of the available computer time. If three activities are running, each gets 33 percent.

Even a multimillion-dollar UNIVAC or IBM mainframe works this way. A hundred or more users might be connected to a single-processor machine. Each appears to have sole control of the computer. The CPU switches between users thousands of times per second, giving each one perhaps 0.001 compute-seconds at a time. The computer is so fast, and humans so slow by comparison, that the users don't notice that they are sharing the machine.

The building block of a multitasking program is the *task*. A task is one logical activity that runs as a whole and that competes for computer time with other tasks. On a large mainframe computer, each task might be a single user's program, or sophisticated users might partition their program into a number of tasks that don't necessarily have to run sequentially.

Since every task competes for computer time, we say that tasks execute "concurrently." A simple multitasking system might alternately run each task in order. For instance, a three-task program would execute task 1, then 2, then 3, then 1 again, in this order, forever.

Of course, computers are never this simple; in most multitasking systems, the time-critical tasks can be commanded to run more often than others. This implies that the tasks run asynchronously with respect to each other. In other words, we really don't know what task is executing at any given time or in what order they'll run. However, since a task is a logically complete processing element and does not depend on the results of other tasks, this isn't a problem.

Many programmers can't envision how one program can be broken into asynchronous, independent activities. A simple example is a low-cost digital thermometer using multiplexed LED displays. A multiplexed display must be constantly refreshed. Only one LED segment is actually turned on at any time, but each segment is cycled so quickly that to the eye, they all appear to be on. Although a hardware-refresh controller is usually used to control the display, a small system can dispense with the extra hardware by using software to control the refresh.

One task would take the data to be displayed and cycle the LED segments as required. Another task would read the thermistor, compute the temperature, and pass the reading to the refresh task for display. Note that each task is completely independent. Each one can run by itself and doesn't care about the others. The only intertask communication is the displayed temperature.

Interrupts: The Key Ingredient

All multitasking systems rely on one critical hardware component: a regular source of interrupts. Whenever an interrupt is detected, the CPU stops executing the current routine and branches to another section of code called an interrupt service routine (ISR). In a multitasking system, an interrupt is applied regularly (say, 100 times per second) to the CPU. Whenever the CPU detects this interrupt, the ISR associated with that interrupt suspends the execution of the current task, preserves the state of the machine at the time of the interrupt, and starts another task going.

This process is called context switching, since the current state, or context, of the machine is preserved before another task is started. Since the entire context of the interrupted task is saved, that task can be restarted exactly as if it had never stopped. In other words, the task itself has no idea it is being interrupted, suspended, and then eventually restarted.

The context-switching interrupt is called the "clock tic," since it resembles the regular tick of a clock. The faster the tics come, the more often each task executes. If a system with 20 tasks has a 20-Hz clock, each task will execute once per second.

It would seem that increasing the frequency of the tics only makes things better, but there is a catch-22. The context task-switching code requires a certain amount of time to decide which task to execute next and to prepare that task for execution. If interrupts come too quickly, the processor spends most of its time deciding what to do next and never actually gets to the task itself.

On most microcomputers, an interrupt rate of 60 Hz tends to work well. This is an ideal number for generating useful timing values, yet it is not so fast that too much time is taken up with task overhead. The 64180 processor includes two separate timers, each of which can be programmed to generate regular interrupts at virtually any rate. BASIC-180 programs timer 0 to

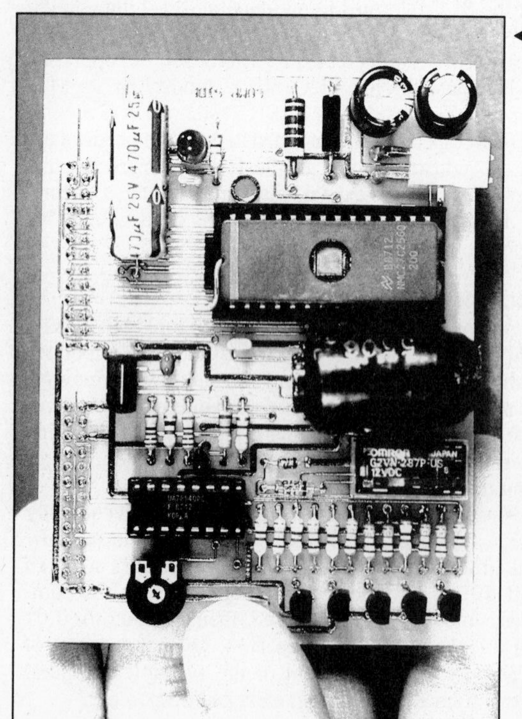

◀**Photo 2:** *The EPROM programmer daughterboard for the BCC180. An EPROM is shown inserted in the zero-insertion-force socket (upper right of board).*

Photo 3: *The EPROM programmer daughterboard (from photo 2) is shown here attached to the BCC180 computer/controller.*
▼

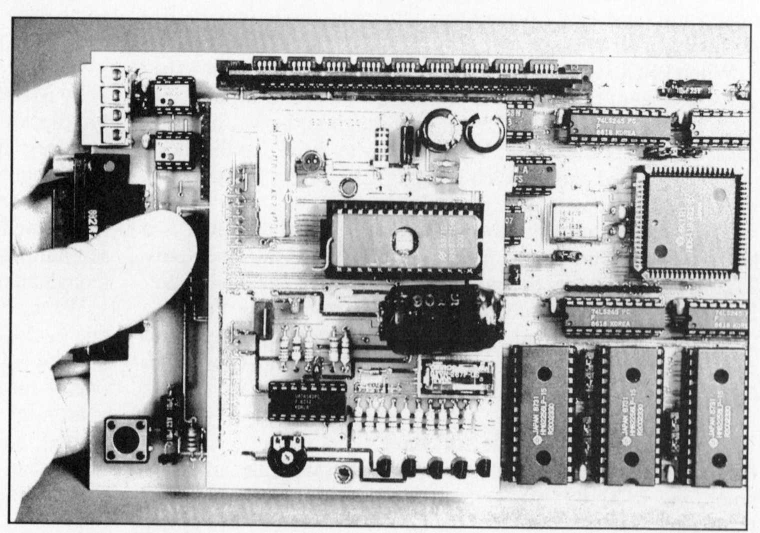

How Fast Is Fast?

BASIC-180 generates true native 64180 code. An intrinsic optimizer is automatically invoked with each compile to ensure fast, efficient code generation. Although BASIC-180 is not as fast as a good C compiler, it compares favorably with other compiled languages. For comparison's sake, we benchmarked it against the MS-DOS version of MTBASIC, GW-BASIC, and CP/M's MBASIC.

BASIC-180 was tested running at 9.216 MHz on a BCC180 board. MBASIC was tested on a 6.144-MHz SB180 board. The MS-DOS BASICs were executed on a 4.77-MHz IBM PC. Comparing the CP/M and MS-DOS BASICs is a little like comparing apples and oranges, but it gives a feel for the differences in speed.

The test program used was the Sieve of Eratosthenes originally described in the September 1981 BYTE. Times listed are for one iteration of the code (see table A).

BASIC-180 has a provision to drastically speed up execution of programs. If the `NOERROR` option is specified before compiling, BASIC-180 generates a smaller file that runs faster. `NOERROR` removes much of the run-time error checking, so it should be used only on debugged programs. For example, when `NOERROR` is specified, no tests are made to see if an array subscript exceeds its maximum allowable value. When the Sieve program was run with `NOERROR`, BASIC-180 ran in 2.1 seconds.

Table A: *Times in seconds for various compilers and interpreters to execute the Sieve program. See the text for a description of each language.*

Language	Time
BASIC-180 (compiler)	7.1
MS-DOS MTBASIC (compiler)	8.6
MS-DOS GWBASIC (interpreter)	101
CP/M MBASIC (interpreter)	141
Whitesmith's C (compiler)	1.6
Microsoft FORTRAN (compiler)	1.7
BDS C (compiler)	4.9

generate mode 2 (vectored) interrupts at a 60-Hz rate (tics are 16.67 ms apart).

BASIC-180's context switcher is started each time the timer interrupts. If a multitasking program is running, this interrupt causes the context switcher to start another task. In the software supplied with the BCC180, a 60-Hz internal interrupt clock results in under 5 percent context-switching overhead (it's less still if we use an external 60-Hz interrupt source).

Tasks can also be controlled by interrupts other than those generated by clock tics. In more sophisticated multitasking systems designed from the start for process control, like the BCC180, tasks can be configured to start on the receipt of an interrupt from an external push button or limit switch, for example (the BCC180 has four external interrupt inputs).

A World of Multitasking Opportunities

Applications for multitasking abound. All large process-control applications involve many tasks that must be handled concurrently. Take a steel mill, for example. The computer controlling a steel-rolling mill's production can't suspend operations when the operator enters data into a keyboard. One task should just handle the keyboard. Another can be assigned to reading steel thickness, generally by measuring the amount of absorption of gamma rays produced by a radioactive source like cesium or americium.

A third task could be responsible for controlling the mill's jack screws to alter the thickness of the steel being rolled. Another task could measure the steel's temperature and compute a correction to the thickness as a function of temperature (2200-degree steel is several percent thicker than room-temperature steel). Other tasks can perform calibrations of the electronics, display computed thickness values on various consoles, and provide financial and historical data on the steel being produced.

Multitasking on the BCC180

On the BCC180, BASIC-180 is both the high-level language and the operating system. It contains all the device drivers and all the multitasking control code. As is the case with an operating system, BASIC-180 provides the entire environment that is seen by the programmer.

Regular Circuit Cellar readers know that I often use BASIC for demonstrating projects. I've found that in the process-control industry, BASIC is the language of choice, too. Let's face it, BASIC is the lowest common denominator in programming languages. Everybody knows BASIC. The same cannot be said of any other language.

BASIC-180 includes a complete set of statements for controlling multitasking. All we have to do is write our multitasking BASIC program, and BASIC-180 will ensure that the tasks sequence properly.

In BASIC-180, all multitasking programs are divided into some number of tasks. A lead task, sometimes referred to as task 0, must always exist. The lead task is the main program. For multitasking to commence, the lead task must start at least one other task running. Let's look at a simple example program:

```
10 RUN 1,60
20 GOTO 20
30 TASK 1
40 PRINT "Task 1"
50 EXIT
```

Although obviously BASIC, this short program contains a number of unfamiliar statements. Lines 10 and 20 comprise the lead task, while lines 30, 40, and 50 define task 1. (Remember that a task is a logically distinct section of code that will be executed concurrently with other tasks.) Every task (except the lead task) starts with a `TASK` statement. This defines the start of the task and assigns a reference to the task for use by the other statements. In this case, the task is defined as task number 1.

Line 50, the `EXIT` statement, defines the termination of the task. Whenever an `EXIT` statement is executed, the task associated with that statement will be terminated.

Line 40 forms the body of the task. In this case, when task 1 runs, it will print `Task 1` once and the task will terminate.

Line 10 is a `RUN` statement, not to be confused with the `RUN` command that starts a program executing in most BASICs. `RUN` always takes two arguments. The first argument is the number of the task to execute (in this case, task 1). The second argument is how often to execute that task.

In any other BASIC, line 20 is a bizarre aberration. It is an infinite loop, so of course no other processing can go on. Not in BASIC-180! Task 1 and line 20 compete for computer time. Processing will be shared between the two activities.

When the program starts, the RUN statement kicks off task 1. It places task 1 in the ready for execution state. When the next tic is detected, task 1 will start running. Although task 1 is short and simple, it is unlikely that the Task 1 message will be printed in less than one tic of the clock. If it is in the middle of printing the message and another tic comes, the loop at line 20 will be executed for the duration of a tic (16.67 ms).

When another tic is detected, task 1 will resume from where it left off. The user will not be able to tell the task was interrupted. After the message is printed, task 1 will exit. When a task exits, it effectively dies and stops competing for processor time.

The second argument of the RUN command (in this case, line 60) tells the context switcher to restart task 1 sixty tics (1 second on the BCC180) after it exits. This is analogous to reincarnation. Although the EXIT statement makes the task die, it will be reborn after a certain period called the schedule interval.

Consider the following program:

```
10 RUN 1,60
20 GOTO 20
30 TASK 1
40 <code>
50 GOTO 40
```

In this case, <code> represents one or more BASIC statements that do whatever the task is responsible for. This case is similar to our previous example, except the task never dies; the GOTO at line 50 keeps the task active forever. It will continue to share time with the line 20 loop, but since an EXIT is never executed, the task never goes away. The reschedule interval in the RUN statement (line 60) is ignored.

The concept of scheduling is an important one in multitasking programs. One of the most important resources in real-time systems is processor time; if it is all used up, the processor will not be able to keep up with the real-world events it is responsible for monitoring.

Although a task that has nothing to do can idle by executing an empty FOR...NEXT loop, this is a terrible waste of computer time. It makes much more sense for the task to execute an EXIT and set a schedule interval so it will be born again when needed. Between the time a task executes the EXIT and the time it is reborn, it uses no processor time.

In the following program, three tasks execute concurrently. Task 1 integrates 10 reads of some asynchronous event. It assumes that some other task is filling variable T with data. Task 1 smooths the data passed in T, returning a filtered floating-point value in AD. Task 2 prints the value in AD once every 2 seconds. Task 3 fills T with data. (In this case, we use random numbers for the data.)

```
10 REAL AD
20 INTEGER T1,T,I
30 RUN 1,1
40 RUN 2,120
50 RUN 3,20
60 GOTO 60
100 TASK 1
110 FOR I=1 TO 10
120 T1=T1 + T
130 NEXT I
140 AD=T1 / 10.0
150 GOTO 110
200 TASK 2
210 PRINT AD
220 EXIT
300 TASK 3
310 T=RND
320 EXIT
```

This example shows that all variables in a BASIC-180 program are global. You can pass data between tasks through the variables. (Integers are all loaded and stored using 16-bit instructions. Since an interrupt can be processed only when an instruction is complete, integers are always stored intact. Floating-point numbers are loaded and stored only with interrupts disabled. BASIC-180 briefly disables the interrupts during these transfers to ensure that the variables will not be corrupted).

Let's look at a program that demonstrates Nyquist's theorem (also known as the sampling theorem). Nyquist said that in order to accurately represent a signal, you must digitize it at a rate of at least twice the highest frequency in the sample. Therefore, to accurately digitize 60-Hz AC, you should sample it at least 120 times per second. If a signal is sampled at too low a rate, the digitized signal may look like something altogether different from the original (an effect known as aliasing).

In the following program, task 1 generates a low-frequency sine wave. Task 2 samples it asynchronously, as would be the case if you constructed an analog-to-digital (A/D) converter to read the AC power's sine wave. You can specify the sample rate to task 2, which is simply how often the task is scheduled. A low number means a high sample rate, and an accurate representation of the sine wave is thus obtained. A large number will cause task 2 to run only occasionally, yielding a distorted picture of the sine wave.

```
100 INTEGER I,J,K,S
110 REAL A
115 I=0
190 PRINT : PRINT : PRINT : PRINT : PRINT
200 PRINT "Sampling Theory demonstration"
220 PRINT
230 PRINT "Enter the sampling rate (1 to
        1000) ";
240 INPUT S
300 RUN 1,20
310 RUN 2,S
330 GOTO 330
500 TASK 1
510 A=SIN(I)
520 I=I + 10
530 IF I < 360 THEN GOTO 550
540 I=0
550 EXIT
600 TASK 2
605 K=A * 30. + 35
610 FOR J=1 TO K
620 PRINT " ";
630 NEXT J
640 PRINT "*"
650 EXIT
```

BASIC-180 supports several other multitasking statements. These statements are designed to give you more control over the operation of each task.

The WAIT statement lets you manually suspend a task for any period of time. When a task issues a WAIT, that task no longer receives access to CPU time until the number of tics given as WAIT's argument have elapsed. WAIT is essentially a free delay mechanism, because the delay requires no CPU time to manage. WAIT takes one argument: the number of tics to delay for. In the null loops in the previous examples, a much more efficient construct is

```
20 WAIT 1000
30 GOTO 1000
```

This uses virtually no computer time, since the program spends most of its time in the WAIT.

CANCEL is a means to stop a task from being rescheduled. Remember that all tasks will restart some time after executing an EXIT statement, the time being determined by the RUN statement's second argument. If you CANCEL a task, once it completes its current execution, it will not be reborn. You can restart scheduling for the task by issuing another RUN command with the appropriate arguments. CANCEL has only one argument: the number of the task to cancel. A task can cancel itself, and any task can cancel any other task.

Finally, PRIORITY is a powerful statement that lets you set a relative importance for each task. All tasks, in the absence of a PRIORITY statement, operate at the same priority (i.e., they all compete for time equally). Any task can raise or lower its priority by issuing a PRIORITY statement, followed by a number indicating relative importance. The number can range from 0 to 63, where 63 is the highest priority and 0 is the lowest. Tasks that don't issue a PRIORITY statement operate at priority level 0.

Normally, whenever a tic interrupt is received, BASIC-180 interrupts the current task and, using the task number that is one greater than the task just interrupted, searches for another task that is ready to execute. In other words, it tries to run task 1, then 2, 3, etc. If a task has issued a WAIT instruction and the wait interval has not elapsed, that task will be skipped. This scheduling technique is called round-robin scheduling.

When tasks execute at different priority levels, every time a tic interrupt is received, BASIC-180 searches for the highest-priority task that is ready to execute. If several tasks are ready, but one has a higher priority than any of the others, that task will execute until it executes an EXIT command or a WAIT command, or lowers its priority.

A task can issue a PRIORITY command at any time. Tasks can dynamically raise and lower their priorities as warranted. BASIC-180 allows up to 32 tasks to be active. With each of 32 tasks raising and lowering priorities, issuing CANCEL and RUN commands at each other, you could construct quite a complex program.

In Conclusion

In Chapter 5, part 3, I'll finish this tutorial on multitasking and the BCC180 with a discussion of BASIC-180's special windowing capability and memory management features.

I'd like to acknowledge and personally thank Ken Davidson and Jack Ganssle for their efforts on the BCC180 project. Ken Davidson's extensive knowledge of the HD64180 helped us avoid the omnipresent hardware design pitfalls, and Jack Ganssle's superb software talents helped explain multitasking in a way that can really be understood.

Editor's Note: Steve often refers to previous Circuit Cellar articles. Most of these past articles are available in book form from BYTE Books, McGraw-Hill Publishing Company, P.O. Box 400, Hightstown, NJ 08250, (1–800–2–MCGRAW).

Ciarcia's Circuit Cellar, Volume I covers articles in BYTE from September 1977 through November 1978. *Volume II* covers December 1978 through June 1980. *Volume III* covers July 1980 through December 1981. *Volume IV* covers January 1982 through June 1983. *Volume V* covers July 1983 through December 1984. *Volume VI* covers January 1985 through June 1986.

It's virtually impossible to provide all the pertinent details of a project or cover all the designs I'd like to in the pages of BYTE. For that reason, I have started a bimonthly supplemental publication called Circuit Cellar Ink, which presents additional information on projects published in BYTE, new projects, and supplemental applications-oriented materials. For a one-year subscription (6 issues), send $14.95 to Circuit Cellar Ink. Credit card orders can call (203) 875-2199.

For more information on MTBASIC for Z80, 64180, or MS-DOS machines, contact Softaid Inc., 8930 Route 108, Columbia, MD 21045, (301) 964-8455.

The following items are available from

Micromint Inc.
4 Park St.
Vernon, CT 06066
For orders: (800) 635-3355
For information: (203) 871-6170
Telex: 643331
Fax: (203) 872-2204

1. A 9-MHz assembled and fully socketed BCC180 computer/controller board with 32K bytes of static RAM, ROM monitor, BASIC-180 development software (same as item 3), and user's manuals.
BCC180-1-20..............$395; for additional 256K DRAM, add $100
2. BCC180 PAK evaluation system. Contains a 9-MHz BCC180 board with 32K-byte static RAM, ROM monitor, BASIC-180 development software, MB08 eight-slot backplane, CC01 10-inch card cage, UPS10 35-watt switching power supply, and user's manuals.
BCC180-PAK ....$595; for additional 256K DRAM, add $100
3. BASIC-180 multitasking BASIC compiler for ROM- or disk-based development. Contains both BASIC-180 EPROM for direct use on BCC180 board and BASIC-180 disk for direct use or software development on SB180. Includes 100-page user's manual. Compiled code may be freely used without further license. BASIC-180 DEV $250
4. An auxiliary 27256 EPROM programmer board for the BCC180.
BCC180PROG5..$89

The following items are available from

Circuit Cellar, Inc.
4 Park St., Suite 12
Vernon, CT 06066
(203) 875-2751
Fax: (203) 872-2204

1. A 9-MHz BCC180 computer/controller complete kit with 32K bytes of static RAM, ROM monitor, BASIC-180 development software, and user's manual. BCC180-KIT-20....................................... $295
2. BCC180 auxiliary 27256 EPROM programmer board full kit.
BCC180PROG5K...$74

For either source above, all payments should be made in U.S. dollars by check, money order, MasterCard, Visa, or American Express. Surface delivery (U.S. and Canada only): add $5 for U.S., $8 for Canada. For delivery to Europe via U.S. airmail, add $14. Three-day air freight delivery: add $10 for U.S. (UPS Blue), $25 for Canada (Purolator overnight), $45 for Europe (Federal Express), or $60 for Asia and elsewhere in the world (Federal Express). Shipping costs are the same for one or two units.

There is an on-line Circuit Cellar bulletin board system that supports past and present projects. You are invited to call and exchange ideas and comments with other Circuit Cellar supporters. The 300/1200/2400-bps BBS is on-line 24 hours a day at (203) 871-1988.

5

THE BCC180 MULTITASKING CONTROLLER

PART 3: MEMORY MANAGEMENT AND WINDOWING

The BCC180's multitasking BASIC compiler can operate on a variety of hardware—even on an IBM PC

In parts 1 and 2 of Chapter 5, I've been discussing a 64180-based multitasking computer, the BCC180 (see photo 1). In common with most of my projects, it is hardware, but because of its unique multitasking capabilities, I have made much of the presentation a tutorial on multitasking. In part 3, I'll conclude with a discussion of the memory management and windowing capabilities of BASIC-180.

Memory and the 64180

The 64180 processor has two great advantages over the Z80. First, it contains a wide variety of on-board peripherals, including timers and universal asynchronous receiver/transmitters (UARTs). Second—and more important—it has an on-board memory management unit (MMU).

The MMU lets the 64180 directly address up to 1 megabyte of memory. The memory connected to the 64180 is called physical memory; it is the entire "universe" of memory available to the processor. The memory that can be addressed with any one map, or configuration, of the MMU is called the logical address space. Every address generated by a user's program is a logical address.

The MMU's role is to translate logical addresses into physical addresses. If, for example, a program tries to reference location 0, this is a reference to logical address 0. The actual physical address will be a function of the mapping of the MMU. It may indeed use physical address 0 (on the power-up, the MMU translates every logical address to exactly the same physical address). Or, if the program had set the MMU, it could use physical address 10000 hexadecimal. (For the remainder of this chapter, all addresses will be in hexadecimal unless otherwise specified.)

Photo 1: *The BCC180 (tall board in the rear) shown here in a typical process-control configuration. In addition to the BCC180's own functions, this system has 32 channels of 12-bit A/D and 16 channels of optoisolated AC I/O.*

The 64180's MMU lets programs segment the memory on any one map into three different areas. For instance, references to logical addresses 0 to 3FFF could access physical addresses 0 to 3FFF, references to logical addresses 4000 to 7FFF could access physical addresses 10000 to 13FFF, and the last half of the logical space could access physical addresses 40000 to 47FFF. Virtually any combination is possible.

Using All Available Memory

Although the MMU provides a fairly convenient way to handle large amounts of memory while still maintaining Z80 compatibility, it is not magic. The MMU is useless unless applications programs are specifically written to use it. None of the thousands of existing CP/M programs can take advantage of the power of the MMU; even running on a 64180, they are still limited to 64K bytes of memory. Handling the MMU is not trivial. Only a handful of programs have been written to take advantage of this powerful feature. Fortunately, BASIC-180 is one of these.

Writing a compiler that generates code to automatically remap the MMU is a pretty tough task. How would you do it? One approach is to start compiling code from address 0 until FFFF is encountered, then (or shortly before) drop in code to remap the MMU so the next 64K-byte region is available, and so on. This approach can be made to work, but you pay a high price in terms of performance. Every FOR...NEXT, GOTO, and GOSUB may require MMU remapping. Worse, every time a program calls built-in run-time routines (of which there are always many), software may have to remap the MMU.

Interestingly enough, the IBM PC's famous lack of performance is partially due to this. Although the PC does not use a conventional MMU, addresses are pretty much limited to 16 bits. Address references outside a 64K-byte region must involve manipulation of the 8088's segment registers, hampering the machine's performance.

BASIC-180 takes a unique approach. In all operational modes except MAPCOMPILE (which I'll describe later), when a BASIC-180 program is running, it ignores the MMU. (Of course, BASIC-180 does initially configure the MMU, but this occurs only at the program's start.) This ensures that BASIC-180 is compatible with all Z80 systems. You can compile programs of any size to disk, even though the compiler uses only around 48K bytes. If you compile a program using the MAPCOMPILE command, BASIC-180 uses the memory map shown in figure 1.

BASIC-180 compiles the "run-time package"—or library of support code required by a program—at logical address 0 to 3FFF. The compiler maps these logical addresses to the same physical addresses (0 to 3FFF).

The variable area consists of both variables defined in the program and temporaries required by the run-time routines. This area must reside in RAM, whereas all the compiled code typically is in EPROM. The temporary area is compiled from the top of logical memory (64K) down for as far as is required. If 16K bytes of RAM is needed, it is compiled to logical address C000 to FFFF. This area is mapped to physical addresses starting at 7FFFF and down.

Finally, BASIC-180 compiles the user's program into logical addresses between the run-time and RAM areas. Every task is compiled to the same range of logical addresses. Thus, the lead task starts at address 4000, as does task 1, task 2, and so on. Although these tasks have the same logical address, they reside in memory at different physical addresses. The upshot is that, since a given task "sees" only its own logical space, multiple tasks executing on the same system are largely unaware of one another.

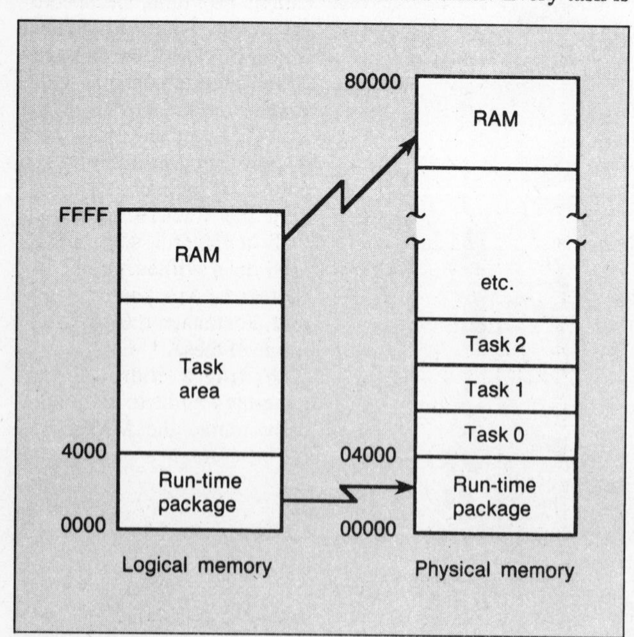

Figure 1: *Memory map for programs compiled using the* MAPCOMPILE *command, showing the relationship between logical memory and physical memory.*

Context Switching

The trick to the success of this approach is context switching. In a multitasking environment, only one task can be executing at any specific moment. BASIC-180 takes advantage of this fact by having the context switcher remap the MMU whenever a tick interrupt is received. The currently executing task's context is saved, and this task is mapped out ("into the ozone," as the experts say) and the next task mapped in. Now, the new task is at the same logical address space as the last one was and can start running.

On the BCC180, a hardware interrupt drives the software context switcher to sequence multitasking. The interrupt is provided by the 64180's internal timer number 0, which BASIC-180 configures to generate 60 interrupts per second. The interrupts from the timer invoke BASIC-180's context switcher, which must then suspend execution of the current task and decide which task to execute next.

The context switcher is structured around a table called the task control block (TCB). Each TCB entry includes the task's current state (e.g., active and suspended), its reschedule interval (specified on the task's RUN statement), how many ticks are left until the interval elapses, the task's priority, the task's memory management map (so the context switcher knows where the task exists in physical memory), and the task's current stack pointer.

Each time the context switcher is invoked, it examines every entry in the table to determine which task must run next, searching in order of priority. The switcher checks tasks' states to find those that are either *suspended* (whose execution was stopped in midstream by an interrupt) or *ready* (they've never been executed but are now ready for execution). It examines tasks of equal priority in a round-robin fashion, giving them all an equal chance to execute.

The switcher starts a ready task at that task's first executable location, and a suspended task at the location it left off when last interrupted. If no task currently requires processor time, the context switcher idles, waiting for the next interrupt. (See table 1 for a further breakdown of task states.)

Context switching is a complex procedure, requiring a substantial number of instructions. The TCB is arranged in a doubly sorted order (by round-robin and by task priority) to speed TCB searches. At 60 ticks per second, less than 5 percent of the processor's time is consumed by context switching.

The process of remapping the MMU occupies a tiny percentage of the context-switching time, so little processing overhead is added. The system can therefore manage large multitasking programs with numerous tasks that sprawl across a megabyte of memory at little expense of processor time.

The 64180's MMU has a minimum mapping resolution of 4K bytes. Therefore, each task compiled by BASIC-180 will take at least 4K bytes, since each task must reside in a separate map.

The revision 1 mask of the 64180 in a plastic leadless chip carrier (PLCC) package supports 20-bit address spaces or 1 megabyte of RAM. BASIC-180 has a somewhat arbitrary limit of 32 tasks per program. In a 1-megabyte address space, each task can therefore take up to 32K bytes (since 32K bytes × 32 tasks = 1 megabyte). This is quite reasonable, since the run-time package uses 16K bytes. If 16K bytes of RAM is required for variables and temporaries, 32K bytes is left for tasks in the logical address space (16K run time + 16K variables/temporaries + 32K program = 64K, the *logical* address space of the 64180).

As you can see, in BASIC-180 large programs are intrinsically tied to multitasking. Nonmultitasking programs are restricted to 64K bytes of total memory use. This encourages programmers to think in terms of segmenting a program into separate, asynchronous activities. A future Circuit Cellar will cover a multiprocessor system, in which this programming style becomes very important.

Table 1: *Task states.*

Active: The task is actually executing. Only one task can be active at any time.

Ready: The task has as yet not been executed but is ready for execution as soon as CPU time becomes available.

Suspended: The task was executed but was preempted by a clock tick. It is ready to resume execution from where it left off as soon as CPU time becomes available.

Waiting: The task was executing but put itself in an idle mode. The CPU will restart the task after a specified amount of time measured in clock ticks.

Dormant: The task has not been made available for execution. It requires no CPU time.

Interactive Compilation

I don't have the patience to use most high-level languages. The scenario usually goes like this: You invoke an editor, enter the program, call up the compiler, which inevitably points out a few misplaced semicolons (I thought the computer was supposed to relieve me of trivia . . .), so you have to go back to the editor, fix the code, and recompile, by now hoping desperately that some other silly syntactical mistake isn't still lurking.

After the compiler is finally happy, you repeat the same process with the linker. (Oh, no! The linker looks at only seven characters of a symbol, but my language recognizes eight! Back to the editor, boys.) Finally, you're ready to test the program. If it doesn't work, you'll often edit lots of print statements into the code to try and localize the problem. No wonder programmers get only 10 lines of good code a day.

BASIC-180 supports two very different development environments. You can write and test code on the BCC180 itself, or you can write and at least partially test code on a host computer like the SB180. This kind of cross-development environment is nice because you can use all the resources of the operating system.

The most important of these is the disk system. (I'm seriously considering adding a floppy disk drive and small computer system interface [SCSI] hard disk interface board to the BCC bus very soon. Circuit Cellar Ink will contain news of that construction project.) You can save various versions of the source and compiled code, adapt portions of old programs for use in a new one, and take advantage of operating-system utilities to ease the development process.

The downside of developing code on a host system is the lack of compatible I/O. In any process-control system, the BCC180 will no doubt be connected to unusual I/O devices not available to the host development system (e.g., parallel I/O lines connected to contact closures, relays, and so forth). For the present, this involves a trade-off.

Most systems use limited access to I/O. If speed is not the overriding consideration, you should use well-designed modular code to segment the I/O drivers into driver routines. When operating in a cross-development mode, you can replace these drivers with code that simulates the behavior of the physical devices. Perhaps only limited simulation is practical, but you can still test much of the code.

BASIC-180 will run on the SB180 just as it does on the BCC180. The BASIC source code is 100 percent compatible. BASIC-180 will even run on an IBM PC under a CP/M simulator like ZSIM from Z-World.

On a cross-development system, you can compile and test BASIC code in one of three ways: interactively, using the `DISKCOMPILE` command, or using the `MAPCOMPILE` command.

When compiling interactively, BASIC-180 appears to operate just like an interpreter. You sign on to the BASIC and edit, run, test, and modify programs, all under the control of BASIC-180. BASIC-180 includes a line editor for easy entry and alteration of programs. Using the `SAVE` and `LOAD` commands, you can save code to and load it from disk.

The `RUN` command starts a BASIC program executing, as in an interpreter. However, when you issue `RUN`, BASIC-180 compiles the program into true native machine code, then executes it. The program—being compiled instead of interpreted—runs faster. Furthermore, the `RUN` command causes BASIC-180 to compile the program to RAM—not disk—so compilation is also speedy. (On the SB180, compilation is over 150 BASIC statements per second.) Best of all, you can test, modify, and retest the program in seconds, just as in an interpreter. As far as I'm concerned, this is the only way to develop code. Only a few other languages now use this approach.

Once you have your program working the way you want it to, you can compile it to an efficient disk file using the `DISKCOMPILE` command. BASIC-180 reads the BASIC source code from a disk file, converts it to native 64180 object code, and saves it to a stand-alone executable file, just like any other CP/M program.

You can use `DISKCOMPILE` to create program files that you ultimately convert to Intel hexadecimal format and burn into EPROMs for stand-alone execution on the BCC180. BASIC-180 always generates ROMable code. This means that programs created with BASIC-180 require only limited operating-system support, if any, and the code and data areas of a BASIC-180 program are kept separate. BASIC-180 lets you define where RAM should be, so you can support the segmented memory typical of ROM-based systems.

To create large multitasking programs, you use BASIC-180's `MAPCOMPILE` command. `MAPCOMPILE` produces a root .COM file that contains a loader and the run-time routines. It then generates a file for each task. These files have the extension .Txx, where xx is the task number. When you invoke the root program (by typing its name, just as you would for any other .COM file), the built-in loader reads each of the task files from disk into the proper area of physical memory, remapping the MMU as required.

Plug and Go

A truly unique feature of the BCC180 version of BASIC-180 is that the compiler itself, and not just the compiled code, can operate from ROM on the BCC180. Most languages need a great deal of operating-system support. A few interpreted BASICs operate from ROM (like the ROM BASIC for the IBM PC), but you'll find few ROM-resident BASIC compilers.

BASIC-180 supports its interactive development environment even in ROM. You'll find nearly all the features available on disk-based systems in the ROM version, including a line editor.

The ROM version of BASIC-180 requires just the BCC180 board, a terminal, and a power supply (you'll need the optional BCC180 EPROM programmer board to generate ROM-based programs). When powered up, the BASIC or a system monitor is available immediately. You can enter, edit, and test programs, just as if you were using an interpreter.

The great advantage of developing code on the BCC180 is that you have all the BCC180's resources available. There is no need to write I/O simulation routines; you can develop critical I/O handlers interactively. Without disks, however, there would seem to be no way to save BASIC source or compiled code. Fortunately, past experience suggested a solution to this problem.

One of the better features of my 8052-based BCC52 is its onboard EPROM programmer that functions like write-once mass storage. I can't stand to rekey programs, so when Softaid's people designed this BASIC-180, I had them add the ability to save programs directly to EPROM in the same manner as the BCC52. When you connect the BCC180's auxiliary EPROM burner, you can use ROMSAVE and ROMLOAD to save and retrieve programs to and from an EPROM (27256). BASIC-180 burns a tokenized (compressed) form of the BASIC source program into the EPROM.

Of course, BASIC-180 doesn't create any sort of file structure on the EPROM. ROMSAVE simply burns each program into the next unused spot in the EPROM, assigning that program a sequential number that ranges from 1 to the number of programs in the EPROM. You load a program from an EPROM using ROMLOAD, which requires as an input argument the ROMSAVE-generated number of the program.

You can also use MAPCOMPILE to compile programs into the EPROM. In this case, MAPCOMPILE burns the programs into the EPROM at the compiled physical addresses, so that you can create a program that executes on power-up by placing the EPROM in the appropriate BCC180 ROM socket(s).

(a)

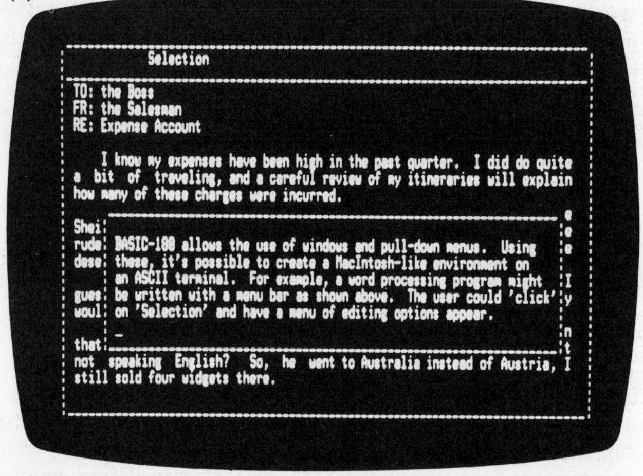

(b)

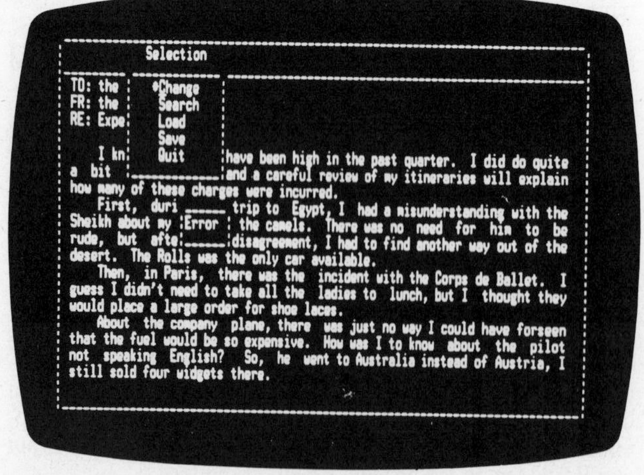

Photos 2a and 2b: *BASIC-180's multitasking windows are sophisticated enough to simulate a Macintosh environment on a serial terminal, as shown in* **(a)** *and* **(b)** *above.*

Windowing

When I first ran into multitasking, I was confused as to how different tasks would handle terminal display I/O. What happens when I program three tasks to simultaneously display results? Do I end up with a jumbled mess of asynchronous PRINT statements going to a common display? Of course not.

BASIC-180 gives you a complete set of window control statements. A window is a subsection of the CRT display terminal with a definable size and position. A window can be any size (up to 24 lines by 80 characters), and you can have up to 10 windows in use at once. When your program selects a window, all output will go to that window until the program selects another. The window's borders act as barriers to PRINT statements (see photos 2a and 2b).

BASIC-180 lets you assign windows to tasks, so that you might assign window 1 to task 4 and window 2 to task 12. All output from task 4 will then go to window 1, and all output from task 12 will go to window 2.

The following BASIC-180 statements control windowing:

WINDOW defines a window's size and location.
WSELECT selects a window for output.
WFRAME draws an outline around a window.
WCLEAR erases a window.
WSAVE saves the contents of a window to an array.
WUPDATE restores to the screen a window that has been saved with WSAVE.

The following program defines a square window 10 characters wide and high, whose upper left corner is at column 10, row 10, on the CRT. It is defined as window number 1. A box is drawn around it using the "_" and "|" characters:

```
10 WSELECT 1
20 WINDOW 10,10,20,20
30 WFRAME "_","|"
```

All subsequent PRINT statements will send their output to this window.

The WSAVE and WUPDATE statements let you write programs with pop-up or pop-down windows, so you can create programs with a Macintosh-like user interface. WSAVE saves on-screen data located behind the pop-up/pop-down window, and WUPDATE restores the screen after the window has been removed.

The following program draws three windows side by side on the screen and assigns a task to each one. All three tasks then run concurrently, each one writing to the window assigned to it:

```
5 INTEGER T1,T2,T3,A,B,C,D,I
10 ERASE
100 DATA 1,1,10,10
105 DATA 1,12,10,22
110 DATA 1,26,10,35
200 FOR I=0 TO 15 STEP 4
210 WSELECT I / 4
220 READ A,B,C,D
```

```
230 WINDOW A,B,C,D
240 WFRAME "_","|"
250 WINDOW A + 1,B + 1,C - 1,D - 1
260 NEXT I
300 RUN 1,30
301 RUN 2,50
302 RUN 3,70
310 GOTO 310
1000 TASK 1
1010 WSELECT 0
1020 WCLEAR
1030 FOR T1=1 TO 9:
     PRINT "TASK 1": WAIT 5:
     NEXT T1
1040 EXIT
1100 TASK 2
1110 WSELECT 1
1120 WCLEAR
1130 FOR T2=1 TO 9:
     PRINT "TASK 2": WAIT 5:
     NEXT T2
1140 EXIT
1200 TASK 3
1210 WSELECT 2
1220 WCLEAR
1230 FOR T3=1 TO 9:
     PRINT "TASK 3": WAIT 5:
     NEXT T3
1240 EXIT
```

Windowing makes extensive use of the cursor-positioning commands of the terminal you're using but, since no standard exists for terminal cursor positioning, the disk-based version of BASIC-180 comes bundled with an installation program that lets you configure it for your terminal or terminal emulator (the ROM-based version has patch points). BASIC-180 can support virtually any terminal's format.

In Conclusion

Finally, I am the quintessential nonprogrammer. I even design hardware to get around programming. The BCC180 and BASIC-180 do serve my purpose, and I hope you find them useful as well. Indeed, they let you work in an easy language, and they support the multitasking that has become so essential as applications become more complex. But, foremost, they raise personal computing performance to a level you would expect only in more expensive microcomputer systems.

I'd like to acknowledge and personally thank Ken Davidson and Jack Ganssle for their efforts on the BCC180 project. Ken Davidson's extensive knowledge of the HD64180 helped us avoid the omnipresent hardware design pitfalls, and Jack Ganssle's superb software talents helped explain multitasking in a way that can really be understood.

Editor's Note: Steve often refers to previous Circuit Cellar articles. Most of these past articles are available in book form from BYTE Books, McGraw-Hill Publishing Company, P.O. Box 400, Hightstown, NJ 08250, (1–800–2–MCGRAW).

Ciarcia's Circuit Cellar, Volume I covers articles in BYTE from September 1977 through November 1978. *Volume II* covers December 1978 through June 1980. *Volume III* covers July 1980 through December 1981. *Volume IV* covers January 1982 through June 1983. *Volume V* covers July 1983 through December 1984. *Volume VI* covers January 1985 through June 1986.

It's virtually impossible to provide all the pertinent details of a project or cover all the designs I'd like to in the pages of BYTE. For that reason, I have started a bimonthly supplemental publication called Circuit Cellar Ink, which presents additional information on projects published in BYTE, new projects, and supplemental applications-oriented materials. For a one-year subscription (6 issues), send $14.95 to Circuit Cellar Ink, P.O. Box 2099, Mahopac, NY 10541–9875. (Canada/Mexico: add $3 surface mail, $12 airmail; all other foreign countries add $12 for airmail. U.S. funds only.) Credit card orders can call (203) 875–2199.

For more information on MTBASIC for Z80, 64180, or MS-DOS machines, contact Softaid Inc., 8930 Route 108, Columbia, MD 21045, (301) 964-8455.

The following items are available from

Micromint Inc.
4 Park St.
Vernon, CT 06066
For orders: (800) 635-3355
For information: (203) 871-6170
Telex: 643331. Fax: 2038722204
Fax: (203) 872-2204

1. A 9-MHz assembled and fully socketed BCC180 computer/controller board with 32K bytes of static RAM, ROM monitor, BASIC-180 development software (same as item 3), and user's manuals.
BCC180-1-20..............$395; for additional 256K DRAM, add $100
2. BCC180 PAK evaluation system. Contains a 9-MHz BCC180 board with 32K-byte static RAM, ROM monitor, BASIC-180 development software, MB08 eight-slot backplane, CC01 10-inch card cage, UPS10 35-watt switching power supply, and user's manuals.
BCC180-PAK ....$595; for additional 256K DRAM, add $100
3. BASIC-180 multitasking BASIC compiler for ROM- or disk-based development. Contains both BASIC-180 EPROM for direct use on BCC180 board and BASIC-180 disk for direct use or software development on SB180. Includes 100-page user's manual. Compiled code may be freely used without further license. BASIC-180 DEV $250
4. An auxiliary 27256 EPROM programmer board for the BCC180.
BCC180PROG5...$89

The following items are available from

Circuit Cellar, Inc.
4 Park St., Suite 12
Vernon, CT 06066
(203) 875-2751

1. A 9-MHz BCC180 computer/controller complete kit with 32K bytes of static RAM, ROM monitor, BASIC-180 development software, and user's manual. BCC180-KIT-20..................................... $295
2. BCC180 auxiliary 27256 EPROM programmer board full kit.
BCC180PROG5K..$74

For either source above, all payments should be made in U.S. dollars by check, money order, MasterCard, Visa, or American Express. Surface delivery (U.S. and Canada only): add $5 for U.S., $8 for Canada. For delivery to Europe via U.S. airmail, add $14. Three-day air freight delivery: add $10 for U.S. (UPS Blue), $25 for Canada (Purolator overnight), $45 for Europe (Federal Express), or $60 for Asia and elsewhere in the world (Federal Express). Shipping costs are the same for one or two units.

There is a multiline Circuit Cellar bulletin board system (running TBBS 2.01M) that supports past and present projects in BYTE and Ink. You are invited to call and exchange ideas and comments with other Circuit Cellar supporters. The 300/1200/2400-bps BBS is on-line 24 hours a day at (203) 871-1988.

6

COMPUTERS ON THE BRAIN

PART 1

Clever signal amplifiers, noise rejection, and A/D conversion are all part of the HAL EEG

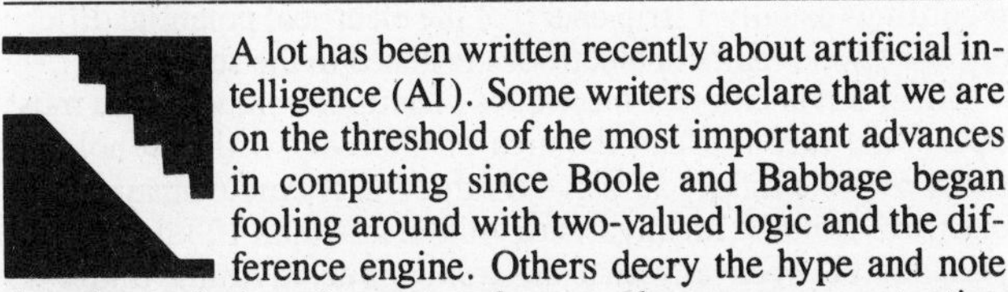

A lot has been written recently about artificial intelligence (AI). Some writers declare that we are on the threshold of the most important advances in computing since Boole and Babbage began fooling around with two-valued logic and the difference engine. Others decry the hype and note that the majority of recent software releases are now touting some form of AI influence in their design or execution, weakening the meaning of the term in order to sell products.

Even after 100 years of study, not all psychologists are in complete agreement as to what constitutes intelligent behavior (look around you—how much have you seen lately?). Intelligence has generally been defined as the global ability to solve problems, to adapt to new situations, to form concepts, and to profit from experience.

However, it is obvious that there are many different types of behavior—many different ways of responding to the same problem—that can be called intelligent. Within the last 20 years, experts have paid much attention to the basic types of intelligence and how they are mediated by the biological substrate of the human brain.

Experts have long supposed that human beings use two major modes of thought: the way of reason and the way of emotion. A commonsense view is that these two ways of thought occasionally conflict. Some writers conceptualize the differences as analytic versus synthetic, successive versus simultaneous, or even digital versus analogical.

Paralleling the conceptualization of two modes of thought have been the results of research on the two hemispheres of the brain. Psychobiologist Roger Sperry of the California Institute of Technology won the 1981 Nobel Prize for Physiology and Medicine for his studies on the functions of the two hemispheres of the brain.

Essentially, Sperry and his colleagues studied individuals who had undergone a commissurotomy, an operation that severs the main bundle of nerve fibers that support the bulk of neural communication between the left and right hemispheres. They found that each hemisphere seems somewhat specialized for different tasks. For approximately 95 percent of the population (right-handed individuals and two-thirds of left-handed individuals), it appears that the left hemisphere of the brain is better organized for executing tasks characterized as:

- Verbal: language skills, speech, reading and writing, recalling names and dates, and spelling.
- Analytical: logical and rational evaluation of factual material.
- Literal: literal interpretation of words.
- Linear: sequential information processing.
- Mathematical: numeric and symbolic processing.
- Contralateral movement: controlling movement on the right side of the body.

The right hemisphere is better organized for tasks characterized as:

- Nonverbal: using imagery rather than words.
- Holistic: processing information simultaneously, in parallel.
- Visuospatial: functions involving perceptions of location and spatial relationships.
- Emotional: experiencing feelings.
- Dreaming: imaginative and metaphoric visual image-making.
- Contralateral movement: controlling movement on the left side of the body.

Hemispheric Activation Level Detector

This chapter's Circuit Cellar project is a brain-wave-monitoring biofeedback device that provides real-time information about predominant hemisphere activation. That is, this Hemispheric Activation Level Detector (HAL, for short) graphically displays the relative amounts of brain-wave activity in each brain hemisphere (see photo 1).

HAL can distinguish among grossly different conscious states, such as between concentrated mental activity and pleasant daydreaming. For example, if you are debugging a program, HAL should show a predominance of left-hemispheric activity. If you are listening to some light music and daydreaming, it should show a predominance of right-hemispheric activity.

HAL is a relatively sophisticated, low-cost, stand-alone, fully isolated four-channel electroencephalogram (EEG) brain-wave monitor. It gathers analog brain-wave voltages from four sets of scalp contacts, filters them, converts them to digital values, and transmits them via an RS-232C port (making HAL compatible with any computer) for recording or analysis.

HAL includes a two-channel fast Fourier transform (FFT) analysis-and-display routine for an IBM PC. (HAL's PC software is intended only as a graphics display demonstration—and there are limitations in processing power when using a straight 4.77-MHz PC—so it displays only two channels, even though HAL sends data on four channels.) If you have a more powerful machine, you should be able to expand the software to display more channels.

When running this special analysis-and-display package, the PC separates out various amplitudes and frequencies of alpha, beta, and theta waves, as well as phase differences between the hemispheres. The result is a graphical representation of what is going on in your brain in real time.

Warning

HAL is presented as an engineering example of the design techniques used in acquiring brain-wave signals. It is not a medically approved device, no medical claims are made for it, and it should not be used for any medical diagnostic purposes. Furthermore, the safe use of HAL requires that the electrical power and communications isolation described in its design not be circumvented. HAL is designed to be battery-operated only. Do *not* substitute plug-in power supplies.

Analyzing HAL's circuitry illustrates practical design techniques, including differential amplifiers for low-level signal detection in a high-background-noise-level environment, a low-frequency band-pass filtering-rectifying-integrating detector, optoisolation for safety, and A/D conversion.

I'm presenting the HAL project in a two-part chapter. In part 1, I'll look at the problems involved in picking up microvolt-level signals, amplifying and digitizing them, and sending them to your computer.

Science and the Brain

As I investigated this area, I found that a great deal of serious research has been going on regarding what we know about how our brain works. Much of this thinking is finding its way into computer science; even the Macintosh and the IBM PC now have neural-network hardware and software available for the experimenter.

In his book *Megabrain*, Michael Hutchison quotes National Institute of Mental Health neurochemist Candace Pert (discoverer of the opiate receptor in the brain and researcher on endorphins—the brain's own painkillers):

> There's a revolution going on. There used to be two systems of knowledge: hard science—chemistry, physics, biophysics—on one hand, and on the other, a system of knowledge that included ethology, psychology, and psychiatry. And now it's as if a lightning bolt had connected the two. It's all one system—neuroscience.... The present era in neuroscience is comparable to the time when Louis Pasteur first found out that germs cause disease.

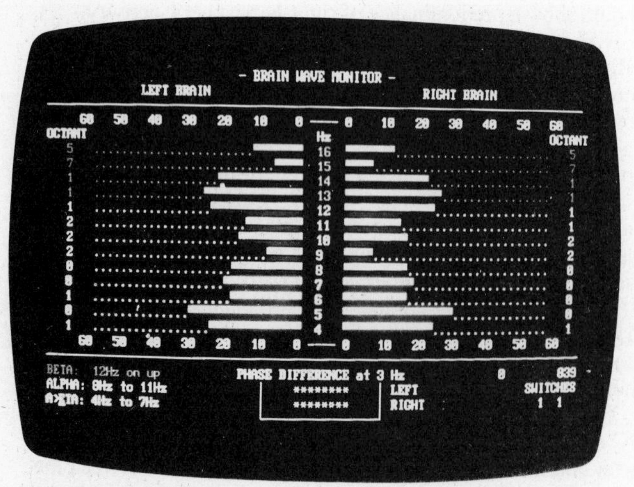

Photo 1: *HAL's output shows energy level by frequency of brain-wave signals for both sides of the brain.*

Hutchison further quotes neuroanatomist Floyd Bloom of the Scripps Clinic in La Jolla, California:

> A neuroscientist used to be like a man in a Goodyear blimp floating over a bowl game: He could hear the crowd roar, and that was about it. But now we are down in the stands. It's not too long before we'll be able to tell why one man gets a hot dog and one man gets a beer.

Much of the activity in this area has centered around the electrical characteristics of the brain. Advances in semiconductor technology have made available inexpensive ICs that let you design physiological monitoring equipment with laboratory quality at experimenter prices. When interest in alpha-wave biofeedback peaked about 15 years ago, a good-quality EEG feedback unit (which provided less information than HAL) cost $1000 for just one channel. Now, you can build four channels for under $200.

Digging into the Waves

The brain is a source of many electrical signals. An EEG is a recording (usually a strip chart) of the electrical potential differences between pairs of electrodes fastened to the scalp.

Silver-silver chloride electrodes pick up the signals. You must take some care to clean the area of the scalp with alcohol and perhaps use a mildly abrasive conducting cream to ensure good electrical contact. Ideally, there will be less than 10 kilohms impedance between any two electrodes, but anything under 25 kohms works (I'll describe placement of the electrodes in part 2 of this chapter).

It takes a trained eye to determine specific information about a person from an EEG. At present, we can only generalize as to what these recordings mean, and we are unable to correlate specific waveforms with intelligence. The observable electrical activity, however, does offer some clues.

According to medical and psychological research, by monitoring this activity, you could, in a gross way, investigate how the brain functions in a variety of circumstances. For example, if you monitor the two hemispheres while a person is solving problems, the type of problem could be indicated by the relative preponderance of one hemisphere's activity as compared to the other's. Sometimes you can even determine the activity (sleep versus reading; relaxed versus agitated).

The electrical signals we are currently able to monitor and identify from the brain are categorized as follows:

Alpha: Research has already indicated that in an awake person, the presence of alpha waves indicates a relaxed person with an absence of problem-oriented brain activity. (Alpha-wave activity describes electrical activity in the range of 8 to 12 Hz, a nearly sinusoidal signal at a voltage level of between 5 and 150 microvolts [μV]—typically 20 to 50 μV.)

Beta: When a person is thinking or attending to some stimulus, alpha-wave activity is replaced by beta-wave activity (14 to 25 Hz, activity of a lower amplitude).

Theta: Theta-wave activity (4 to 8 Hz, 20 μV and higher) usually appears during sleep, but it has been associated with deep reverie, mental imagery, creativity, dreaming, and enhanced learning ability.

Delta: Delta-wave activity (from 0.5 to 4 Hz) is seen in the deepest stages of sleep.

In addition, you must remember that I am describing an attempt to correlate cerebral electrical activity with subjectively observed events (types of cognitive tasks). While brain waves may be varying tens of times per second, our subjective experi-

To eliminate unwanted noise, HAL incorporates a band-pass filter that rejects frequencies under 4 Hz and over 20 Hz. While this compromises delta-wave acquisition, it does filter out most of the undesired signals.

ence varies more slowly. It may take a second or two to change concentration and to focus on a new task. Hence, you need to integrate the readings over a short period of time. Previous research in this area suggests that ¼ second to ½ second is reasonable.

A Noisy Environment

It's possible for HAL to "hear" more than we want. HAL is sensitive enough to detect artifactual signals: muscle activity from the forehead, eye and head movements, heart-rate activity, brain-wave "spikes" or irregular slow-wave activity, and—if you're not careful—60-cycle power-line hum. To eliminate this noise, HAL incorporates a band-pass filter that rejects frequencies under 4 Hz and over 20 Hz. While this compromises delta-wave acquisition, the benefit is that it filters out most of the undesired signals.

Photo 2: *HAL in the prototype stage. The input jacks are arranged along the top; HAL's battery is near the bottom.*

Photo 3: *HAL's optoisolated RS-232C circuitry.*

Detecting 4- to 20-Hz signals with a minimum amplitude of 5 μV from a source with approximately 10- to 20-kohm impedance is not an insignificant task. Ideally, the band-pass filter section should have a flat (±1-decibel [dB]) response across the passband; it should provide at least −18 dB per octave attenuation of signals outside the passband. The frequency response of the amplifier should be at least 50 to 60 dB down at 60 Hz. An equivalent input noise level of 0.5 μV or less would be good. Finally, input DC current should be less than 50 nanoamperes.

HAL's Circuitry

HAL's hardware circuitry is divided into two sections: preamplifier/filter and digitizer/control (see photo 2). The preamplifiers and filters acquire and boost the microvolt-level analog signals to useful levels. The digitizer section does the signal conditioning and A/D conversion and sends the data through an optocoupler to the host computer for analysis (see photo 3).

Several factors contributed to the evolution of the analog section of the circuitry. Initially, I planned to use narrow passband hardware filters to detect and measure only the alpha waves for each channel. Such an approach would discard a significant amount of information coming from the brain, essentially making the monitor capable of only simple "digital" discrimination—the presence or absence of alpha waves. This hardly seemed an achievement, since it merely duplicated the simple alpha biofeedback units available for the last 15 years.

Discussions with hardware and software experts eventually led to the conception of a more sophisticated system, one in which I considered the slowness of the EEG waveforms, the speed of the A/D conversion, and the analyzing power of an IBM PC. Ultimately, I decided that the HAL EEG monitor would function as a raw data accumulator and transmitter. The host computer would perform all signal analysis and display the results. (HAL's data output is RS-232C serial and can be analyzed and displayed on any computer. I chose to use an IBM PC here only for convenience.)

I expanded the bandwidth to allow the possibility of analyzing beta and theta waves. Even though these amplitudes are much lower than alpha waves, they are associated with some interesting phenomena.

To accomplish this task, I had to develop a special preamplifier/filter that would amplify only the specific EEG signals picked up from the scalp of the subject and amplify them to a level that is high enough for A/D conversion. Each HAL preamplifier/filter channel takes six operational amplifiers (op amps). Four of them provide amplification and impedance matching, and two others provide 60-Hz rejection filtering.

I designed the amplifiers and active filters in figure 1a around the TL-084 quad op amp and used as many common values as possible. The TL-084 provides junction-field-effect-transistor inputs with picoampere bias currents, low power consumption, and adequate input noise level. (If you are building this project, you should not substitute another type of op amp.) The bandwidth of the analog section is about 16 Hz (−3 dB at 20 Hz).

You can calculate the equivalent input noise by integrating the noise voltage as a function of frequency over the bandwidth. This 180-nanovolt equivalent noise, combined with the noise from the differential input stage multiplied by the system gain, yields a calculated output noise level of approximately 2.5 millivolts (mV).

Actual measurements of the noise output of the four-channel prototype were 3.5 mV root mean square, with a source impedance of 13-kohm impedance per input. I decided this was acceptable for the system with a 10-mV per bit A/D sensitivity.

I used three sections of IC1A to make a differential input instrumentation preamplifier. (Note that all six op amps associated with channel A are labeled IC1A and IC2A. Channel B's op amps are labeled IC1B and IC2B, respectively, and so forth.) An ideal difference amplifier will amplify only the voltage difference between the two inputs. Voltages that appear on both inputs when referenced to the ground lead are called common mode voltages.

For example, if the voltage on one input is $+50\ \mu V$ and the other input is $+15\ \mu V$, the difference signal would be $35\ \mu V$ and the common mode signal would be $15\ \mu V$. HAL measures the difference signal between the two electrode positions. The difference amplifier measures this difference by applying one signal to the inverting input of the op amp and the other signal to the noninverting input.

The ability of the op amp to amplify only the difference is specified as the common mode rejection ratio. In HAL, I measured this experimentally by shorting the inputs, applying an input signal between the shorted inputs and ground, and comparing the output with that obtained by applying the same signal across the two inputs. The common mode output was 43 dB down below the differential output. (You would correctly suspect that the major component of common mode voltages in HAL will be induced by the 60-Hz power line. I'll discuss how HAL rejects the 60-Hz signals later.)

I set the voltage gain of the preamplifier to 5800 and incorporated AC coupling between the stages to eliminate DC offset voltages and provide some low-frequency roll-off. Feedback capacitor C3 provides high-frequency roll-off, with the gain down 9 dB at 60 Hz. The third-order active filter stage has a −3-dB frequency of 22 Hz and is 30 dB down at 60 Hz. You'll find the same third-order filter at the input of each final amplifier to the A/D converter (ADC), thus providing another 30 dB, for a total of 69 dB attenuation at 60 Hz.

The interstage coupling capacitors set the low-frequency passband of the amplifiers. The low-frequency roll-off is 24 dB

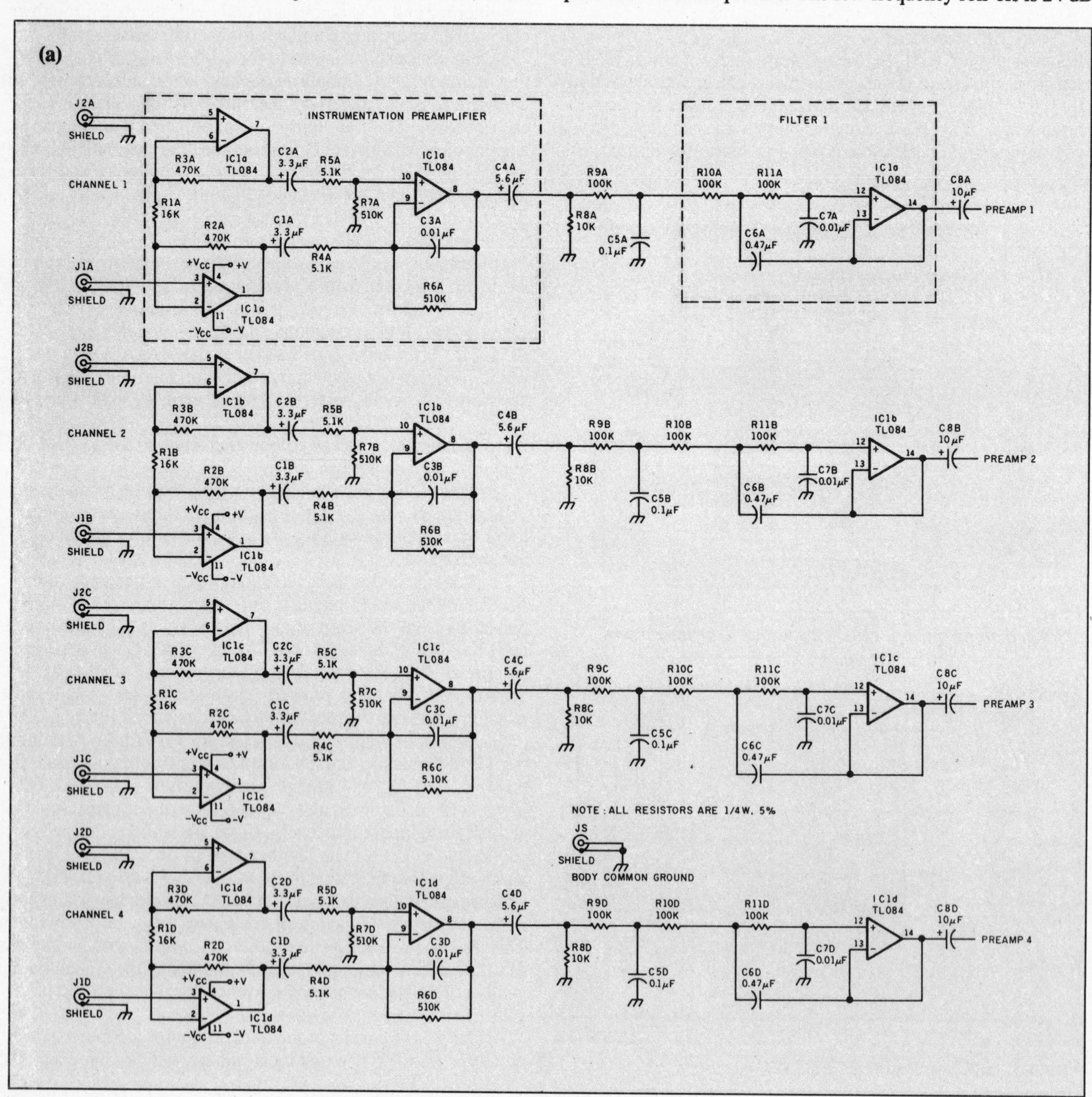

per octave, with the −3-dB point at 6 Hz. This is well above muscle activity and other noise.

A 2.5-volt reference diode sets the analog references to the ADC at 1.75 V and 3.25 V, or ±1.25 V of half the power supply. The last amplifier stage is DC offset to one half the power supply voltage, with the AC signals having a permissible peak value of 1.25 V. I set the overall gain of the amplifier stage to 12,500 so that a 100-μV signal would be the maximum input. This amounts to about 0.8 μV per bit sensitivity.

Since the ADC0808 is generally thought of as a DC converter and HAL measures AC signals, offsetting the reference to the ADC lets it measure signals that swing above and below some point designated as "zero" (offset binary converter). When you apply 0 V to the ADC, its output will be 80 hexadecimal.

A voltage gain of 12,500 corresponds to 82 dB (20×log Av). The 60-Hz rejection of 69 dB results in a 60-Hz gain of 13 dB (82 dB − 69 dB). The common mode rejection of 43 dB reduces the 60-Hz gain to a loss of −30 dB (13 dB − 43 dB). This all means that a 60-Hz common mode signal at the inputs is reduced by a factor of 0.03 in getting to the ADC.

To show up as a ±1-bit ripple on the data, the common mode input signal would have to have an amplitude of 300 mV peak to peak. This 300 mV would be reduced by a factor of 0.03 to become 10 mV at the ADC. When I connected a 1-inch unshielded lead to HAL's input, it picked up about 100 mV peak to peak of noise. This seems adequate, but all the same, don't use HAL while standing directly beneath a neon sign transformer!

The Digitizer and Control Section

The signals from the four preamplifier/filter channels go to four of the eight analog inputs of the ADC0808. An 80C31 CPU performs channel selection and transmission to the host CPU. (While it is possible to duplicate the preamplifier/filter section to ultimately produce an eight-channel version of HAL, the current level of software for the 80C31 is designed for only four channels.)

Figure 1c shows the microcontroller part of the headset circuit

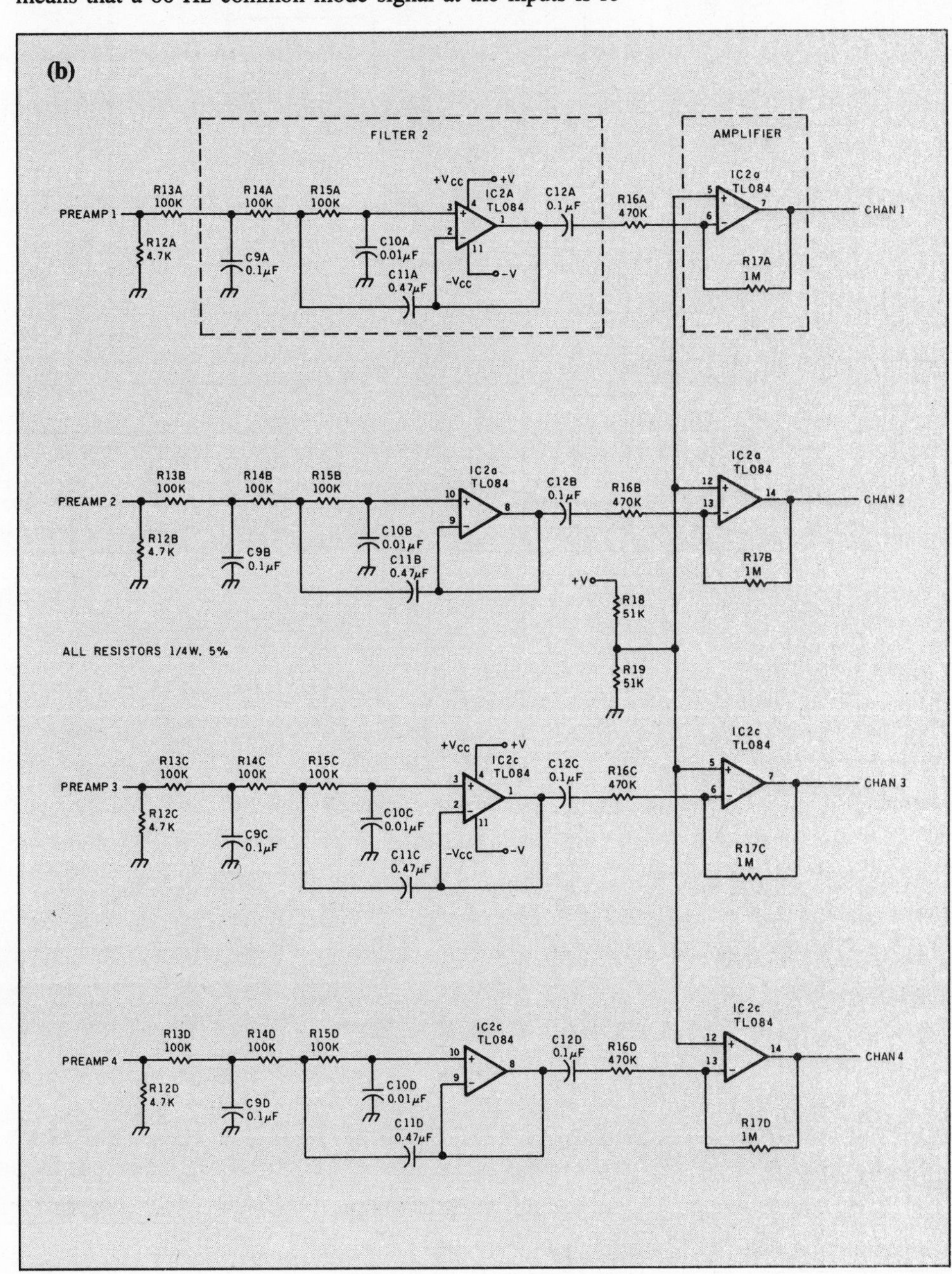

Figure 1: *The schematic for HAL.* **(a)** *Electrode pickup, preamplifier, and part of the filter stage.* **(b)** *More filtering and the final amplifier.*

cuitry. The 80C31's port 1 connects to the ADC0808's data outputs, with all the control and status bits handled by port 3. Bits from port 1 also drive the serial output line and the two event marker switch inputs. Because port 2 is dedicated to the upper half of the program address and port 0 is the EPROM data bus, no port bits are left for anything else.

The timing requirements are so simple that the code doesn't even need interrupts. It samples the two switches, reads the left and right hemisphere voltages from the ADC0808, and sends the results out serially. Each data sample consists of a 5-byte

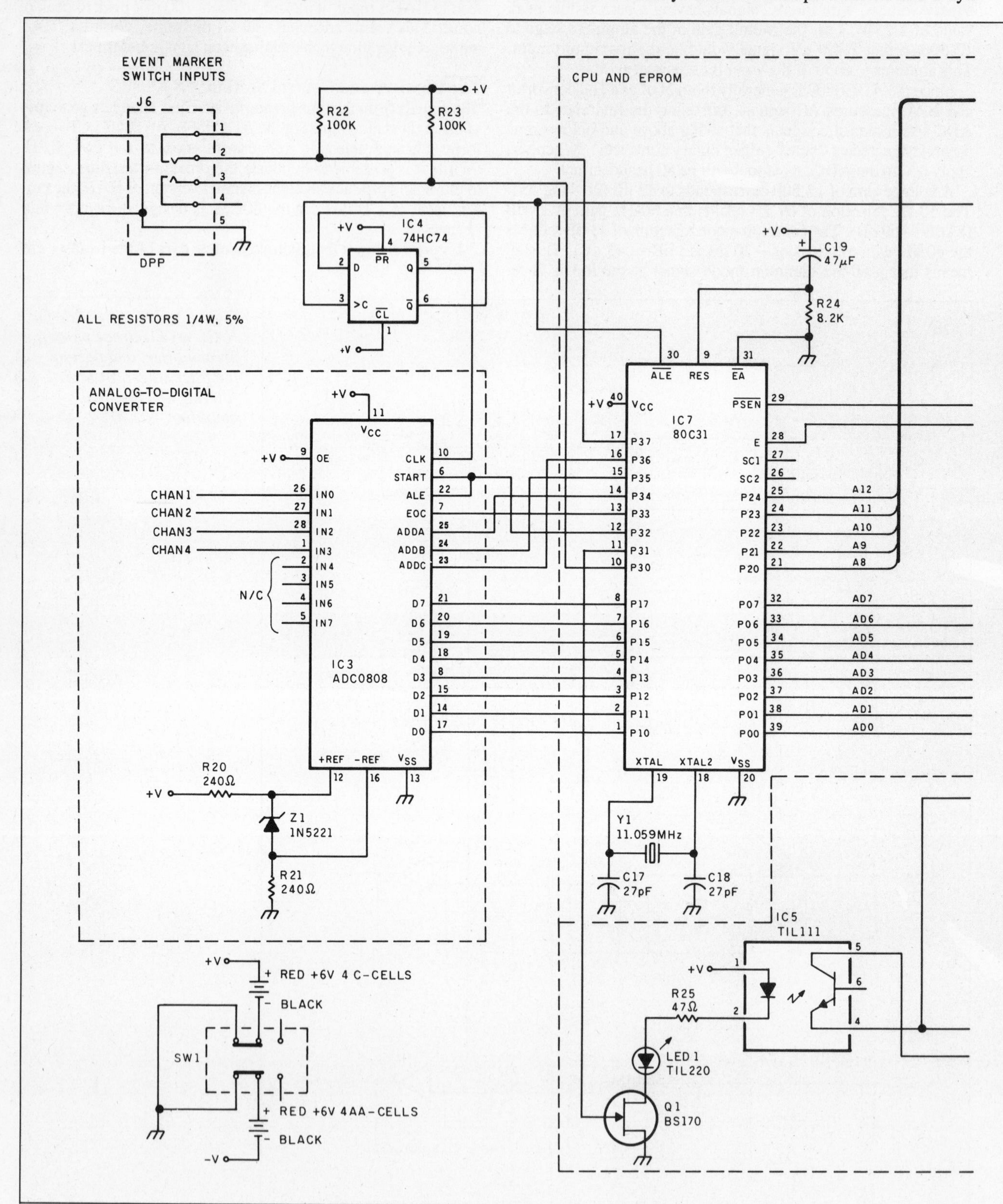

Figure 1c: *HAL's A/D converter, on-board CPU, and RS-232C port.*

transmission. The data sequence is first byte, two switch position codes with 6 bits of leading zeros, followed by 4 bytes of sequentially sampled A/D channels.

HAL is battery-powered. Four alkaline C cells provide +6 V, and 4 AA cells provide −6 V for the op amps. The CMOS digital circuitry runs from the 6-V supply. Current drain with all CMOS components is 225 milliamperes (mA) at 6 V and 50 mA at −6 V. (OK, I know that most chips like 5 V, but CMOS digital chips will work fine in this application at 6 V.) An addi-

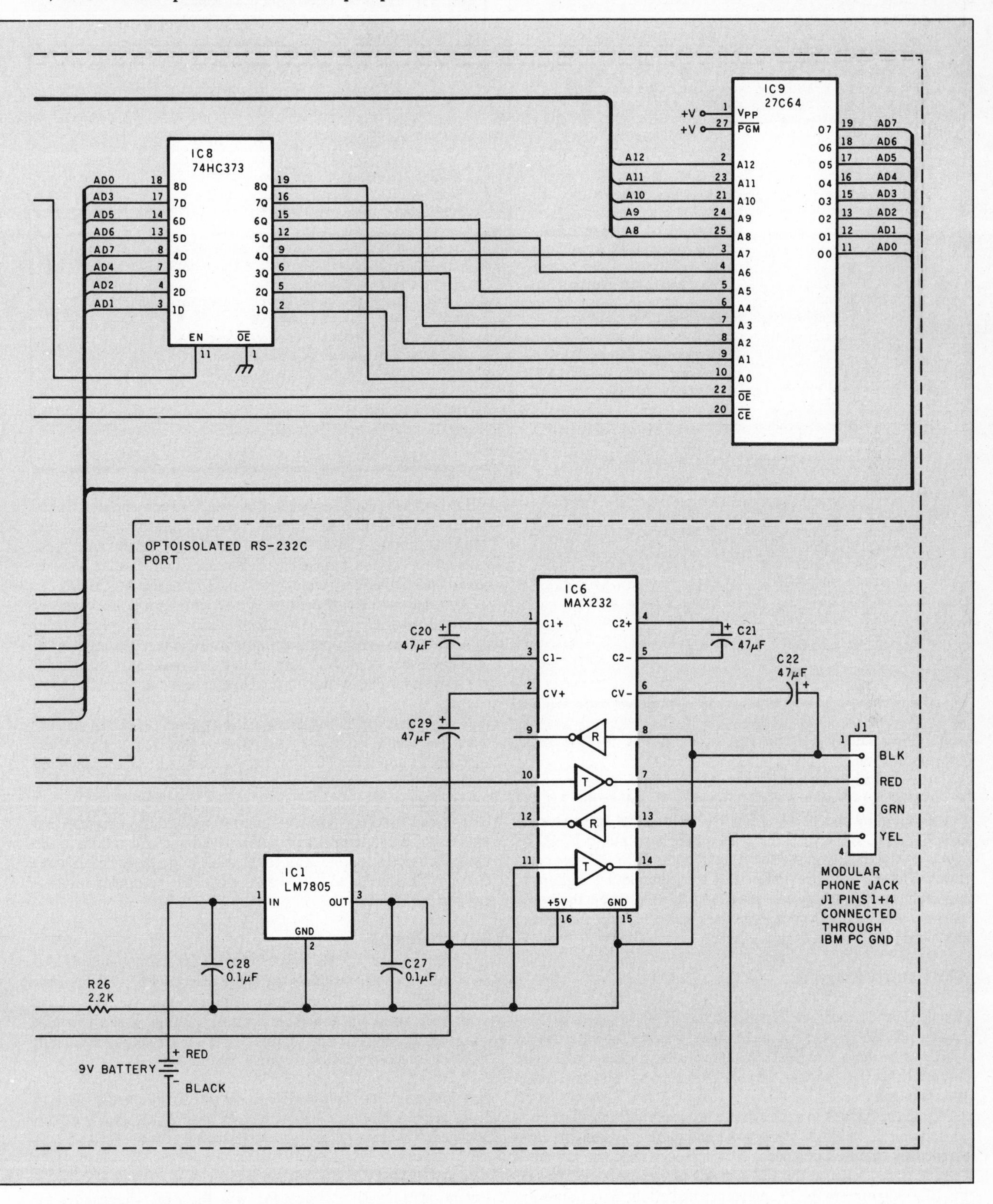

Listing 1: *Source code for HAL's main loop.*

```
; -- 80C31 main loop

ADCdata  EQU  P1    ; data inputs
ADCctls  EQU  P3    ; control I/O
start    EQU  2     ; + to start conversion
ready    EQU  3     ; + on end of conversion
switches EQU  81H   ; switch bit locations
ADC0     EQU  10H   ; low order address bit
ADCaddr  EQU  70H   ; address bit location
sync     EQU  P3.0  ; sync out of switch input!

;>> setup code omitted

;wait for Timer 0 to run out, then reload it
again EQU  $
      JNB   TF0,again     ; loop until timer tick
      CLR   sync          ; blip scope sync down
      NOP
      SETB  sync          ; ... and back up again
      CLR   TR0
      MOV   TH0,#HIGH -T0period ; reload
      MOV   TL0,#LOW  -T0period
      CLR   TF0           ; clear end flag
      SETB  TR0           ; restart counter
;--- read the channels in a great rush
      CALL  getADC
;--- read the switches and send them out
      MOV   A,ADCctls     ; grab input bits
      ANL   A,#switches   ; strip switches
      RL    A             ; move to bits 1-0
Lsw   JNB   TI,Lsw        ; will set every time!
      MOV   SBUF,A        ; send byte
      CLR   TI            ; reset ready flag
;--- send analog data
L0    JNB   TI,L0         ; wait for trans ready
      MOV   SBUF,R0       ; drop in the value
      CLR   TI            ; reset trans ready
L1    JNB   TI,L1         ; repeat for channel 1
      MOV   SBUF,R0
      CLR   TI
      JMP   again
```

tional 9-V battery (10 mA) provides power for the serial communication.

The two-push-button switch inputs (J6) allow operator signaling to the host computer. Serial data output drives the TIL111 optocoupler by means of a BS-170 field effect transistor. A MAX232 (IC6) converts the optocoupler's output to RS-232C levels compatible with the serial input of the IBM PC host computer. The MAX232 is powered by the separate 9-V battery to maintain isolation between HAL and the PC when the serial port is connected. (Do not try to use the 6-V C cells that power the main HAL circuit to power IC6.)

The Control Program

HAL's firmware control program (the main loop is shown in listing 1), contained in a 2764 EPROM (IC9), is called BIO31. Nearly all BIO31's time is spent in line waiting for timer flag 0 (TF0) to become a logic 1. Whenever that happens, the code reloads timer 0 to produce the next 1/64-second delay and clears the flag again.

While the 5 bytes in each sample take only 6.25 milliseconds to transmit at 4800 bits per second and there's lots of idle time on the link (the PC code needs 64 samples per second; we don't send it faster because the PC analysis program would choke), the FFT software in the PC presumes that all the data points are sampled at the same instant in time. As a requirement, then,

Listing 2: *The code HAL uses to read its A/D converter.*

```
;--------------------------------
; Get channels from the ADC input
; Values are stashed in registers
getADC PROC
       MOV  A,ADCctls  ; reset address
       ANL  A,#NOT ADCaddr
       SETB ACC.ready  ; ensure this bit is a 1
       MOV  ADCctls,A
;--- grab channels
       SETB ADCctls.start ; blip start line
       CLR  ADCctls.start ;  with 1 us pulse
; Wait for EOC to go away
Lw0r   JB   ADCctls.ready,Lw0r
; Now wait for EOC active
Lw0e   JNB  ADCctls.ready,Lw0e
       MOV  R0,ADCdata ; save data in reg
       ADD  A,#ADC0    ; tick channel number
       MOV  ADCctls,A
       SETB ADCctls.start ; repeat for chan. 1
       CLR  ADCctls.start
Lw1r   JB   ADCctls.ready,Lw1r
Lw1e   JNB  ADCctls.ready,Lw1e
       MOV  R1,ADCdata
       RET
getADC ENDPROC
       END
```

BIO31 runs the ADC as fast as possible between samples. I used in-line code to eliminate the overhead of subroutine calls and returns, although I'll be the first to admit that the few microseconds probably don't make any difference at all. Listing 2 shows what's needed to grab channels 0 and 1 from the ADC0808.

Throughout the conversions, the accumulator holds a copy of port 3, so changing the ADC channel address is simply a matter of adding 1 to the proper accumulator bit and reloading port 3. The code sets the ADC ready bit to a 1 to make sure that the bit is always an input; writing a zero to that bit would turn it into an output.

Toggling the ADC's start bit using a pair of CLR/SETB instructions provides a 1-microsecond pulse on that output. One of the nice things about the 8031 is that you can tell exactly how long each instruction will take, so generating precise time intervals is quite simple.

The ADC0808 takes a few microseconds to drop the line that signals the end of conversion before starting the next one, so the code includes a loop to wait for that bit to go away before continuing. This is one of those cases where the computer can outrun the peripheral!

In Conclusion

In Chapter 6, part 2, I'll examine the software components of HAL, including an 8088 machine language discrete FFT callable from BASIC. I'll provide BASIC source code so that you can design your own software and reconfigure HAL into a sophisticated brain-wave biofeedback monitor or a continuously recording EEG, or so you can add additional channels.

Special thanks for help provided on this article to Dr. Robert Stek, David Schulze, Rob Schenck, Jeff Bachiochi, and Ed Nisley.

BIBLIOGRAPHY

Hutchison, Michael. *Megabrain: New Tools and Techniques for Brain Growth and Mind Expansion*. New York: Morrow, 1986.

This book contains the results and speculations of some pioneers of brain-machine technology.

Meikson, Z. H., and Philip C. Thackray. *Electronic Design with Off-the-Shelf Integrated Circuits*, 2nd. ed. Englewood Cliffs, NJ: Prentice Hall, 1984.

1983-84 Opto-electronics Data Book. Texas Instruments.

Editor's Note: Steve often refers to previous Circuit Cellar articles. Most of these past articles are available in book form from BYTE Books, McGraw-Hill Publishing Company, P.O. Box 400, Hightstown, NJ 08250, (1–800–2–MCGRAW).

Ciarcia's Circuit Cellar, Volume I covers articles in BYTE from September 1977 through November 1978. *Volume II* covers December 1978 through June 1980. *Volume III* covers July 1980 through December 1981. *Volume IV* covers January 1982 through June 1983. *Volume V* covers July 1983 through December 1984. *Volume VI* covers January 1985 through June 1986.

It's virtually impossible to provide all the pertinent details of a project or cover all the designs I'd like to in the pages of BYTE. For that reason, I have started a bimonthly supplemental publication called Circuit Cellar Ink, which presents additional information on projects published in BYTE, new projects, and supplemental applications-oriented materials. For a one-year subscription (6 issues), send $14.95 to Circuit Cellar Ink. Credit card orders can call (203) 875-2199.

The following item is available from

CCI
4 Park St.
Suite 12
Vernon, CT 06066

Circuit Cellar Hemispheric Activation Level Detector (HAL) complete printed circuit board kit. Comes with printed circuit board and all chips and board-mounted components, including 80C31, IC sockets, 27C64 EPROM, user's manual, and IBM PC demonstration display on PC format disk. Does not include battery holders, batteries, scalp electrodes, or connecting cables. HAL-4-KIT.................................. $179; add $19 if you want an IBM PC cable.

The Circuit Cellar HAL design is available for licensing to qualified manufacturers. Call for information.

All payments should be made in U.S. dollars by check, money order, MasterCard, or Visa. Surface delivery (U.S. and Canada only): add $5 for U.S., $8 for Canada. For delivery to Europe via U.S. airmail, add $14. Three-day air freight delivery: add $7 for U.S. (second-day Federal Express), $15 for Canada (DHL), $22 for Europe (DHL), or $30 for Asia and elsewhere in the world (DHL).

There is a multiline Circuit Cellar bulletin board system (running TBBS 2.0M) that supports past and present projects in BYTE and Ink. You are invited to call and exchange ideas and comments with other Circuit Cellar supporters. The 300-/1200-/2400-bps BBS is on-line 24 hours a day at (203) 871-1988.

6

COMPUTERS ON THE BRAIN

PART 2

Making sense out of the data

In Chapter 6, part 1, I introduced you to the Circuit Cellar electroencephalogram (EEG) monitor called HAL (Hemispheric Activation Level Detector) and discussed its hardware design. In part 2, I'll look at the system software and provide some directions for HAL's use.

Software Overview

The overall operation of HAL is relatively straightforward. It samples four channels of analog brain-wave data 64 times per second and transmits this data serially to a host computer. The host analyzes two channels of the data to determine frequency components, and it then displays the results on a monitor as a series of continuously changing bar charts.

As I mentioned in part 1, the frequency range of interest in brain-wave detection is 4 Hz to 20 Hz, which includes alpha, beta, and theta waves. I could have designed HAL with many independent band-pass filters to separate each channel's complex waveform into discrete sinusoidal components, but that would have involved much more hardware. Instead, HAL's raw digitized analog data is analyzed on the host computer using a fast Fourier transform (FFT).

An FFT is like a mathematical prism. Just as a prism breaks down light into its component colors, the FFT breaks down a waveform into the sinusoidal components that make up the signal. In this way, I can extract amplitude and phase values for any given frequency in the target spectrum of 4 Hz to 16 Hz (I didn't display frequencies from 16 Hz to 20 Hz because of speed constraints on the IBM PC).

The system software coordinates four major tasks, which are graphically outlined in figure 1. The Intel 8031 microcontroller's task is precise sampling of hemispherical data and transmission via an optocoupled RS-232C link to the host IBM PC (the 8031 firmware is called BIO31). An IBM PC task called `COMMO` is an interrupt-driven communications routine that receives HAL's incoming data. Being interrupt-driven, `COMMO` operates independently of the activities of the FFT analysis-and-display routine (and any other tasks that happen to be running on the IBM PC).

As the data comes in, the host loads it into 1024-byte sample queues for left brain and right brain data. The FFT algorithm reads this incoming data, calculates amplitude and phase values per frequency, and updates integer arrays. Finally, the BIO program reads these arrays to continually update the bar graphs on the video display.

Simple—right? Just like using a hammer to make gravel out of a boulder. Simple in theory, at least.

While HAL is technically independent of the host computer (by virtue of its serial interface), I chose to present the demonstration display on an IBM PC for the sake of convenience. (I invite conversion to other computers and will make such conversions available through the Circuit Cellar bulletin board system.)

Because HAL is supposed to be both an educational and a functional presentation, my fundamental goal for the software was to keep it simple and fast. To that end, I used BASIC wherever possible. However, to maintain processing speed, I incorporated certain assembly language routines that are called from BASIC (i.e., the QuickBASIC compiler).

Refer again to figure 1, the overview block diagram of the system. The `COMMO` and FFT routines are written in assembly language. `COMMO` is really a group of subroutines; BASIC merely calls the initialization code.

I wrote the FFT block in assembly language for maximum speed in the FFT calculations. Using a BASIC program, a 64-point FFT took about 20 seconds to do. The existing FFT algorithm in assembly language will calculate 64-point transforms for both channels in about 240 milliseconds (ms). More about this later.

HAL's Brain

The 8031 is a remarkably capable processor that I have been using for many stand-alone projects lately. I described the inner workings of BIO31 last month with the hardware description, since BIO31 works so closely with the hardware. Basically, BIO31 is a simple read-transmit-wait loop (see figure 2). The read portion of the loop samples the switches and the A/D converter. It then places this information in the output buffer and transmits it out the serial port at 4800 bits per second (bps) to the host computer. The transmission is a 5-byte sequential string: switch position/sync, channel 1, channel 2, channel 3, and channel 4.

Once transmission is completed, the program waits for the 8031's internal timer to time out. Since proper sampling of a continuous waveform requires sampling at a fixed rate, this precise time delay is what provides the host program with data samples that the FFT can successfully analyze.

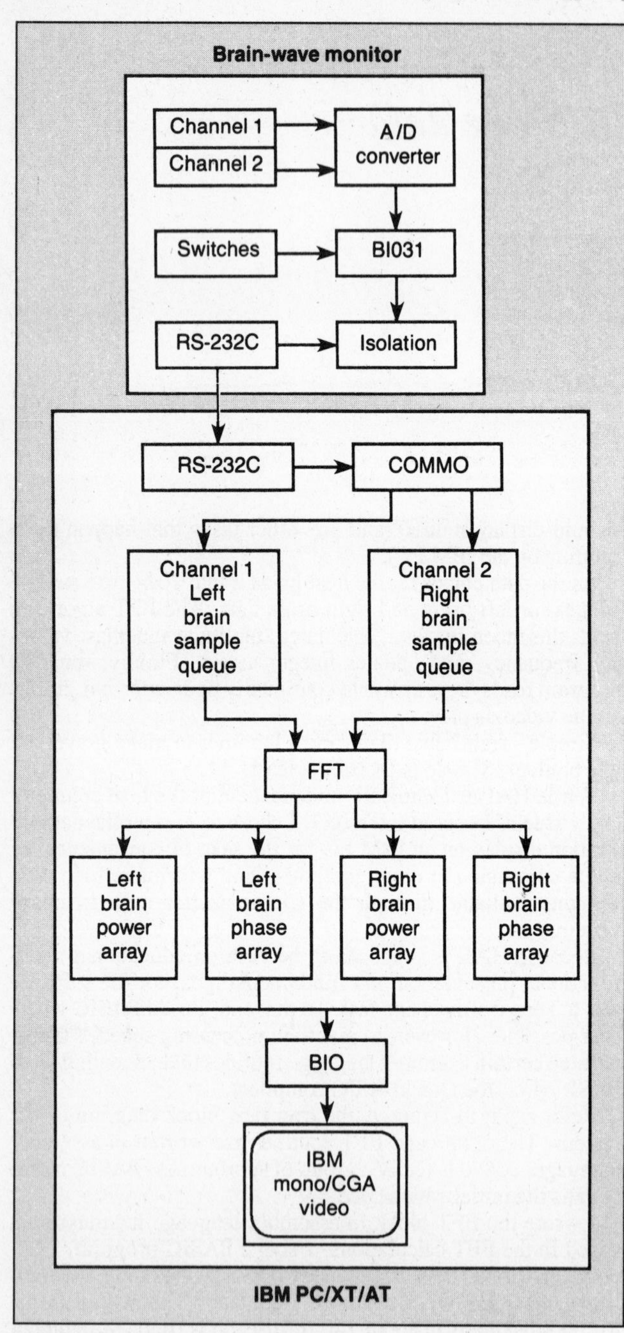

Figure 1: *A block diagram of the complete HAL system, showing major components of the brain-wave monitor and the host machine.*

The IBM PC Communications Interface

The IBM PC's COM1 port is configured as the headset interface port. Recall that the data coming from HAL arrives at 4800 bps in packets of 5 bytes, transmitted 64 times a second. The first byte indicates HAL's switch configuration, and the host computer uses this byte for synchronization purposes.

A switch position/sync byte value of 00000011 binary indicates that both the left and right switches are closed; a value of 00000000 indicates they are both open. The maximum value the switch position/sync byte can have is 3. BIO31, as one of its functions, tests all outgoing data and limits the lowest A/D channel data to a value of 4. Since any incoming data with a value of 3 or less must therefore be the switch position/sync byte, it is possible for `COMMO` to synchronize easily with the incoming data stream.

`COMMO`'s service routine uses a state machine to synchronize itself with the incoming data (see figure 3). It reads the data byte from the Data Receive register and stores it according to which state the routine is in. If state equals 0, it saves the switch bits and advances to the next state. If state equals 1, it loads the sample data into the left brain sample queue. If state equals 2, it loads the sample data into the right brain sample queue. If a value of 3 or less is encountered, the state will automatically change to 0, and it will save the data as switch parameters.

The Software Prism

After the computer receives HAL's data, the FFT takes over, transforming the collected information into something useful. In this application, I wanted to see what frequencies are active in whatever brain is connected to the EEG and how powerful those frequencies are.

My first impression of the FFT was stated earlier: a mathematical prism. I've read articles that have used the FFT on everything from spectral analysis and digital filtering to biorhythms and random numbers. But what does the FFT do here?

Consider figure 4. The four sine waves on the bottom represent energy levels at four different frequencies plotted with respect to time. The vertical bars represent a 1-second sample window. (The 64 dots on each graph between the two vertical lines represent $\frac{1}{64}$-second intervals.) The top squiggle is an integrated waveform, the sum of the four waveforms below it at any given point in time.

This integrated waveform is similar to what we find in the real world. I used values generated from this integrated waveform to test HAL's software. The results of the FFT on this waveform appear on the screen shown in photo 1. The bars are read side to side, with their respective frequencies labeled in the middle. Energy levels are defined both above and below the bar charts. Notice that I applied the same test data to both the left brain and right brain sample queues. I will discuss the display in greater detail later.

There is a little more to structuring an FFT than just using an algorithm out of a book. Without getting into any mathematical formulas, I will explain the impact of modifying FFT parameters. The text box "Understanding the Fast Fourier Transform" on page 90 explains the FFT, using a version written in BASIC. The initial lines of code are configuration data, and only the last 11 lines perform the actual transform on the data in the array RAWDATA.

You must address three major factors when building a simple FFT algorithm: the duration of the sample window, the sample

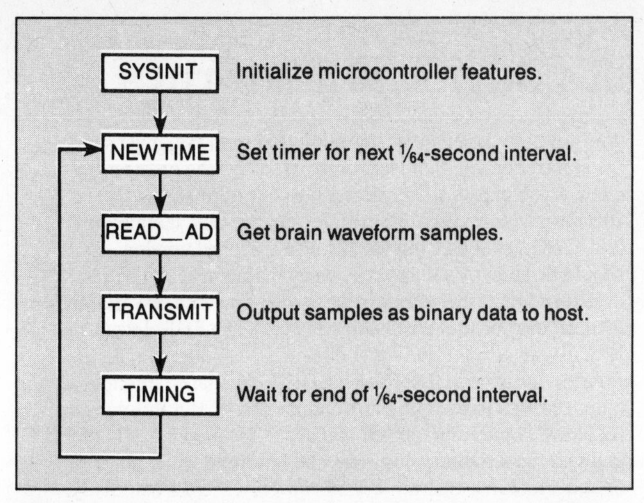

Figure 2: *A flowchart for HAL's firmware program, BIO31.*

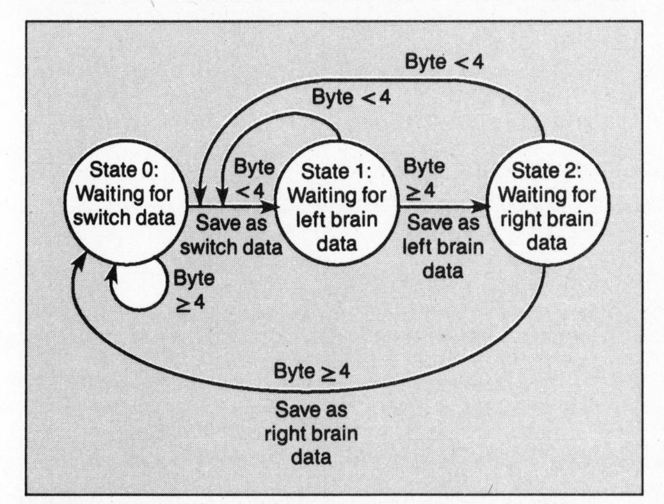

Figure 3: *The host computer's receiver routine executes a finite-state machine, shown here graphically. Each circular node of the graph represents a state, and transitions from state to state occur when the system receives a byte.*

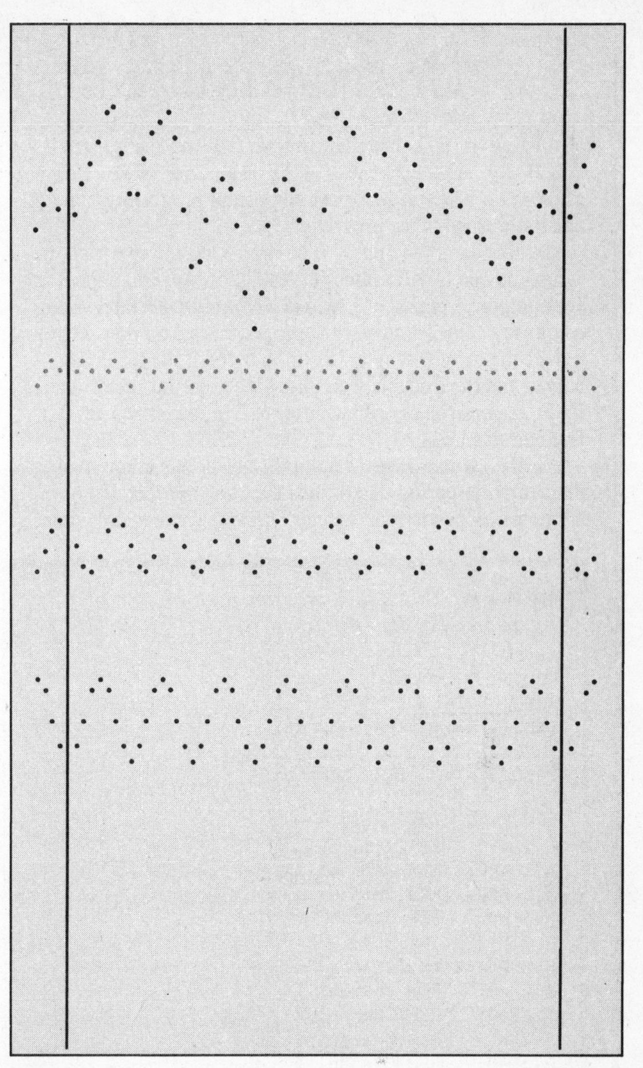

Figure 4: *If you add together the lower four sine waves, you get the topmost, random-looking waveform. The FFT does just the reverse: It takes the composite waveform and extracts its components.*

rate, and the actual frequencies desired for output. These factors are identified as:

- `WDW`: The sample window in seconds.
- `PT`: The number of samples to be taken in the sample window.
- `MIN` and `MAX`: Lowest and highest frequencies to be evaluated.

The `FREQ` array represents the sinusoidal frequencies to be integrated into the waveform I talked about earlier, `A` represents the peak amplitude of these sine waves, and `P` serves to introduce a phase offset in a given channel. The variable `SM` specifies the starting offset of the points of data to be analyzed by the FFT. Graphically, `SM` locates the vertical bars on the display seen in figure 5.

Faster FFT

Speed was a key issue in this project. For a visual display to be of any value to a user, the software has to analyze and display the brain-wave data as quickly as possible. Ideally, you should

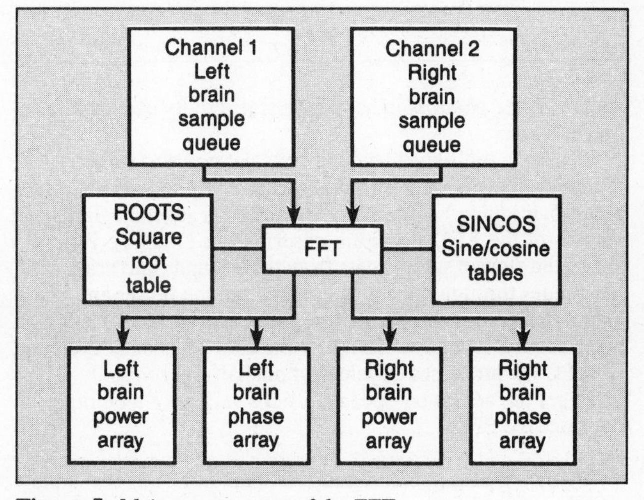

Figure 5: *Main components of the FFT.*

Understanding the Fast Fourier Transform

Listing A is a heavily commented code fragment in BASIC that performs the FFT on an array of data points (RAWDATA). The following are important program variables and their usage in the program.

WDW, the sample window: The time period over which the sample is taken will affect the resolution you can expect in your output. A period of 1 second will yield 1-Hz intervals of frequency. This is the way the program is set up now. If you reduce the sample window to half a second, the resolution is halved and the frequency output will be in 2-Hz increments. By increasing the sample window to 2 seconds, you get 0.5-Hz intervals.

PT, the sample rate: The number of points taken during the sampling period determines the range of frequency increments (dependent on window period) you can calculate. According to the Nyquist sampling theorem, you must sample at a frequency of at least twice as fast as the greatest frequency you expect to encounter. If your high end is 40 Hz, you must take at least 81 samples during that sample window. HAL takes 64 samples during a 1-second window, so it should be able to evaluate frequency output up to 30 Hz.

MIN and MAX, the output frequencies: The countervariable HZ need only be the frequencies desired. The FFT algorithm is really an accelerated version of what is called a discrete Fourier transform (DFT). The DFT calculates frequency output from 0 to the number of points taken. But this output includes something called negative frequencies. These mathematical monkeyshines can be ignored for the purpose of understanding this project. However, FFT.BAS can easily be tweaked to display this graphically, also.

Listing A: *This BASIC program performs the FFT on an array of data points.*

```
REM FREQUENCY(Hz)
FREQ(0)=16:  REM  Channel 1
FREQ(1)=9:   REM  Channel 2
FREQ(2)=8:   REM  Channel 3
FREQ(3)=2:   REM  Channel 4

REM AMPLITUDE
A(0)=10:  REM Channel 1
A(1)=20:  REM Channel 2
A(2)=30:  REM Channel 3
A(3)=40   REM Channel 4

REM PHASE (Radians)
P(0)=0:   REM Channel 1
P(1)=0:   REM Channel 2
P(2)=0:   REM Channel 3
P(3)=0:   REM Channel 4

WDW=1.0: REM Sample period in seconds
PT=64:    REM Number of sample points
MIN=2:    REM Min freq to evaluate (1 - 25)
MAX=16:    REM Max freq to evaluate (MIN - 25)

SM=040:    REM Offset to start of sample period

FOR HZ = MIN TO MAX
REM ** Clear the real and imaginary
REM ** components.
  REAL = 0
  IMAG = 0
REM ** Begin the frequency analysis
REM ** loop.
  FOR X=0 TO PT-1
REM ** Calculate reference angle
    G = 2*PI*HZ*X/PT
REM ** Calculate real components
    REAL = REAL + RAWDATA(X)*COS(G)/PT
REM ** Calculate imaginary component
    IMAG = IMAG - RAWDATA(X)*SIN(G)/PT
  NEXT X
REM ** Power calculation
  FFT(HZ,0) = 2*SQR(IMAG*IMAG + REAL*REAL)
REM ** Phase calculation
  FFT(HZ,1) = ATN(IMAG/REAL)
NEXT HZ
```

be able to see changes in hemispherical activity every half second or sooner.

Clearly, a high-level language was not going to do the trick. (Running the FFT algorithm above in compiled BASIC takes about 5 minutes.) Ultimately, I created a machine language version of the FFT algorithm described in FFT.BAS. Figure 5 shows the flow of sample data from the left and right brain sample queues through the algorithm and into the power and phase arrays defined in the BASIC program. On an IBM PC XT, it takes about 2 seconds to complete the entire feedback cycle; an IBM PC AT processes a cycle in about half a second.

To get the necessary speed, I used a variety of techniques to optimize the FFT:

- *Number of data points*: I chose 64 data points as the sample size for a few reasons. First, you can perform integer division by 64 using shift operations instead of a divide instruction. When I used the 8088's integer-division instruction, it took about 340 ms to calculate a single 64-point FFT. In contrast, division using shift operations takes about 120 ms to accomplish the same thing.

  Second, taking the division by shifting into account, 64 points per second will give the resolution and the range needed to evaluate frequencies from 4 Hz to 20 Hz. The Nyquist sampling theorem dictates that you must sample a waveform at a frequency of at least twice its highest frequency component. Consequently, 32 points would not be enough samples to analyze anything above 15 Hz. On the other end, 128 points would give you a greater frequency range for analysis, but it would take considerably longer to calculate.
- *Table lookup*: The FFT algorithm requires only a fixed number of sine and cosine values. By building these 64 values into a table rather than calculating them on the fly, you can save a lot

You can display HAL's output in either CGA or monochrome.

of time. Using a table of square roots is similarly much faster than calculating them.

- *Scaling*: When I built the tables, I took care to scale the integer values so power calculations can divide by 256. This magic number lets the software shift bytes instead of bits, giving considerable savings in time.
- *Phase octants*: Because the accuracy of this device is limited to integer calculations—and again because of speed—the program reports phase results in 45-degree increments. Instead of using the arctangent to calculate phase results, it is quicker in machine language to determine them using the sign and absolute amplitude of both real and imaginary components. The result is in phase octants.

Finally, the FFT is a much deeper subject than I can possibly cover in this chapter. I recommend a book entitled *The FFT: Fundamentals and Concepts* by Robert W. Ramirez (Prentice Hall, 1985). It provides an excellent look at the FFT and the theory behind it.

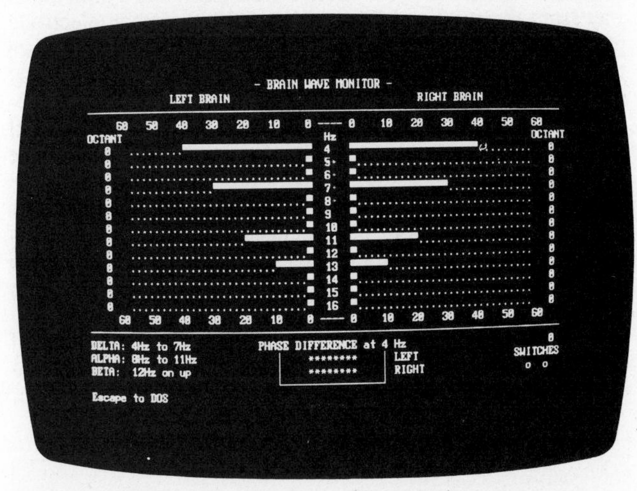

Photo 1: *HAL properly unravels the complex waveform shown in figure 4. Note the four long horizontal bars; each corresponds to one of the four component sinusoidal waves.*

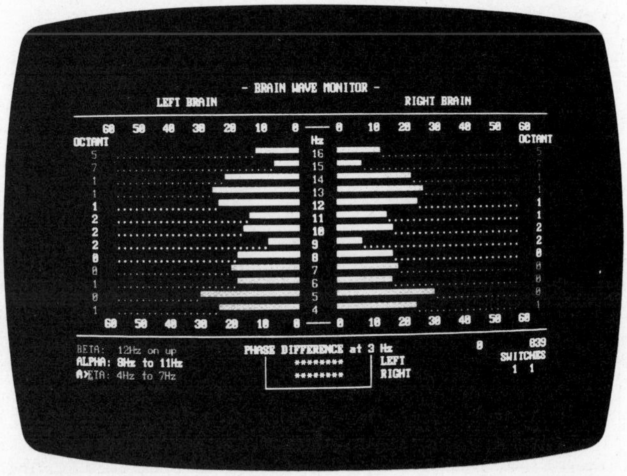

Photo 2: *HAL's output can be displayed on an IBM PC using the CGA.*

Driving the IBM Video Display

The last major block in the system diagram is called BIO. This module takes calculated output from the power and phase arrays just loaded by the FFT, converts this output into character strings, and writes them to the display. This part of the software displays the window opened by HAL into your brain.

Photo 2 shows a typical HAL display. You can display HAL's output in either CGA or monochrome. The horizontal bars represent the relative energy levels found in the frequencies evaluated. The bars that lead to the left represent the energy in the left hemisphere (channel 1); the bars to the right of center are the right hemisphere (channel 2).

Respective frequencies are labeled down the middle of the display under the "Hz" heading. The numbers 0 to 60 located in the rows above and below the energy bars represent relative energy levels only (pressing F3 changes the scale). They do not necessarily indicate absolute voltage levels that a much more expensive EEG machine might provide.

The phase angle of any given frequency is shown under the "OCTANT" heading for its respective hemisphere. An octant is nothing more than an eighth of a cycle. Octant 0 means the waveform is within the first 45 degrees of its cycle. A value of 1 puts it between 45 and 90 degrees, and so on.

Relative phase is indicated in a box at the bottom center of the screen under the "PHASE DIFFERENCE" heading. This feature selects the frequency from the left hemisphere with the greatest energy level and compares its phase angle with the phase angle of the corresponding frequency in the right hemisphere. The two rows of asterisks represent the position of the right hemisphere relative to the left hemisphere. If the right

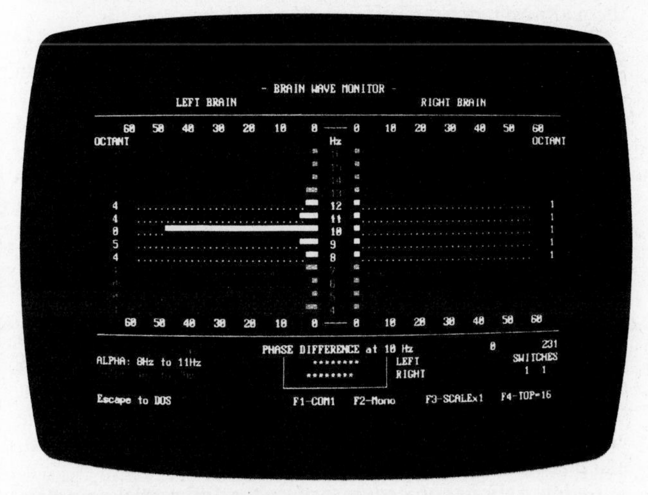

Photo 3: *While testing HAL, I injected a 10-Hz sine wave through the headset electronics. The resulting output, shown here, indicates that the system works properly.*

hemisphere is one octant ahead of the left hemisphere, the right hemisphere indicator will be advanced one column in front of the left hemisphere indicator.

Yes, But Does It Work?

How do we know that what we see on the display is really what is going on in our head? What makes data from the headset electronics different from a string of random data? Testing, of course.

I've already discussed the first test: Taking a waveform built of known sinusoidal components, passing it through the FFT algorithm, and verifying that the frequency and energy levels coming out are what I put in. The next level of confidence comes from using a sine-wave generator to force-feed the hardware with real data. Photo 3 is a sample of a 10-Hz sine wave being introduced into the headset electronics channel 1 input. Note that all other frequency bands are null.

Once we know how the FFT algorithm works—and we have the confidence that the hardware will deliver the data correctly—the rest is up to interpretation by the users. A sample session using the complete system seems to indicate that many frequencies are active. There also appears to be a difference between the right and left hemispheres in both amplitude and phase. But just what that means will have to be left to a person more involved with biofeedback.

Using HAL

HAL is one of the few Circuit Cellar projects in which the circuitry and software are relatively simple in comparison to the application. While the basic EEG apparatus has been in use for 60 years, we are just beginning to understand the "circuitry" and "software" operating within our own heads. Hundreds of volumes are devoted to clinical and research electroencephalography. I have included some possibilities for further reading at the end of the article; you may find them interesting as you explore this complex and fascinating field.

HAL provides an effective demonstration of how our gross behavior is in some way mirrored in the electrical activity of the brain. To view this behavior, however, you must learn how to connect HAL to your head.

You can purchase disposable EEG electrodes with adhesive pads and conductive gel from medical supply houses. Reusable silver/silver-chloride electrodes are also available. The reusable electrodes are more expensive initially, but you can use them almost indefinitely if you care for them properly.

Five wires are involved in a two-channel HAL connection: common reference, left-channel differential input pair (J1A and J2A), and right-channel differential input pair (J1B and J2B). You should construct the differential pair wires with shielded cable. I found that standard shielded microphone cable—with a male RCA connector at one end and a snap to mate with the electrode at the other end—is quite sufficient. The shield is, of course, grounded only at the RCA jack on HAL. Be careful not to accidentally short the shield to the differential input electrodes or HAL will produce erroneous results.

You can place the common reference electrode on the mastoid, the bony projection just behind the ear. You should first clean the area with soap and water to remove oils from the hair and scalp. Next, rub the area lightly with a piece of alcohol-dampened gauze. Finally, peel the adhesive from a disposable electrode, fill the well with conductive gel (but do not overfill), and place the electrode on the selected spot.

Follow a similar procedure to place the differential-pair electrodes on each hemisphere. One wire (J2A or J2B) goes over the frontal lobe—directly above the eye and just below where your hairline was before it started receding! The second electrode (J1A or J1B) is more difficult to place (unless you happen to be completely bald). You should put it over the occipital lobe, which in most people is covered with hair.

On the back of your head is a ridge, where your skull begins to bend inward toward your neck. Find a spot about a third of the way from the midline of your head to the common electrode placed on the mastoid and just on or below the ridge. Prepare the scalp as before, taking care to hold the hair carefully away from the site. You might try using a sweatband to hold the electrode in better contact with the scalp, making several small holes in the sweatband to allow easier access to the electrode connector.

Figure 6 shows the general location and nomenclature for commonly used electrode placements. If you build a multichannel model of HAL, you should refer to more detailed literature about electrode placement, monopolar and bipolar placement, and other esoteric subjects.

Assuming that you have successfully placed both channels of electrodes appropriately, it is time to fire up the software supplied with HAL. Remember, to retain its electrical isolation,

Warning

HAL is presented as an engineering example of the design techniques used in acquiring brain-wave signals. It is not a medically approved device, no medical claims are made for it, and it should not be used for any medical diagnostic purposes. Furthermore, the safe use of HAL requires that the electrical power and communications isolation described in its design not be circumvented. HAL is designed to be battery-operated only. Do *not* substitute plug-in power supplies.

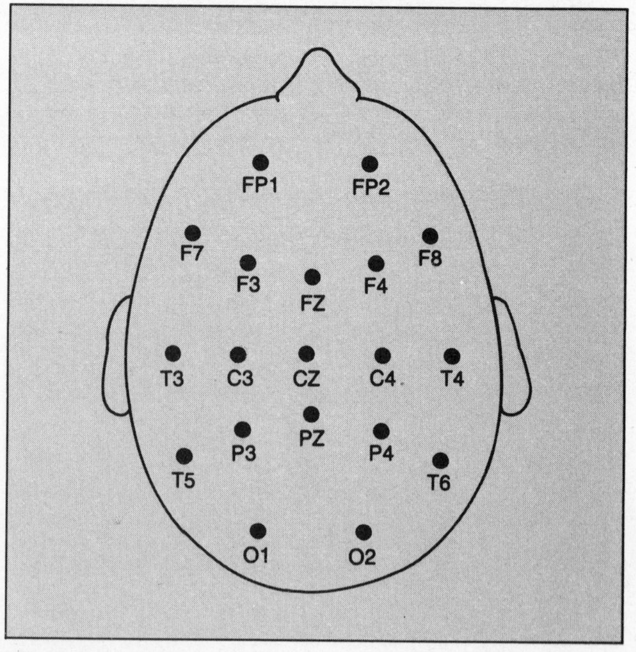

Figure 6: *Commonly used locations and nomenclature for electrode placement.*

you should operate HAL only on batteries, *exactly* as presented.

Attach the electrode wires to the electrode connectors, connect HAL to the computer's COM1 port, turn on HAL, load `BIO.EXE` into the computer, and watch your monitor. The sample display illustrated previously will let you see your own brain activity as you produce it. Actually, due to sampling and processing time, you will see it delayed by about half a second.

You may see little high-amplitude activity, and the dominant frequency may drift quite a bit (pressing F3 will change the amplitude range). But if you close your eyes and make your mind less blank than usual, your "lab partner" should see a dominant frequency appear somewhere in the alpha-wave range. Of course, if you open your eyes to see it, it will immediately disappear. Somehow this is reminiscent of the old question about a tree falling in a forest making a sound or not. Since HAL's data is serial, you could store a session to disk and view it "off-line" later.

The display also shows the condition of the two optional switches that can be attached to HAL. The intent here was to provide a way for you to monitor either some external condition or internal state, so you can correlate it with your brain-wave activity. You could use two momentary contact switches, one for each hand perhaps, to see if you can correctly identify from internal cues some aspect of brain functioning (e.g., what band you are producing a dominant frequency in, whether the left or right hemisphere is showing greater power output for different mental tasks, and so forth).

Alternatively, some external source—the presence or absence of music, a strobe light, or other stimulus—could trigger the switches. Since you have two switches, you could test for four different conditions.

You can also modify the BIO code to turn HAL into a standard biofeedback device. In this case, you would want to add threshold controls. That is, the program could provide audio feedback only when you had achieved certain minimum values of relative amplitude within a certain frequency band. You could provide different feedback for each channel, or feedback could be tied into both channels meeting the same (or different) criteria.

A multichannel HAL would need a completely different display to provide meaningful feedback about the brain's activity. An EGA display could be mighty useful here, in some way showing activity in terms of color on a map of the brain.

A Final Thought

As I said earlier, we are only starting to understand the relationship between our brain-wave activity and our mental states. HAL is not intended to take the place of a $100,000 EEG analyzer in a modern neurologist's lab, but it certainly can provide the serious experimenter a valid vehicle for entry into the fascinating world of neuroscience.

In Future Chapters

In Chapter 8, I'll present a development system for Intel's 8051 family of microcontrollers using the IBM PC as a base.

BIBLIOGRAPHY

Brown, Barbara. *New Mind, New Body: Bio-Feedback; New Directions for the Mind*. New York: Harper & Row, 1974.

Kooi, Kenneth. *Fundamentals of Electroencephalography*. New York: Harper & Row, 1971.

Scientific American Editors. *Altered States of Awareness*. San Francisco: W. H. Freeman & Co., 1972.

Segalowitz, Sid J. *Two Sides of the Brain: Brain Lateralization Explored*. Englewood Cliffs, NJ: Prentice Hall, 1983.

Special thanks for help provided on this article to Dr. Robert Stek, David Schulze, Rob Schenck, Jeff Bachiochi, and Ed Nisley.

Editor's Note: Steve often refers to previous Circuit Cellar articles. Most of these past articles are available in book form from BYTE Books, McGraw-Hill Publishing Company, P.O. Box 400, Hightstown, NJ 08250; (1–800–2–MCGRAW).

Ciarcia's Circuit Cellar, Volume I covers articles in BYTE from September 1977 through November 1978. *Volume II* covers December 1978 through June 1980. *Volume III* covers July 1980 through December 1981. *Volume IV* covers January 1982 through June 1983. *Volume V* covers July 1983 through December 1984. *Volume VI* covers January 1985 through June 1986.

Circuit Cellar Ink

It's virtually impossible to provide all the pertinent details of a project or cover all the designs I'd like to in the pages of BYTE. For that reason, I have started a bimonthly supplemental publication called Circuit Cellar Ink, which presents additional information on projects published in BYTE, new projects, and supplemental applications-oriented materials. For a one-year subscription (6 issues), send $14.95 to Circuit Cellar Ink. Credit card orders can call (203) 875-2199.

The following item is available from

Circuit Cellar, Inc.
4 Park St., Suite 12
Vernon, CT 06066
(203) 875-2751

Circuit Cellar Hemispheric Activation Level Detector (HAL) complete printed circuit board kit. Comes with printed circuit board and all chips and board-mounted components, including 80C31, IC sockets, 27C64 EPROM, user's manual, and IBM PC demonstration display on PC format disk. Does not include battery holders, batteries, scalp electrodes, or connecting cables. HAL-4-KIT .... $179; add $19 if you want an IBM PC cable.

The Circuit Cellar HAL design is available for licensing to qualified manufacturers. Call for information.

All payments should be made in U.S. dollars by check, money order, MasterCard, or Visa. Surface delivery (U.S. and Canada only): add $5 for U.S., $8 for Canada. For delivery to Europe via U.S. airmail, add $14. Three-day air freight delivery: add $7 for U.S. (second-day Federal Express), $15 for Canada (DHL), $22 for Europe (DHL), or $30 for Asia and elsewhere in the world (DHL).

There is a multiline Circuit Cellar bulletin board system (running TBBS 2.0M) that supports past and present projects in BYTE and Ink. You are invited to call and exchange ideas and comments with other Circuit Cellar supporters. The 300-/1200-/2400-bps BBS is on-line 24 hours a day at (203) 871-1988.

7

BUILD THE CIRCUIT CELLAR IC TESTER

PART 1: HARDWARE

This versatile tester can save you hours of troubleshooting when building and debugging electronic systems

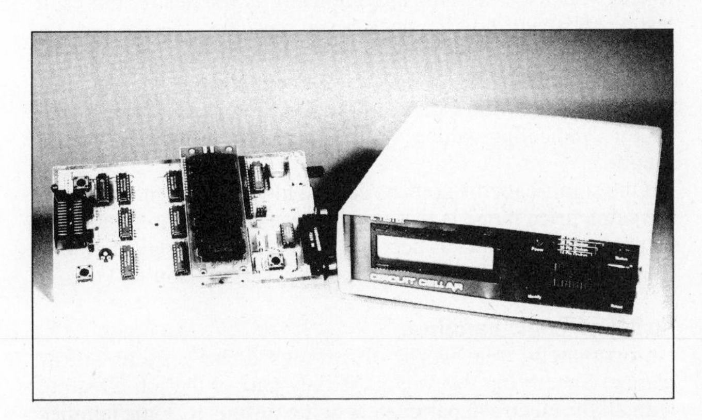

Having designed and debugged many electronic systems, I have seen more than my share of defective ICs. I have also wasted more time than I care to remember discovering that my latest creation was not deficient, but that one of the factory-fresh ICs I put in it was in fact defective. You'd think they'd test them, wouldn't you?

An IC tester can provide both time savings and increased confidence when building and debugging electronic systems. In fact, finding defective ICs before manufacturing an electronic product can also save a considerable amount of money by minimizing the labor and board damage costs involved with reworking electronic boards.

For the most part though, IC testers are used for repairing failed electronic circuits. My latest example was my home: While I was preparing this project, lightning struck my house and practically everything got blitzed. If it were not for my IC tester's help in finding the 29 blown chips in my home-control and automatic-lighting system, I'd still be sitting in a dark, dead house (I thought I had added every preventive measure I could, but I can see we'll need another project on transient protection). I was especially thankful that it could successfully test open-collector driver chips—a problem for most economical testers.

Having an IC tester saved my day, and it may be something you have always needed, too. In this part of Chapter 7, I will describe the design and construction of a digital IC tester with tutorial emphasis on the thinking I had to go through in the process of building it. In Chapter 7, part 2, I will conclude with a discussion of the specific operation of this tester and its advanced software.

Design Considerations

The first step in designing any project is to carefully consider and define what the device is to do. For the IC tester, I first looked at units already on the market and noted their features, prices, deficiencies, and benefits.

I found a price range that varies from less than $200 to several thousand dollars. They also vary considerably in their operation and capability. The low-cost units are generally bus-specific—plugged into a computer slot (Apple II or Commodore 64)—and include operating software. Up the scale from those are the stand-alone—but relatively "dumb"—IC identifiers. With these, if you put a good chip into the socket, a two-, three-, or four-digit number indicating its identification appears on the seven-segment LED display.

The low-end (less than $1000) testers I found have fixed device libraries and perform only simple digital tests (i.e., no AC-parametric tests and no logic-threshold tests). Most, however, indicate that they do provide "periodic" library updates as new standard parts become available.

The high-end testers, costing several thousand dollars, allow some AC-parametric testing, threshold testing, and testing of analog ICs. While they are probably incapable of verifying complete compliance to manufacturers' data sheets, they certainly come close. They can help identify chips with marginal timing specifications. The cost of these devices (including the cost of maintenance, special adapters, and new device support) makes them prohibitive to ordinary users; such devices typically find their home in large corporations with special testing requirements (often those involved with military or aerospace applications).

Flexibility at an Economical Price

My goal in developing the IC tester was to provide as much capability and flexibility as possible in an affordable device that can be used by small businesses and electronic experimenters.

Certainly, economics played its part in requiring compromises in the design. I decided that AC-parametric testing and threshold-level testing would put the device into a higher-price category than I was targeting, so these features were the first to go. Then, I needed to determine what the user interface should be like.

One possibility was to design a card that plugged into an IBM PC slot, with an external test box connected by a cable. This approach would let me develop and include PC software permitting users to develop tests for their own devices. This would include standard devices not yet in the master library and custom devices, like programmable array logics. Unfortunately, this limited the use of the tester to owners of PCs or compatibles (with a free backplane slot and a long extension cord), and the tester would hardly be portable.

Another possibility was to configure the tester to connect to a dumb terminal, or to any computer with terminal-emulation capability, via RS-232C. While this would broaden the number of potential users of the tester, and would give the tester a little more flexibility, it would also take away the flexibility of user-generated device tests unless that extra (and I might add, very intensive) software capability was provided within the tester.

Finally, I could choose the pure stand-alone approach. Such a configuration would be a self-contained portable tester with its own display and some form of entry panel. Even though it's an easier concept, a stand-alone unit would be more expensive to build and would potentially have the same limitations as terminal-based testers unless it also contained the "smarts" of a larger computer.

Three Units in One

After considering the various circuit possibilities, I concluded that my IC tester should support all three modes of operation. With only a slight increase in hardware complexity, I could present a single design that operates in different ways depending upon which peripheral components and software you install (see photo 1). The operating configurations are called PC-host mode, terminal mode, and stand-alone LCD mode.

The PC-host and terminal modes simply require a serial port for operation. In terminal mode, the tester presents all statements regarding test functions and results on the video terminal's display. The PC-host mode is similar, with the exception that it has the added flexibility of letting you directly modify and extend the device library.

In the stand-alone LCD mode, the tester shows device parameters and data on a 2-line by 20-character LCD. (It should be noted that the LCD is optional; you can operate the tester in the other two modes without it.)

In essence, the stand-alone LCD mode provides a portable (i.e., battery-operated) IC tester suitable for testing any chips that are precoded within its extensive EPROM-resident device library. (The Revision 1.0 library currently contains about 600 74xx00-series and CMOS 4000-series chips.)

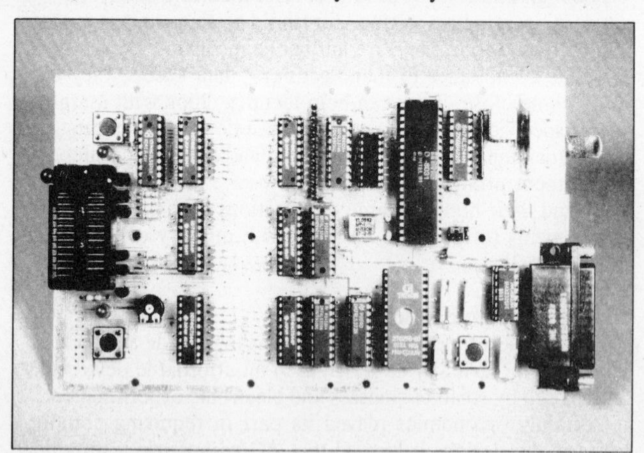

Photo 1: *The prototype IC tester printed circuit board configured for terminal operation. The IC under test is inserted into a special zero-insertion-force socket on the left side, and test information and menu selections are displayed on a terminal connected through the DB-25 connector on the right side.*

The terminal mode provides a menu format intended to maximize the information displayed, while the PC-host mode converts this otherwise stand-alone piece of hardware into an interactive and configurable diagnostic tool with the intelligence of a full computer.

Testing Logic ICs

How do you go about testing ICs? Certainly, I had to answer this question before I could design the tester.

Testing 7400-series logic devices appears relatively straightforward (I didn't consider AC and voltage threshold checking for reasons of economics). To test a two-input NAND gate, for example, you merely set specific logic levels on the gate inputs and check that the outputs are what they are supposed to be.

The process involves a series of test vectors. A test vector is a pattern of bits (0s and 1s) applied to the inputs of the device under test (DUT), to which the DUT responds with a response vector (a pattern of bits on the DUT's outputs). You then compare the response vector from the DUT to the expected response vector, with bit differences indicating pin failures.

You can specify any number of test vectors for a device, allowing you to test the chip as completely as you desire. For each test vector specified for a device, you must also specify a corresponding expected response vector. Since there are cases when some outputs of a device may be in an unknown state, you must also provide a "don't care" mask for each expected response vector, indicating which bit comparisons the tester should ignore.

One significant difference between my IC tester and others in the same price range is that mine does a full-function logic test using as many vectors as necessary to exercise all logic possibilities on the test device. Most inexpensive testers don't do this.

So Many Logic Families

Unfortunately, real-world electronics doesn't quite follow theory. Specifying test vectors is only part of the job. Dealing with all the electrical parameters of the various IC logic families is the real problem.

Since its initial development and introduction by Texas Instruments, the 7400 series of ICs has become an industry standard—at least in terms of device functions and pin-outs. These chips are composed of a large variety of SSI-, MSI-, and LSI-logic building blocks, which designers put together to produce the desired functions.

The original 7400-series family consisted primarily of simple functions, like gates and flip-flops. These were adequate for many applications, but designers kept demanding devices with increasingly greater complexity and functionality.

IC technology did not stand still as designers needed more devices with higher speed and lower power. These requirements led to the introduction of the 74H00-series (high-speed) and 74L00-series (low-power) devices. For the most part, these new series maintained the device pin-outs established by the standard-TTL predecessors (the 7400 series). However, the 74H00-series devices consumed substantially more power than, and the 74L00-series devices were slower than, the standard 7400-series devices.

As the technology improved, even more families appeared. A faster family using Schottky technology was established, the 74S00 series, along with a popular low-power Schottky family, the 74LS00 series.

Eventually, the very-low-power CMOS devices that had been manufactured with 4000-series numbering shifted over to the more popular 74xx00-series pin-out and numbering scheme with the introduction of the 74C00-series family of devices. These devices were slow and had low-current-drive outputs, but they filled a niche in designs requiring extremely low power consumption.

Other families include 74ALS (advanced low-power Schottky), 74AS (advanced Schottky), 74HC (high-speed CMOS), 74HCT (high-speed CMOS, TTL-compatible), 74AC (advanced CMOS), 74ACT/74AHCT (advanced CMOS, TTL-compatible), and 74F (Fairchild advanced Schottky).

Simple Concept, Tough Trade-offs

Digitally speaking, the logical parameters of a 74xx00 are the same regardless of its family, and you could easily be misled into thinking that we are designing a digital tester. However, each of these families has analog characteristics that differ from the other families. The IC tester is actually more an exercise in analog design. Let me explain.

Typical differences between logic families are power consumption, speed, output current drive, input current loading, input transition thresholds, and output voltage swings. Comparisons of some of these parameters for a 74x00 quad NAND gate from several families are shown in table 1. (While the parameters specified in table 1 for the 74x00 devices do not apply to all devices within the respective families, they are representative of the majority of the devices).

In effect, table 1 shows the wide variations of input and output parameters that the ideal IC tester must support. Low-level input currents range from 1 microampere to 2 milliamperes (and much higher on some device inputs), and low-level output currents range from 360 μA to 20 mA.

The tester's ability to identify a device presents an important consideration. If the tester is designed for 74ALS or 7400 "straight" TTL, you might smoke a 74C chip if you inserted it into the tester operated at the current levels of those devices.

Any truly general purpose (read usable) tester must accommodate the wide ranging voltage and current parameters of all the families. Since the tester may not know at the outset what device is installed in the ZIF (zero insertion force) socket (remember, one of the modes is to identify unmarked chips), it cannot make any assumptions as to which pins are inputs and which are outputs.

The tester requires a certain amount of trial and error to identify an unknown device, and it must employ current-limiting resistors between the DUT (in the ZIF socket) and the IC tester's vector-generation circuitry (for when a DUT and tester output are connected together).

Also, while most devices have totem-pole outputs, some have tristate, open-collector, or open-drain outputs. The tester must be able to pull tristate outputs high and low when they are in the high-impedance state to verify the state, and it must also be able to pull open-collector and open-drain device outputs high and low to verify proper operation.

The catch-22 is to determine a resistor value that will support the input and output current specifications for all the device families to be tested, yet not overstress the DUT. If you go strictly by the book, no single current-limiting resistor value works for both inputs and outputs in all families.

The device specifications provided in table 1 are the manufacturer's recommended operating conditions (ROCs). Looking further into the data sheets, however, we find more information regarding what the chips can do if they have to, such as limited-duration short-circuit output current.

In effect, if we take advantage of our regulated testing environment, we can stretch the ROC a little to choose a resistor that presents the best compromise for handling all the logic families. Think of it as the electronic equivalent of poetic license.

All things considered, I found that the resistor value should be in the 390- to 421-ohm range. Since 390 ohms is the nearest standard resistor value (5 percent tolerance), I chose it for the tester. (After I built the tester, I substituted all standard resistor values between 300 and 430 ohms, inclusive, and verified that the 390-ohm choice provides the best overall performance.)

How It Works

After determining the above, I had one more hurdle. The tester needed to be able to apply virtually any number of test vectors to the DUT without losing the device's state from the previous vector—and without causing undo stress on the DUT (i.e., without keeping any of the DUT outputs in a high-current output mode for an extended period of time). I solved this with what I like to refer to as a combinatorial-latch circuit.

Each ZIF-socket pin typically has three circuit connections to the IC tester (see figure 1). One connection (connection A) is to

Table 1: *Comparison of specifications for various 74xx00 devices. (Subscript identifiers are IL—input low, IH—input high, OL—output low, and OH—output high.)*

Device name	I_{IL} max	I_{IH} max (μA)	V_{IL} max (V)	V_{IH} min (V)	V_{OL} max (V)	V_{OH} min (V)	I_{OL} max (mA)	I_{OH} max (mA)
74LS00	−0.4 mA	20	0.8	2.0	0.5	2.7	8.0	−0.4
74H00	−2.0 mA	50	0.8	2.0	0.4	2.4	20	−0.5
74L00	−0.18 mA	10	0.7	2.0	0.4	2.4	3.6	−0.2
74S00	−2.0 mA	50	0.8	2.0	0.5	2.7	20	−1.0
74AS00	−0.5 mA	20	0.8	2.0	0.5	2.5	20	−2.0
74ALS00	−0.1 mA	20	0.8	2.0	0.5	2.5	8.0	−0.4
74HC00	−1.0 μA	1.0	1.2	3.15	0.33	3.84	4.0	−4.0
74HCT00	−1.0 μA	1.0	0.8	2.0	0.33	3.84	4.0	−4.0
74F00	−0.6 mA	20	0.8	2.0	0.5	2.7	20	−0.36
74C00	−1.0 μA	1.0	1.5	3.5	0.4	2.4	0.36	−0.36
7400	−1.6 mA	40	0.8	2.0	0.4	2.4	16	−0.4

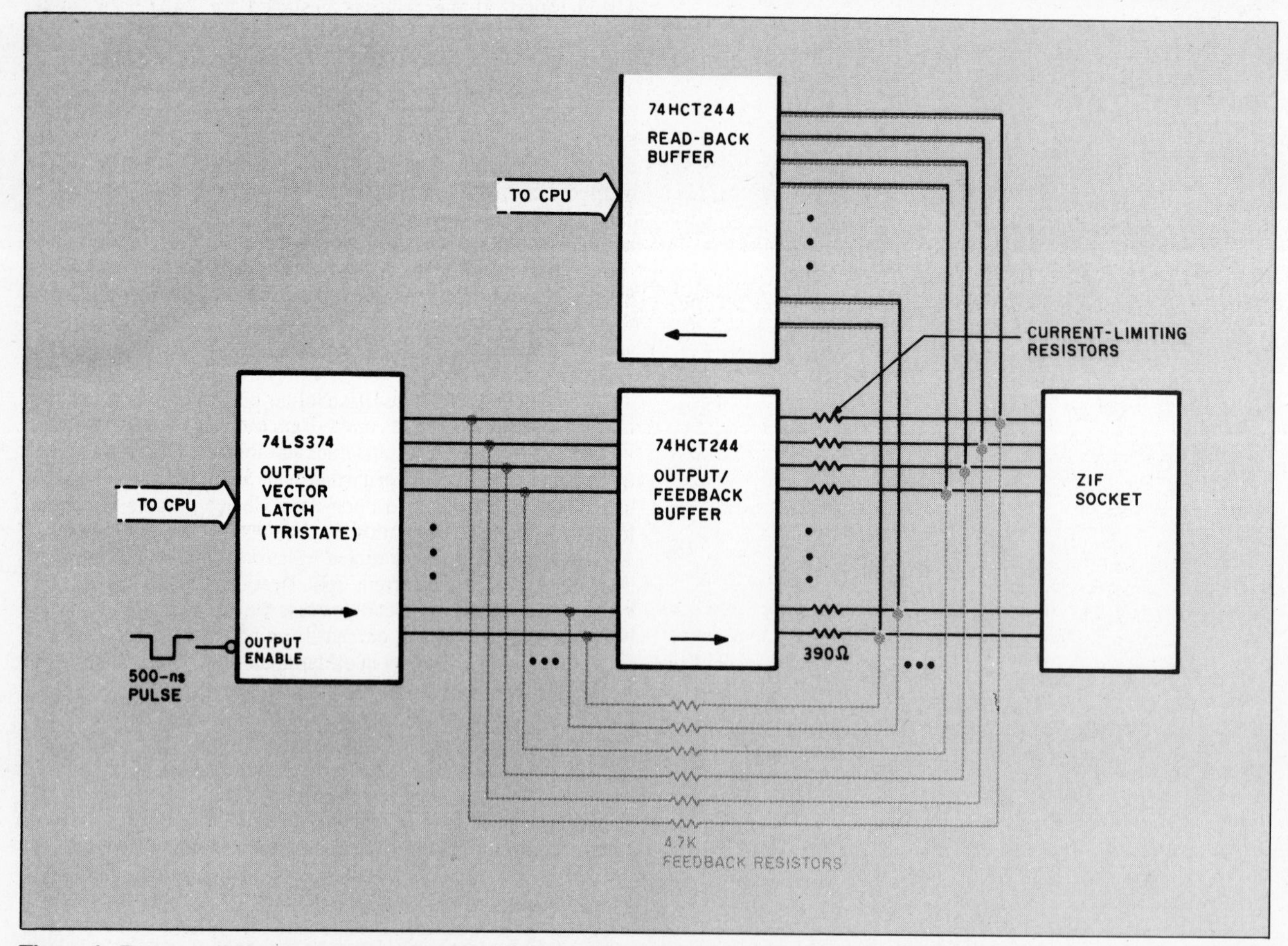

Figure 1: *Diagram of the IC tester's combinatorial-latch circuit. The zero-insertion-force socket holds the device under test.*

The DB-25S connector provides the RS-232C interface connection to an IBM PC or any dumb terminal.

an output of a 74HCT244 buffer—the feedback buffer—through a series 390-ohm current-limiting/load resistor. Another connection (connection B) is to the corresponding input of the same 74HCT244, through a 4.7-kilohm series feedback resistor. The 74HCT244 input is also connected to an output of a 74LS374 tristate latch.

The final ZIF-socket-pin connection (connection C) is directly to an input of another 74HCT244 tristate buffer—the read-back buffer. By reading the 74HCT244 read-back buffer, the processor can determine the logic levels of the DUT pins (the ZIF-socket pins).

The IC tester sends a test vector to the DUT by writing the desired bit pattern into the 74LS374 latch, while the latch's outputs remain in the high-impedance state. The system then enables the outputs of the 74LS374 (i.e., they are allowed to go active) for a period of 500 nanoseconds, applying the test-vector bit pattern to the inputs of the feedback 74HCT244 buffer.

During the 500-ns 74LS374-enable period, the relatively high value of the feedback resistors (4.7 kilohms) ensure that the 74HCT244 inputs will see the test-vector logic levels from the 74LS374, regardless of the logic levels present at the DUT pins.

Within a few nanoseconds (i.e., propagation time) of the time the feedback 74HCT244 first sees the new logic levels from the 74LS374, the same logic levels will appear on the outputs of the 74HCT244; these logic levels will remain on the 74HCT244 outputs for the duration of the 500-ns pulse.

If a DUT output in the ZIF socket is in the opposite logic state as the corresponding 74HCT244 output, the resistor between the 74HCT244 output and the DUT pin will present a load to the DUT output, possibly causing it to go into its "overdrive" mode in an attempt to retain its desired output logic level. The overdrive operation will continue until the end of the 500-ns pulse, when the 74LS374 outputs are finally disabled, returning to their high-impedance state.

When the 74LS374 outputs are disabled, the only inputs to the feedback 74HCT244 will be from the DUT feedback resistors. Since the feedback buffer is a 74HCT-series device, it presents negligible input current loading (about 1 μA), so the voltage levels reaching the 74HCT244 inputs through the feedback resistors will be nearly the same as those at the corresponding DUT pins.

If the voltage coming through a feedback resistor to the 74HCT244 is the same logic level as that presented previously by the enabled 74LS374 output (the case when the DUT pin is an output of the same logic level or when the DUT pin is an input), the 74HCT244 output will remain unchanged. Thus, the logic level is combinatorially latched by the 74HCT244.

If the voltage appearing at the 74HCT244 input from the feedback resistor is the opposite logic level of that presented previously by the 74LS374 (which is the case when the DUT pin is an output of the opposite logic level), the 74HCT244 will see the new logic level at its input and change its output to match. When this occurs, the 74HCT244 output then matches the output of the DUT pin, eliminating the loading that was present. Again, the new logic value will be combinatorially latched by the 74HCT244 using the feedback loop.

You can see that the loading duration on a DUT output will essentially be the duration of the enable pulse—only 500 ns. This keeps potential chip stress to a minimum, while verifying the ability of device outputs to operate properly under load conditions.

The IC Tester Hardware

The schematic for the IC tester is shown in figure 2. The 8031 single-chip microcontroller (IC1) is the brains of the tester. The firmware to run the tester is provided in an EPROM at IC6. The current standard device library (version 1.0) is supplied on a 27256, but IC6 can accommodate several EPROM types, including 2764, 27128, and 27512 devices. The type you would use is determined by the JP1's jumper configuration.

The ZIF socket (IC17) is an Aries universal socket. This specific socket supports devices up to 24 pins, having either 0.3- or 0.6-inch DIP-package widths. When you insert devices into the ZIF socket, you bottom-justify them.

Unfortunately, one problem with using a single ZIF socket on a tester is configuring the power pins for the DUT. Most ICs conform to the standard diagonally opposite corner-pin power/ground configuration: pins 24/12, 16/8, and 14/7. However, a number of devices have oddball power and ground pin-outs. These include 14-pin ICs with ground on pin 11 and power on pin 4, 16-pin ICs with ground on pin 12 and power on pin 5, and 16-pin ICs with ground on pin 13 and power on pin 5, among others (there are also devices with two power pins to support voltage-level conversion).

After reviewing the devices in each oddball pin-out category, I chose to support the two categories with the most devices: 14-pin devices with ground on pin 11 and power on pin 4 and 16-pin chips with ground on pin 12 and power on pin 5. This is, of course, in addition to supporting devices having corner power and ground pins. (In the stand-alone identify-unmarked-chip operating mode, the tester will successfully identify only corner-pin-powered chips.)

The DB-25S connector provides the RS-232C interface connection to an IBM PC or any dumb terminal. The connector is configured as a DCE (data communication equipment) device, allowing you to use a straight-through cable. You need only three pins on the connector (pins 2, 3, and 7—receive, transmit, and signal ground, respectively), but I've hard-wired the DTR (pin 6) handshaking line to a logic high for terminals that need it.

The IC tester has push buttons and some switch-selectable options. A four-position DIP switch (SW1) is used for several purposes, including data-transfer-rate selection, PC-host/terminal mode selection, and 74Cx mode selection (to be described next month). Push buttons PB1 and PB2 are for supporting stand-alone mode operation. PB1 is the identify button, and PB2 is the retest button.

J3 is the connector for the optional LCD, which uses the 8031's P1 connector as its data bus. I chose the P1 bus as the LCD's driver to meet the LCD's (relatively slow) timing requirements. The 74LS139 (IC7) is the address-decoding circuit for accessing several devices on the tester. It decodes the ZIF tristate latches (IC8 through IC10) and read-back buffers (IC14 through IC16), as well as the power/ground transistor latch (IC19).

The 74LS139 also provides a special signal that enables the outputs of the 74LS374 tristate latches for approximately 500 ns (the 8031 WR\ strobe duration), transferring the latched 74LS374 bits to the combinatorial latches formed by the 74HCT244s (IC11 through IC13) and their associated feedback resistors.

For the tester's buffers (IC11 through IC13), I chose 74HCT devices instead of 74LS (or other family) devices. Members of this family drive their outputs close to the power and ground rails, can source a lot of current, and provide negligible load on the resistor-feedback circuit. Similarly, the read-back buffers (IC14 through IC16) are 74HCT devices to keep loading to an

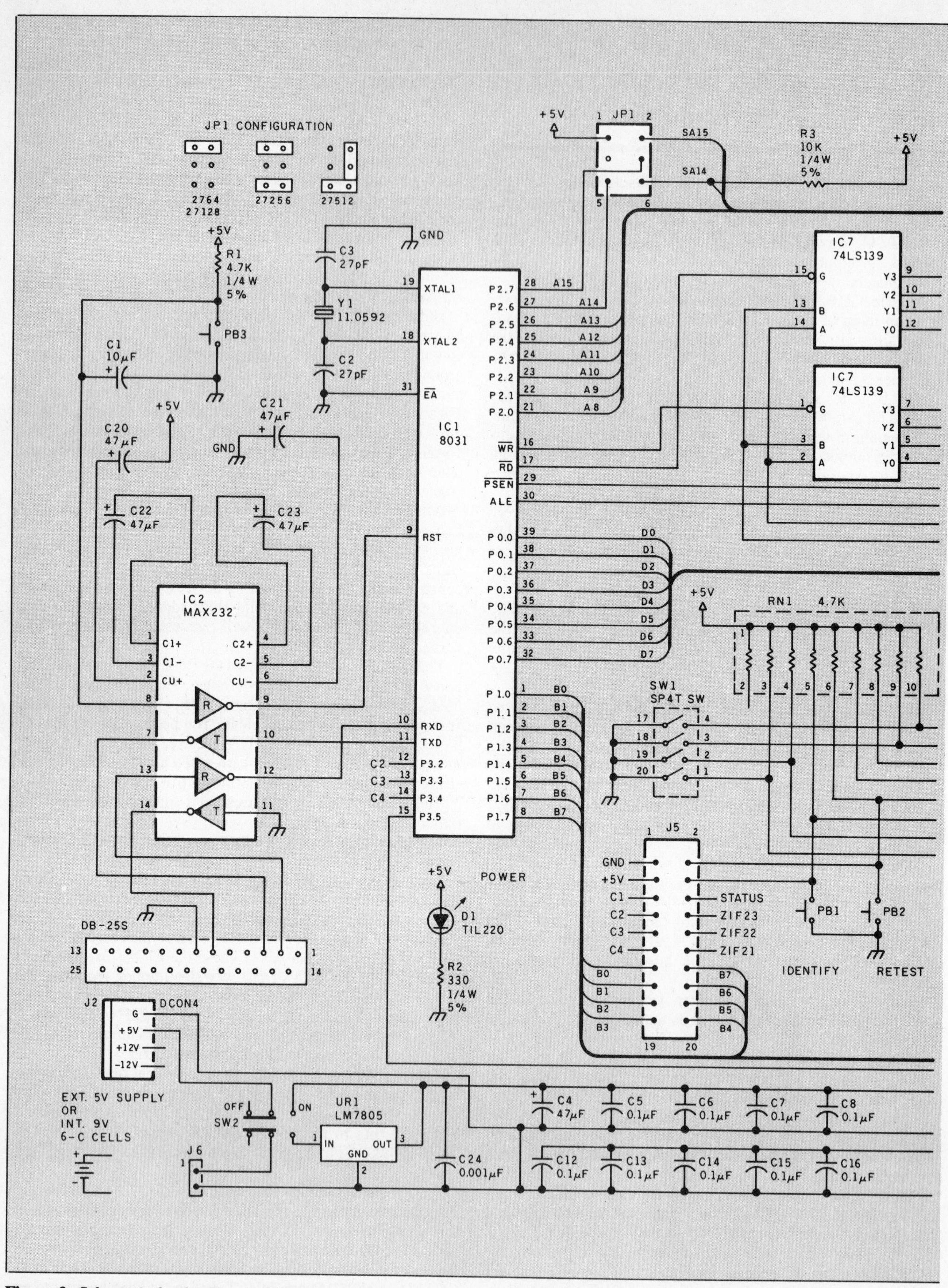

Figure 2: *Schematic for the Circuit Cellar IC tester.*

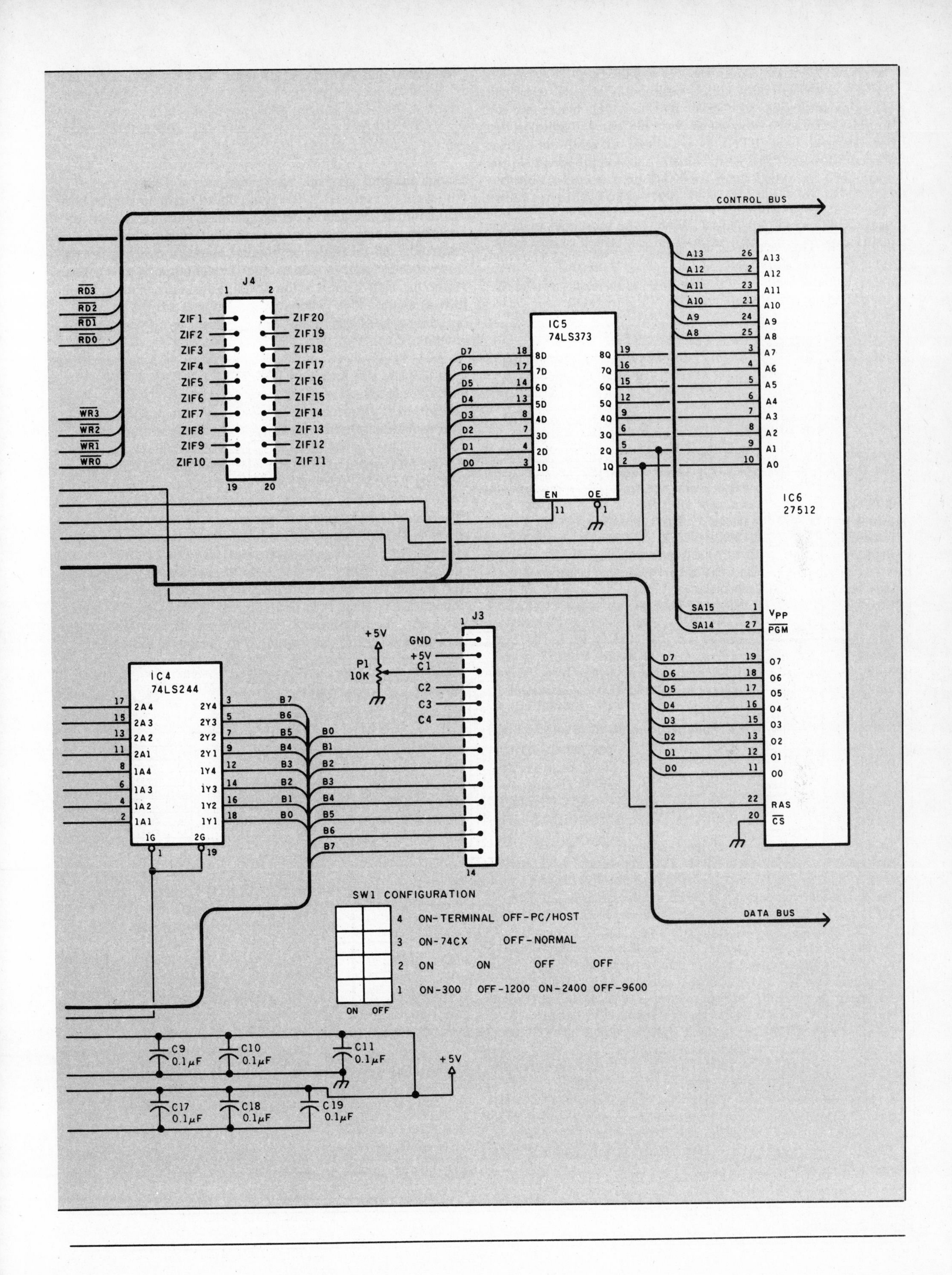
CONTROL BUS
DATA BUS
J4
ZIF1
ZIF2
ZIF3
ZIF4
ZIF5
ZIF6
ZIF7
ZIF8
ZIF9
ZIF10
ZIF11
ZIF12
ZIF13
ZIF14
ZIF15
ZIF16
ZIF17
ZIF18
ZIF19
ZIF20
IC5
74LS373
IC6
27512
IC4
74LS244
J3
GND
+5V
C1
C2
C3
C4
P1
10K
SW1 CONFIGURATION
4 ON-TERMINAL OFF-PC/HOST
3 ON-74CX OFF-NORMAL
2 ON ON OFF OFF
1 ON-300 OFF-1200 ON-2400 OFF-9600
ON OFF
C9 0.1μF
C10 0.1μF
C11 0.1μF
C17 0.1μF
C18 0.1μF
C19 0.1μF
RAS
SA15
SA14

absolute minimum (do *not* substitute 74LS devices).

The discrete transistors (Q1 through Q6) provide the power and ground switching for the ZIF socket (IC17). Pin 24 of the ZIF socket is connected directly to +5 volts, eliminating the need for an additional transistor. The PN2907s (Q3 through Q6) are for turning on power (+5 V) to various ZIF-socket pins (9, 19, 20, and 22), while the PN2222s (Q1 and Q2) are for turning on ground to two of the ZIF-socket pins (12 and 16).

The 74HCT374 latch (IC19) controls the transistors. As mentioned earlier, 74HCT devices can source and sink current equally well. This fact made the 74HCT374 a good choice for driving the transistors, since it can handle the transistor base currents equally well for the ground switches (high 74LS374 outputs) and the +5-V switches (low 74LS374 outputs).

The tester has two LEDs. D1 is merely a power-on indicator that lights whenever power is applied. D2 is a software-controlled status LED used to indicate when the device is operating in an RS-232C mode (PC-host or terminal, LED on) or a stand-alone mode (LED off).

Experimenters

While you can order printed circuit boards and kits for the Circuit Cellar IC tester, I encourage you to build your own. If you don't mind doing a little work, I will again support your efforts. A hexadecimal file of the executable code for the 8031 Revision 1.0 system EPROM code, suitable for stand-alone or terminal operation, is available for downloading from my bulletin board at (203) 871-1988.

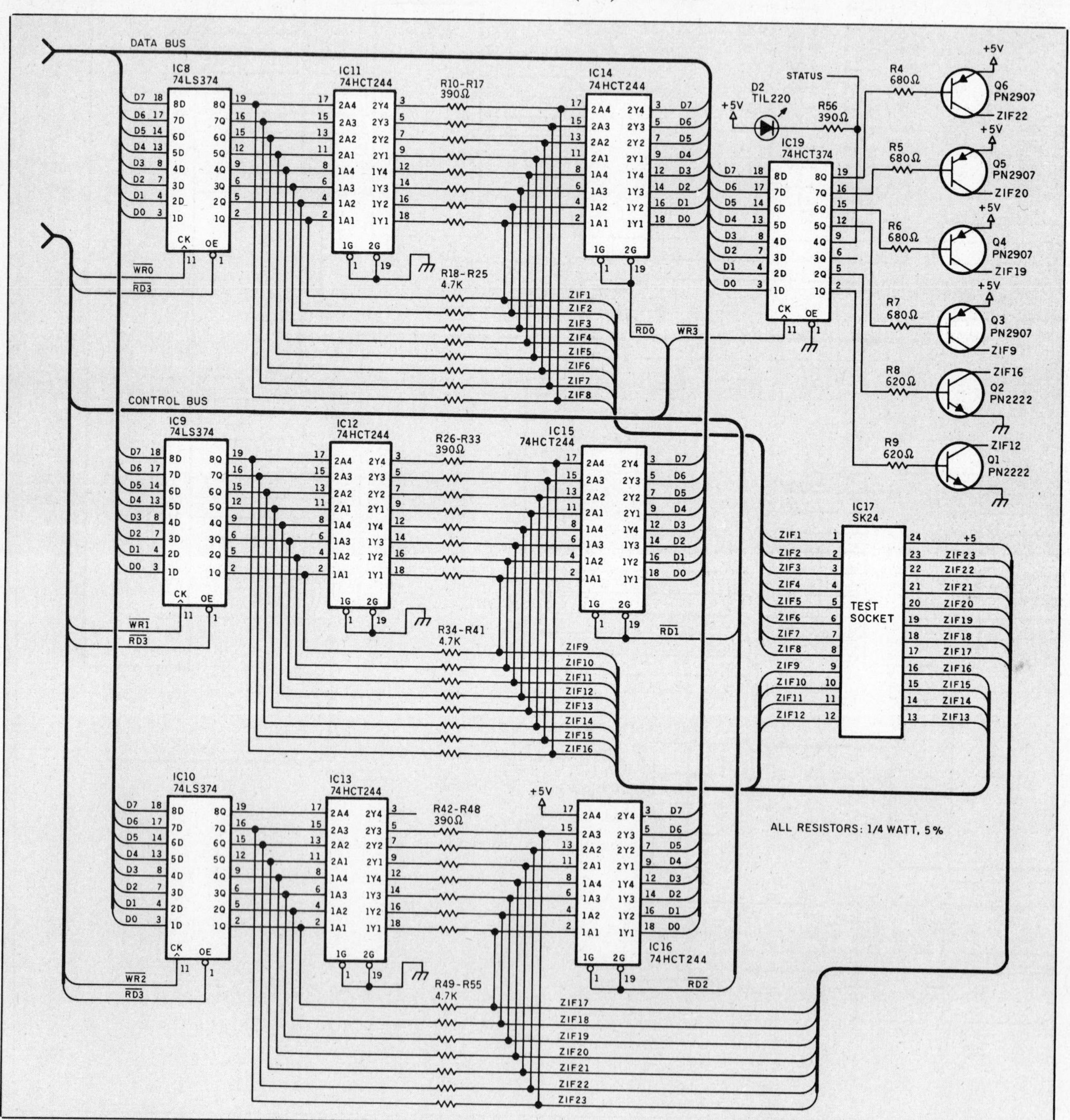

Figure 2: *Continued.*

Alternatively, you can send me a preformatted IBM PC 5¼-inch disk with return postage, and I'll put the file on it for you (the hexadecimal file could be used with my CCSEP serial EPROM programmer, for example). Of course, this free software is limited to noncommercial personal use.

In Conclusion

In Chapter 7, part 2, I will present the tester's software, which lets you develop and debug your own test vectors and device libraries.

Special thanks to Roger Alford, Jeff Bachiochi, and William Potter for their work on this project.

Editor's Note: Steve often refers to previous Circuit Cellar articles. Most of these past articles are available in book form from BYTE Books, McGraw-Hill Publishing Company, P.O. Box 400, Hightstown, NJ 08250, (1–800–2–MCGRAW).

Ciarcia's Circuit Cellar, Volume I covers articles in BYTE from September 1977 through November 1978. *Volume II* covers December 1978 through June 1980. *Volume III* covers July 1980 through December 1981. *Volume IV* covers January 1982 through June 1983. *Volume V* covers July 1983 through December 1984.

The following items are available from

Circuit Cellar, Inc.
4 Park St., Suite 12
Vernon, CT 06066
(203) 875-2751

1. Circuit Cellar IC tester experimenter's kit for stand-alone or terminal operation. Contains IC tester printed circuit board, 11.0592-megahertz crystal, programmed 27256 EPROM with Revision 1.0 device library, MAX232 level shifter, Aries 24-pin narrow-format ZIF socket, and manual with complete parts list.
ICT01-EXP $99
2. Circuit Cellar IC tester full printed circuit board kit for stand-alone, terminal, or PC-host operation. Contains IC tester printed circuit board, 8031 processor and crystal, programmed 27256 EPROM with Revision 1.0 device library, Aries 24-pin narrow-format ZIF socket, IC sockets, all board-mounted components and ICs, PC-host software on PC format disk, power supply, and manual.
ICT01-FULL $179
3. Complete Circuit Cellar IC tester kit with stylish enclosure. Full printed circuit board kit with all components, right-angle-mounted enclosure adapter board with ZIF socket and LCD, software on PC format disk, power supply, and manual.
ICT02 .... $349
4. Two-line by 20-character LCD and 14-pin Berg connector for either item 1 or 2.
2x20 LCD $32

All payments should be made in U.S. dollars by check, money order, MasterCard, Visa, or American Express. Surface delivery (U.S. and Canada only): add $5 for U.S., $8 for Canada. For delivery to Europe via U.S. airmail, add $14. Three-day air freight delivery: add $10 for U.S. (UPS Blue), $25 for Canada (Purolator overnight), $45 for Europe (Federal Express), or $60 for Asia and elsewhere in the world (Federal Express). Shipping costs are the same for one or two units.

There is an on-line Circuit Cellar bulletin board system that supports past and present projects. You are invited to call and exchange ideas and comments with other Circuit Cellar supporters. The 300/1200/2400-bps BBS is on-line 24 hours a day at (203) 871-1988.

7

BUILD THE CIRCUIT CELLAR IC TESTER

PART 2: SOFTWARE AND OPERATION

Steve guides us on a tour of the software that makes his inexpensive IC tester possible

In part 1 of this chapter, I talked about the design of my IC tester. In this part, I'll talk about its software and operation.

Three in One

To refresh your memory, the IC tester supports three modes of operation: PC-host mode, terminal mode, and stand-alone LCD mode.

PC-host mode requires that you connect the tester to a serial port on an IBM PC or compatible. In this mode, the PC handles all test-vector transfers and comparisons and provides the highest level of flexibility and power.

To operate the tester in terminal mode, you connect it to a dumb terminal or any microcomputer that emulates a terminal (see photo 1). The options are essentially the same as those offered in PC-host mode, although you can use only a fixed, ROM-resident device library.

The stand-alone mode of operation lets you operate the tester with only two push-button switches and a 2-line by 20-character LCD. As in terminal mode, this mode operates only with a fixed, ROM-resident device library. It lacks some features of the other two modes, but it permits device identification (using the Identify push button) and specified-device testing (using the Retest push button). The latter lets you determine specific pin failures on a bad IC and display this information on the LCD.

Much of the flexibility of the IC tester comes from its modifiable and expandable device library. While an IBM PC (or clone) is essential for PC-host mode operation, it is required if you're going to make any system software changes, like adding new chips to the library.

With the exception of a single assembly language serial-port driver, all the software was written in Turbo Pascal on an IBM PC. (While the programs do take advantage of some PC-specific features of Turbo Pascal, you shouldn't have much trouble converting them to other Pascal compilers.)

The Definition of a Test Vector

In order to define test vectors, it is important to develop a straightforward means of describing the vector information. What information do we need to define a device and its test vectors?

The device definition consists of the device name (e.g., 7400), the specific package size (e.g., 14 pins), the locations of the power and ground pins (e.g., 14 and 7), and which pins are inputs, outputs, or tri-state.

A test vector merely specifies the high (1) and low (0) logic levels to be written to the pins of the device under test (DUT). A test vector written to the DUT pins is referred to as an output vector.

To determine if the DUT responded properly to the output vector (i.e., to make sure outputs switched as expected and to verify that no inputs are shorted), the tester must read a corresponding read-back vector from the DUT and compare this to an expected read-back vector. Each complete test vector consists of an output vector and an expected read-back vector.

The format for specifying the vector-definition modules is shown in table 1. The order of the different line types is important, though you may freely intersperse comment lines. (Like many assemblers, all characters on a line following an asterisk are ignored by the test-vector compiler.)

The best way to understand the vector-definition module format is by example. Table 2 shows the vector-definition module for a 7400 quad two-input NAND gate. As its name implies, this device contains four two-input NAND gates (the pin-out is shown in figure 1).

Photo 1: *The Circuit Cellar IC tester shown here is operating in terminal mode, connected to a Tandy DT-100 terminal via the RS-232C port on the top of the tester.*

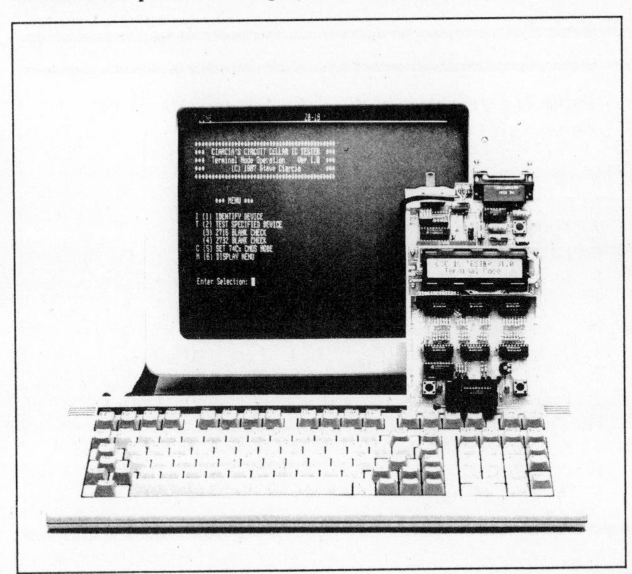

In table 2, the first line is the device name. The name can appear anywhere on the line after the pound sign (preceding and following spaces are ignored). As a general rule, you should keep device names as generic as possible. Instead of using the name "74LS00" use "7400," and so on. Since the tester will logically identify both a 74LS00 and 74HC00 as the same chip, it is better to display "7400," or perhaps "74xx00." There are, of course, cases where you can make exceptions.

The second noncomment line of the vector-definition module is the setup line, which has an *S* in the first column. Three numbers with delimiting spaces must follow the *S*; the first number indicates the number of pins the device has, the second indicates the ground pin number, and the third indicates the power pin number. These numbers tell the compiler (and the tester) what the chip's device is. As I described in Chapter 7, part 1, the tester supports six device types (see table 3).

Following the next comment line is the pin-function line, which has an *F* in the first column. This line specifies a pin-function identifier for each pin, with the identifiers being separated by one or more spaces. Valid identifiers are *I* for input pins, *O* for output pins, and *T* for tri-state pins.

The pin-function line also determines the columnization for the remainder of the device vector definition. All 1s and 0s in the test vectors must be aligned under these columns, and the pin numbers in the pin-number line (the next line in the definition) must also be aligned under these columns.

The next line in the vector-definition module is the pin-number line. It has the letter *P* in the first column. This line specifies the device pin numbers used in testing. The numbers must correspond to the pin-function identifiers specified in the pin-function line and must fall in the columns defined by the function identifiers. If the pin number for a column has two digits (e.g., pin 14), either of the two digits can fall in the column.

The next several lines in the vector-definition module are the actual test vectors. The lines beginning with *I* are initial vectors (output vectors), and the lines beginning with *R* are the expected read-back vectors.

For *I* vectors, the acceptable identifiers are 1 and 0, corresponding to high and low digital values, respectively. For *R* vectors, acceptable identifiers are 1, 0, and *X*, with *X* indicating "don't care." (*X* indicates that the tester should ignore the specified pin when comparing the actual read-back vector to the expected read-back vector. If the 1 or 0 bit value of a column does not change from one line to the next, leaving the column blank in the subsequent line[s] implies that the value should be the same as the last value explicitly stated for that column.)

The last line in the vector-definition module is the end line,

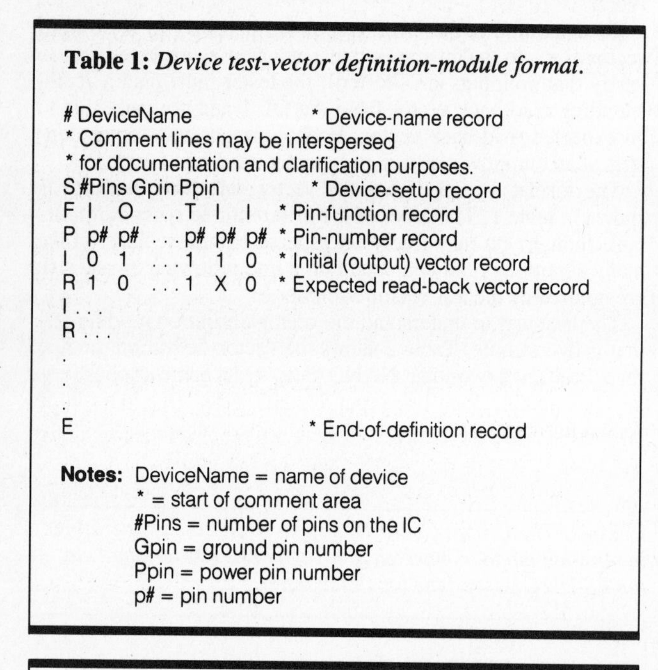

Table 1: *Device test-vector definition-module format.*

```
# DeviceName              * Device-name record
* Comment lines may be interspersed
* for documentation and clarification purposes.
S #Pins Gpin Ppin         * Device-setup record
F  I   O  ...T   I   I    * Pin-function record
P  p#  p# ...p#  p#  p#   * Pin-number record
I  0   1  ...1   1   0    * Initial (output) vector record
R  1   0  ...1   X   0    * Expected read-back vector record
I...
R...
.
.
.
E                         * End-of-definition record
```

Notes: DeviceName = name of device
* = start of comment area
#Pins = number of pins on the IC
Gpin = ground pin number
Ppin = power pin number
p# = pin number

Table 2: *Device test-vector definition module for the 7400.*

```
# 7400          * Quad two-input NAND
S 14 7 14
*  NAND 1     NAND 2    NAND 3     NAND 4
F  I  I  O    I  I  0    I  I  O    I  I  O
P  1  2  3    4  5  6    9  10 8    12 13 11
I  0  0  0    0  1  0    1  0  0    1  1  1
R        1          1          1          0
I  0  1  0    1  0  0    1  1  1    0  0  0
R        1          1          0          1
I  1  0  0    1  1  1    0  0  0    0  1  0
R        1          0          1          1
I  1  1  1    0  0  0    0  1  0    1  0  0
R        0          1          1          1
E  *  end of 7400
```

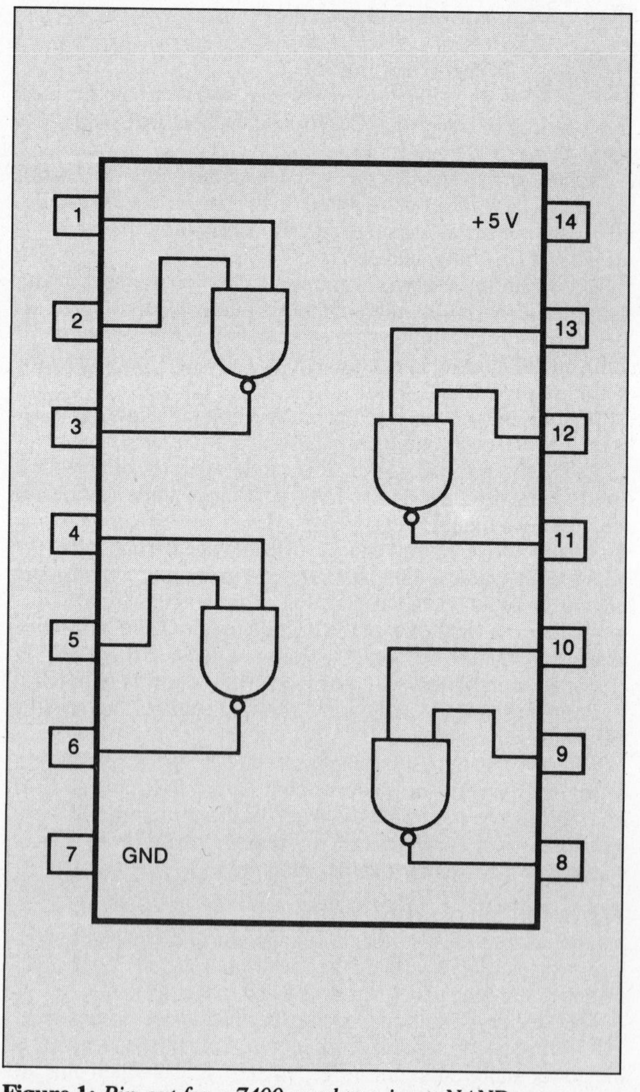

Figure 1: *Pin-out for a 7400 quad two-input NAND gate.*

which begins with an *E*. This is the only letter required to specify the end of the vector-definition module.

Finally, there is the issue of logically identical devices that have different part numbers (like 74LS04 and 74LS14). Differences typically lie in some of the special operational parameters, like Schmitt-trigger inputs or improved current drive capability, which cannot be detected by this IC tester.

Functionally identical devices (i.e., the same test vectors would pass on both devices) are declared to be clones of a specific device. An example of this is shown in table 4. The device-name and end lines are the same as the standard vector-definition module, but only the *C* line is found between them, indicating which device it is supposed to be cloned from.

Compiling Test Vectors for Use

The test-vector compaction compiler, VECCPT.PAS, is a Turbo Pascal program that accepts files conforming to the device test-vector definition format described above. It converts the device and vector information into a single compact module that the computer and tester use to test the devices.

VECCPT.PAS uses seven primary arrays to store the compacted vector information. The primary array, `VectorTable`, holds the actual test-vector information, including the device pin-function information (i.e., which pins are inputs, which are outputs, and which are tri-state), the output vector bytes, the input vector bytes, and the "don't care" mask bytes.

Because the ZIF (zero insertion force) socket has 24 pins, tester software uses 3 bytes for pin and vector information for every device, regardless of size. Consequently, the pin-function and test-vector information is stored as if a 24-pin device were being tested.

The "don't care" mask generated by VECCPT.PAS automatically masks the read-back vector pins not associated with the device being tested. For example, if a 20-pin device is being tested, the bits of the 3-byte read-back vector associated with ZIF-socket pins 1, 2, 23, and 24 will be masked by the "don't care" mask (power and ground pins are automatically masked).

Each of the six device types supported by the IC tester has its own associated array for storing device names and pointers into the `VectorTable` array. These arrays are called `DeviceType` arrays.

While the `VectorTable` array uses variable-length records, with each record being the information to support one device, the `DeviceType` arrays use fixed-length records, with each record containing a 9-byte field for the device name (8 bytes for the name and 1 byte for the string size) and an integer (2-byte) field for the `VectorTable` pointer.

Figure 2 illustrates the information stored in the various arrays and how the arrays interact. As shown, device names are stored in the appropriate `DeviceType` array, and the device pin-function and test-vector information is stored in the `VectorTable` array. A pointer in the `DeviceType` array indicates the start of the corresponding vector-information record in `VectorTable`.

The `VectorTable` device record begins with a 2-byte field indicating the number of bytes in the record. The next 3 bytes specify which pins are inputs and which are outputs (set bits are inputs, and cleared bits are outputs).

The following 3 bytes indicate which pins are tri-state (set bits are tri-state). If a pin is indicated as being tri-state, the 1/0 value in the corresponding bit position of the previous I/O definition bytes is irrelevant. By default, VECCPT.PAS specifies unused ZIF (zero insertion force)-socket pins as being tri-state.

Following the 2 record-size bytes and 6 device pin-function definition bytes, the actual test-vector information begins. Each complete test vector consists of 9 bytes in the record. The first 3 specify the output vector, the next 3 specify the expected read-back vector, and the last 3 specify the "don't care" mask.

As VECCPT.PAS executes, it stores device-name, pin-function, and test-vector information into the appropriate arrays. Notice that the program does not need to store device-type information, since a device's type is determined by which `DeviceType` array it is placed in.

Device clones are handled somewhat differently. When a device is specified as a clone of another device (the "original" device), the name of the clone is placed into the next available record of the appropriate `DeviceType` array. The record number of the original device (in the same array) is then determined, and the value 32,767 is subtracted from the record number; this value (always negative) is then stored in the pointer field of the clone record.

Thus, when the operating software finds a negative integer value in the pointer field of a device record, it will know the device is a clone of another device. It then adds 32,767 to the pointer value to get the record number of the original device.

I should point out that when VECCPT.PAS processes a clone, it looks through its arrays to find the named original device. If the specified original device is not found in any of the six `DeviceType` arrays, the software generates an error, and the clone device will not be stored in any array (the compiler would not even know which array should get the clone record). Thus, it is essential that you specify clone devices only after the corresponding original device.

When compaction of the test-vector files is complete, the compacted information is stored in a binary file. (The format of the data stored in the compacted file is shown in figure 3.)

Operating Software

Once the device test vectors have been developed and compiled into a compacted file, we are ready to use the tester for testing and identifying devices. This involves the cooperation of several programs.

First, there's a ROM-resident program on the IC tester. This program is written in 8031 assembly language and handles the three operating modes from the tester's vantage point. Then there's a Turbo Pascal program that executes on the IBM PC (or XT or AT) for operating the tester in PC-host mode.

Finally, another Turbo Pascal program converts the information in the output file produced by VECCPT.PAS into Intel hexadecimal ASCII format. This permits you to download to an EPROM burner. This lets you put new device vector information into the IC tester's ROM for operation in the terminal and stand-alone LCD modes.

Table 3: *The six device types supported by the tester.*

Device type	Number of pins	Gnd pin	+5-V pin
1	14	7	14
2	14	11	4
3	16	8	16
4	16	12	5
5	20	10	20
6	24	12	24

Table 4: *Definition module for the 7437, a "clone" of the 7400.*

```
#7437     * Quad two-input NAND buffers
C 7400    * Clone of 7400 (Dev. type=1)
E         * End of 7437 definition
```

Explaining all the software for the IC tester would involve considerably more space than I have available here (see the Circuit Cellar Ink applications publication for additional support materials). While my description here is tailored to the application and use of the IC tester, the user's manual and distribution software contain much source code and go into significant detail describing the process for creating a new device library and testing custom devices.

PC-Host Mode

The PC-host mode of operation is the most powerful of the three modes. I'll start with its description, because the basic testing technique is the same for all three modes.

The PC-host mode provides flexibility in letting you download and use different device libraries and offers test-vector debugging features not available in the other two modes. Functions like `Identify` and `Test Specified Device` differ only in the information displayed and are the same in all modes.

Once you give the PC-host mode operating program the name of the compacted test-vector file and the serial-port number (1 or 2), the software attempts to establish a communication link with the IC tester. If the tester does not respond, the PC will perform two retries (three tries total) before printing an error message and sounding a beep.

Once communication is established, the PC reads the specified compacted test-vector file, downloads it to the IC tester, and displays the version number and a formatted operation menu on the screen. The typical menu offers four device-testing options and two mode-selection options.

The display also shows three status/information lines. The first line, `Device:`, indicates the name of the current or most recent device being tested, or the name of an identified device. The second line, `Message:`, displays messages like `Device Passed` and `Device Not Found`. The third line, `Pin Failures:`,

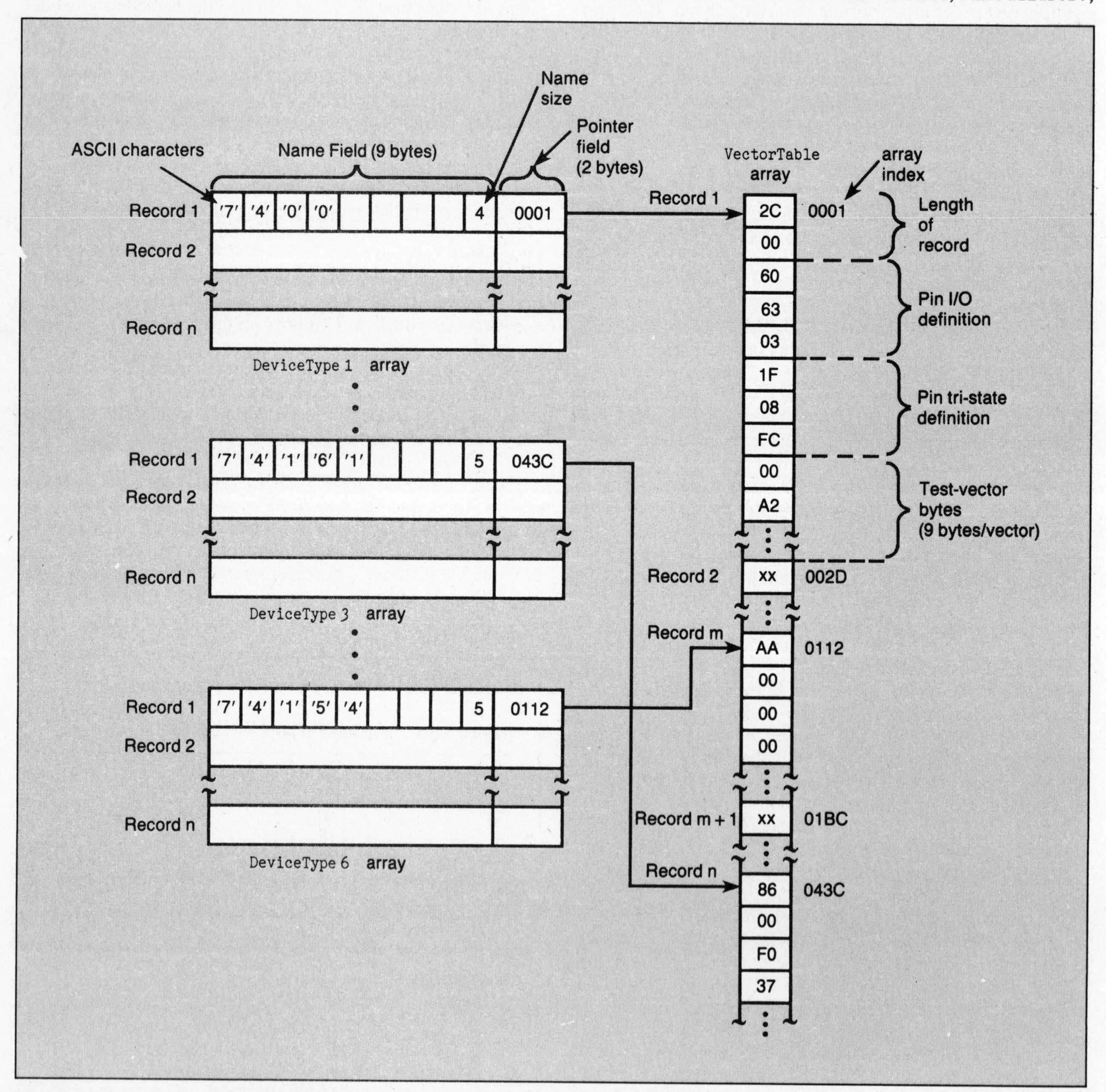

Figure 2: *Illustration of the information storage in VECCPT.PAS's primary arrays.*

displays pin numbers that failed vector tests when testing a specified device or an EPROM.

The first menu item, `Identify Device`, tells the tester to attempt to identify the device in the ZIF socket (the device-identification algorithm supports only devices having the corner power and ground pins). To identify a device, the system powers the ZIF socket for a 24-pin device and then applies the first 24-pin device test vector (if any) in the device library to the DUT.

If the read-back vector compares favorably to the expected read-back vector (along with the "don't care" mask), the next vector for the same device is applied, and so on. This continues until the DUT passes all the test vectors—indicating proper device identity—or until a vector failure occurs. If a vector failure occurs, a check is made to see which bits in the read-back vector, if any, are different from those sent out in the output vector. These bits represent pins that must be either output or tri-state pins, and the pin values are noted in an accuracy array.

If the DUT passes all the test vectors, the tester has identified the device; its name is displayed, and control returns to the menu. If the DUT fails a vector, the next device in the 24-pin library is checked.

Testing continues until the DUT is identified or no more 24-pin devices are left to test. If the program runs out of 24-pin devices, it clears the accuracy test array and repeats the same procedure with the 20-pin, then 16-pin, and finally 14-pin devices. Inability to finally identify the part is only the result of the device not being in the library, or because it is defective.

The second menu item, `Test Specified Device`, moves the cursor to the `Device` line. If any devices have already been tested or identified, the name of the last device tested is automatically displayed on the line. If you desire to retest the same part type, press Return (or Enter). If you wish to choose a different device, enter the new device name and press Return to test the DUT.

By telling the IC tester what type of chip is in the ZIF socket, all the test vectors for that device will be applied to the device and checked, regardless of whether they pass or fail. If vector failures do occur, you'll see the pin numbers on the `Pin Failures:` line.

The first two menu items represent the operations you will probably want to do 99 percent of the time and can be done in all three operating modes. Sometimes, however, you may have 2716 or 2732 EPROMs that you would like to verify are blank. Menu items 3 and 4 provide this capability.

In addition to performing a blank check on the EPROM, the EPROM tests also check for shorts on the EPROM input pins. If shorted pins are detected, an error message is displayed and the failed pins are displayed on the `Pin Failures:` line. Since the ZIF socket is only 24 pins, the tester cannot accommodate larger EPROMs.

The third menu selection deals with CMOS logic devices only. As I discussed in part 1 of this chapter, all the standard 74xx00-series logic families except the 74C00 series (and some specific devices within other families) are capable of sourcing and sinking enough current on their outputs for proper operation of the tester.

The 74C00-series devices (and the similar 4000-series CMOS devices) have a problem sinking enough current to switch logic states when an output is pulled up to +5 volts. Most of the tests for the 74xx00-series families attempt to load the device outputs in the direction opposite the expected state (if an output is expected to go low, it is loaded with a pull-up resistor), causing particular problems when testing the 74C00-series family devices when reading outputs that are expected low, but are being pulled up.

The remedy for the 4000-series devices is simple: Write all test vectors for these devices always using a pull-down load on all outputs. In order to keep the 74xx00-series tests the same for all families, however, I had to use a different approach. Menu item 5 lets you `Set 74Cx Mode`.

In this mode, regardless of the original output vector-bit levels, all output vector bits that correspond to device output (non-tri-state) pins are changed to low (pull-down). This allows the 74C00-series devices to pass the generic 74xx00-series tests. You can also select this mode for identifying 74C00-series devices.

The final menu option is `Set Diagnostic Mode`. This option is available only when operating the tester in PC-host mode. It adds an extra line to the bottom of the display, `Vector Failures:`, to indicate which test vectors failed when testing a device.

When testing a specified device (not when identifying a device) in diagnostic mode, the `Device:` line indicates the number of pins the device has, as well as the ground pin number and the power pin number. If the device is a clone of another device, this is also indicated, along with the device name of the original device.

If the device being tested fails, the `Message:` line indicates how many vectors failed (along with the normal failure message), and the `Vector Failures:` line indicates the vector numbers of the first 10 failed vectors (or all failed vector numbers, if fewer than 10 failed). The extra information can prove helpful when debugging new test vectors.

ROM-Resident Control Program

The IC tester's 8031 assembly language control program provides local support for all three modes. A software-readable, four-position DIP switch selects mode and data transfer rate, while a status LED indicates the tester's current operating disposition (a second LED acts as a power-on indicator).

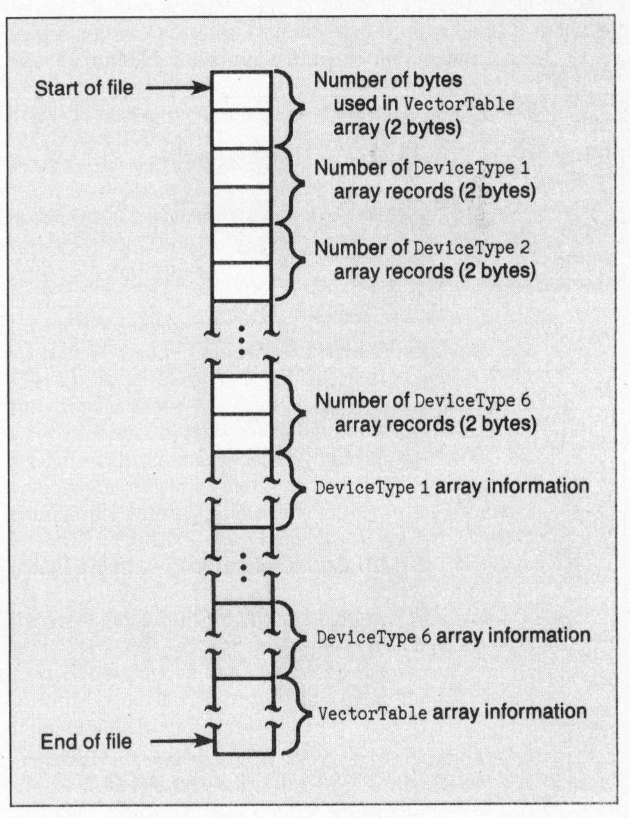

Figure 3: *Format of the vector-compaction file output by VECCPT.PAS.*

Upon power-up or reset (using the on-board reset button), the program initializes the 8031's on-chip ports to turn off all power and ground transistors to the ZIF socket and to place the LCD interface lines in their appropriate default states.

It then generates a brief delay (nominally, 1 second) to provide time for power to stabilize for all devices on the board. Software then checks two of the DIP switches to determine the desired data transfer rate and configures the 8031's on-chip UART to handle serial communications at the specified data transfer rate.

Once initialization is completed, the program checks another DIP switch to see if the user has selected PC-host mode or terminal mode. For terminal mode, the system turns on the status LED (to indicate that a serial operating mode, as opposed to stand-alone mode, is currently enabled) and sends a sign-on message and menu out the serial port to the attached terminal. For PC-host mode, no sign-on message is sent.

In either case, the tester also displays a sign-on message on the optional LCD, if present. In order to select the stand-alone mode, you merely press the "Identify" push button—which is constantly polled during both serial operating modes—and the system will turn off the status LED to indicate stand-alone mode operation. The only way to return to serial mode operation is by pressing Reset.

When operating in PC-host mode, the IC tester's ROM program merely responds to commands from the host. Various commands allow "reset" (power and ground transistors turned off), software version request, power and ground switch setup, and DUT output vector application and read-back vector reading. Terminal mode operation is similar to PC-host mode operation, with the exception that you are restricted to the device library stored in ROM, and the diagnostic mode described earlier is unavailable.

Stand-alone operation requires no connection to the serial port, but it does require that you have the LCD installed (see photo 2). All interaction is via the on-board "Identify" and "Retest" push-button switches and the LCD. A DIP switch enables or disables "74Cx" mode.

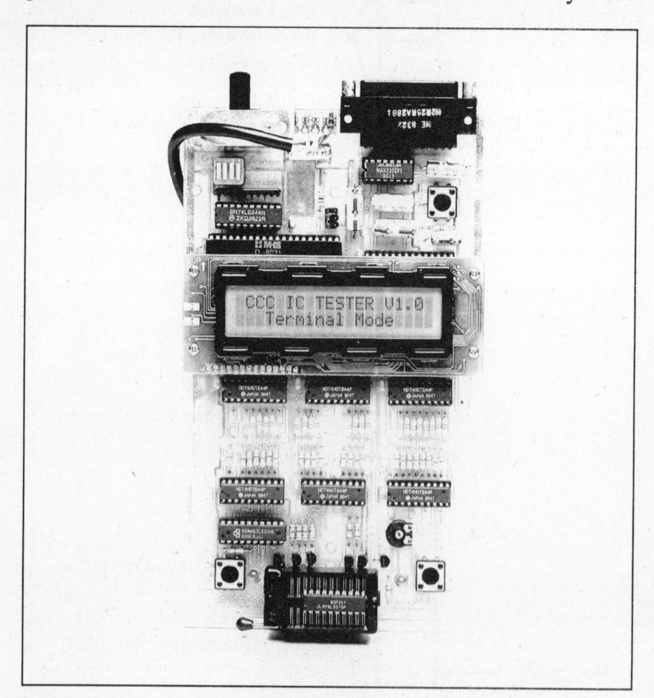

Photo 2: *You can use the IC tester in stand-alone mode, provided you have attached the tester's optional LCD. The push-button switches in the upper right, lower left, and lower right control the tester's operation.*

Pressing the "Identify" push button causes the tester to attempt to identify the device in the ZIF socket. If the identification is successful, the device name is displayed on the LCD; otherwise, an identification failure message is displayed.

Once a device has been identified, you can test other devices of the same type using the "Retest" push button. The test vectors for the identified device are then applied to the DUT, and detected pin failures, if any, are displayed on the LCD.

Flexibility

While the Circuit Cellar IC tester represents hundreds of hours of hardware and software development, the end result is something that was designed to be simple to operate. It clearly offers a great deal of flexibility for testing common devices, but it is also useful for developing tests for custom or proprietary devices like programmable array logic.

In order to test a PAL, you must develop a series of test vectors that apply bit patterns to the device inputs and watch for expected output values just like those from any standard 74xx logic device. The PAL test vectors are based on the logic-transfer functions (the logic equations) of the device.

You compile and name the test vectors and then add them to the device library. To test PALs, you run the IC tester in the normal way: Just insert the PAL to be tested in the ZIF socket (bottom-justified) and specify either the `Identify Device` option (the easier choice) or the `Test Specified Device` option, giving the device's name, "PAL1," for example.

In Conclusion

The powerful, yet easy-to-use, Circuit Cellar IC tester can provide testing and identification for innumerable standard and custom IC devices, in packages ranging from 14 to 24 pins. It's a tool that can save you time and money by catching potential problems during production, helping debug problem boards, and by identifying and/or verifying unknown devices or devices with uncertain operation. The flexibility and capability offered by this tester were previously available only to those willing to spend thousands of dollars.

In all honesty, I have to admit that the hardware for this project was trivial compared to the enormous software task involved in creating the operating system and device library. The initial Revision 1.0 ROM-resident library contains more than 200 generic entries. Considering that a generic entry of "7400" can cover 10 clone entries, the library physically covers about 800 chips. I owe a special debt of gratitude to those who helped put this project together and saved me from having to deal with all this software.

Experimenters

While you can order printed circuit boards and kits for the Circuit Cellar IC tester, I encourage you to build your own. If you don't mind doing a little work, I will again support your efforts. A hexadecimal file of the executable code for the IC tester's 8031 EPROM (a 27256) is available free for downloading from my bulletin board at (203) 871-1988. It contains the complete Revision 1.0 ROM-resident device library and software for complete stand-alone and terminal mode operation.

Alternatively, you can send me a preformatted IBM PC 5¼-inch disk (2.0 or higher) with return postage, and I'll put the file on it for you. Please add $5 for a printed copy of the user's manual. Of course, as always, this free software is limited to noncommercial personal use.

I would like to personally thank Roger Alford and Bill Potter for their collaborative efforts on this project. Bill Potter's tireless dedication creating the test-vector library and Roger Alford's clever programming expertise served to make the Circuit Cellar IC tester a true performer.

Editor's Note: Steve often refers to previous Circuit Cellar articles. Most of these past articles are available in book form from BYTE Books, McGraw-Hill Publishing Company, P.O. Box 400, Hightstown, NJ 08250, (1-800-2-MCGRAW).

Ciarcia's Circuit Cellar, Volume I covers articles in BYTE from September 1977 through November 1978. *Volume II* covers December 1978 through through June 1980. *Volume III* covers July 1980 through December 1981. *Volume IV* covers January 1982 through June 1983. *Volume V* covers July 1983 through December 1894.

It's virtually impossible to provide all the pertinent details of a project or cover all the designs I'd like to in the pages of BYTE. For that reason, I have started a 24-page bimonthly supplemental publication (with no advertising) called Circuit Cellar Ink, which presents additional information on projects published in BYTE, new projects, and supplemental applications-oriented materials. For a one-year subscription, send $14.95 to Circuit Cellar Ink, or call(203) 875-2199.

The following items are available from

Circuit Cellar, Inc.
4 Park St., Suite 12
Vernon, CT 06066
(203) 875-2751

1. Circuit Cellar IC tester experimenter's kit for stand-alone or terminal operation. Contains IC tester printed circuit board, 11.0592-megahertz crystal, programmed 27256 EPROM with Revision 1.0 device library, MAX232 level shifter, Aries 24-pin narrow-format ZIF socket, and manual with complete parts list. ICT01-EXP...$99
2. Circuit Cellar IC tester full printed circuit board kit for stand-alone, terminal, or PC-host operation. Contains IC tester printed circuit board, 8031 processor and crystal, programmed 27256 EPROM with revision 1.0 device library, Aries 24-pin narrow-format ZIF socket, IC sockets, all board-mounted components and IC's, PC-host software on PC format disk, power supply, and manual. ICT01-FULL......$179
3. Complete Circuit Cellar IC tester kit with stylish enclosure. Full printed circuit board kit with all components, right-angle-mounted evclosure adapter board with ZIF socket and LCD, software on PC format disk, power supply, and manual. ICT02.............................$349
4. Two-line by 20-character LCD and 14-pin Berg connector for either item 1 or 2. 2x20 LCD..$32

All payments should be made in U.S. dollars by check, money order, MasterCard, Visa, or American Express. Surface delivery (U.S. and Canada only): add $5 for U.S., $8 for Canada. For delivery to Europe via U.S. airmail, add $14. Three-day air freight delivery: add $10 for U.S. (UPS Blue), $25 for Canada (Purolator overnight), $45 for Europe (Federal Express), or $60 for Asia and elsewhere in the world (Federal Express). Shipping costs are the same for one or two units.

There is an on-line Circuit Cellar bulletin board system that supports past and present projects. You are invited to call and exchange ideas and comments with other Circuit Cellar supporters. The 300/1200/2400-bps BBS is on-line 24 hours a day at (203) 871-1988.

8

WHY MICROCONTROLLERS?

PART 1

Photo 1: *Two versions of the Intel MCS-51 family: the 8031 ROMless microcontroller and the 8751 programmable (EPROM on-chip through the quartz window) microcontroller.*

Microcontrollers are used in keyboard and disk interfaces and in numerous other devices. Here's a tutorial on the 8031/8051 microcontroller family.

Gone are the days when a complex project required a suitcase full of TTL ICs and a wire-wrap gun. As with most of the recent Circuit Cellar designs, under the hood there are a few carefully chosen discrete ICs controlled by a microprocessor. The trade-off is simple: Hardware is expensive and software is cheap—once you get it right.

Many of you have noticed that I have been using two popular microcontrollers from Intel—the 8031 and the 8051—as the control elements in Circuit Cellar designs. I've received many requests for general information about these microprocessors and for guidance in applying them elsewhere. With that as an incentive, I decided to present this combination tutorial/development system project.

In this first part of a two-part chapter, I'll introduce the members of the 8031/8051 chip family and describe using them as embedded microcontrollers. The second part will present all the elements of a useful development system intended to simplify the process of designing systems using the 8031/8051 family.

The Difference in a Name

It's important to be clear on the distinction between microcomputers, microprocessors, and microcontrollers. That way, you'll understand where the 8031/8051 family fits in the big picture.

A microprocessor is just the CPU part of a computer, without the memory, I/O, and peripherals needed for a complete system. For example, 8088 and 80286 chips are microprocessors (the "micro" prefix designates that this CPU element is at the chip level). All other chips in an IBM PC are there to add features not found within the microprocessor chip itself. The hardware designer can choose different chips to implement those features in different ways, although a designer has little room for choice if the end result is supposed to be an IBM PC clone.

When a microprocessor is combined with I/O and memory peripheral functions, the combination is called a microcomputer. Of course, vendors anxious to designate that their computer is more powerful than others often shed the "micro" prefix, but it's still a microcomputer given today's definition. Ultimately, good economic sense suggests that all computers, including minicomputers and mainframes, will utilize the same basic elements, just differentiated by quantity.

The fact that combining a CPU with memory and I/O produces a microcomputer also holds true at the chip level. Many companies add these peripheral functions onto the same substrate with the CPU to make a complete microcomputer. These devices are called single-chip microcomputers to differentiate them from their big-cousin desktop microcomputers.

Generally speaking, microcomputer chips are designed for very small computer-based devices that don't need all the functions of a full computer system. In cost-sensitive control applications, even the few chips needed to support a CPU like an 8088 or Z80 are too many. Instead, designers often employ a single-chip microcomputer (or a slightly expanded circuit using one) to handle control-specific activities. When single-chip microcomputers are designed or used in industrial control systems, they are often called single-chip microcontrollers. Basically, there is no difference between microcomputers and microcontrollers; the name depends on how we use them.

Frequently, microcontrollers are used to replace circuit functions that ordinarily require many low-level chips or need the main CPU's attention each time the circuit is active. The IBM PC keyboard-interface circuit is a prime example of the use of a microcontroller chip. In the PC, a half dozen chips (excluding I/O addressing and decoding) are necessary to receive and decode the serial clock and data bit stream from the keyboard. In the IBM PC AT (and the CCAT I present in Chapter 11), this low-level circuitry is replaced with an 8742 microcontroller that completely simulates the old circuit and incorporates additional features in one chip.

The Intel 8051 Family

The Intel 8051 is a classic microcontroller (a generation more advanced than the 8742) and a true single-chip microcomputer containing parallel I/O, counter/timers, serial I/O, RAM, and EPROM or ROM (depending on the part type). The 8051 family contains several members (Intel refers to it as the MCS-51 family), each adapted for a specific type of system.

The different versions are outlined in table 1, and a block diagram of the 8051 is shown in figure 1. The 8051 has two close relatives, the 8751 and the 8031 (see photo 1), and a cousin, the 8052. All versions contain the same CPU, RAM, counter/timers, parallel ports, and serial I/O. The 8051 contains 4K bytes of ROM, which must be custom-masked when the chip is manufactured. In the 8751, the ROM is replaced with EPROM that you can program (the schematic for an 8751 programming adapter for the Circuit Cellar serial EPROM programmer presented in October 1986 is available by writing to me).

The 8031 is meant for expanded applications and uses external memory. The 8031 uses three of the four on-chip parallel ports to make a conventional address and data bus with appropriate control lines.

You might wonder why you'd choose a single-chip microcomputer in the first place if you end up converting it back to function as a CPU with other peripheral chips. Basically, it depends on the degree of expansion required. Since the 8031 still contains RAM, a parallel port, and a serial I/O port—even when functioning as the CPU core of an expanded circuit—the eventual number of chips necessary to expand the I/O or memory is still considerably less for the same ultimate capability than with a straight

Table 1: *Members of the 8051 microcontroller family tree.*

Device name	ROMless version	EPROM version	ROM bytes	RAM bytes	16-bit timers
8051	8031	(8751)	4K	128	2
8051AH	8031AH	8751H	4K	128	2
8052AH	8032AH	8752BH	8K	256	3

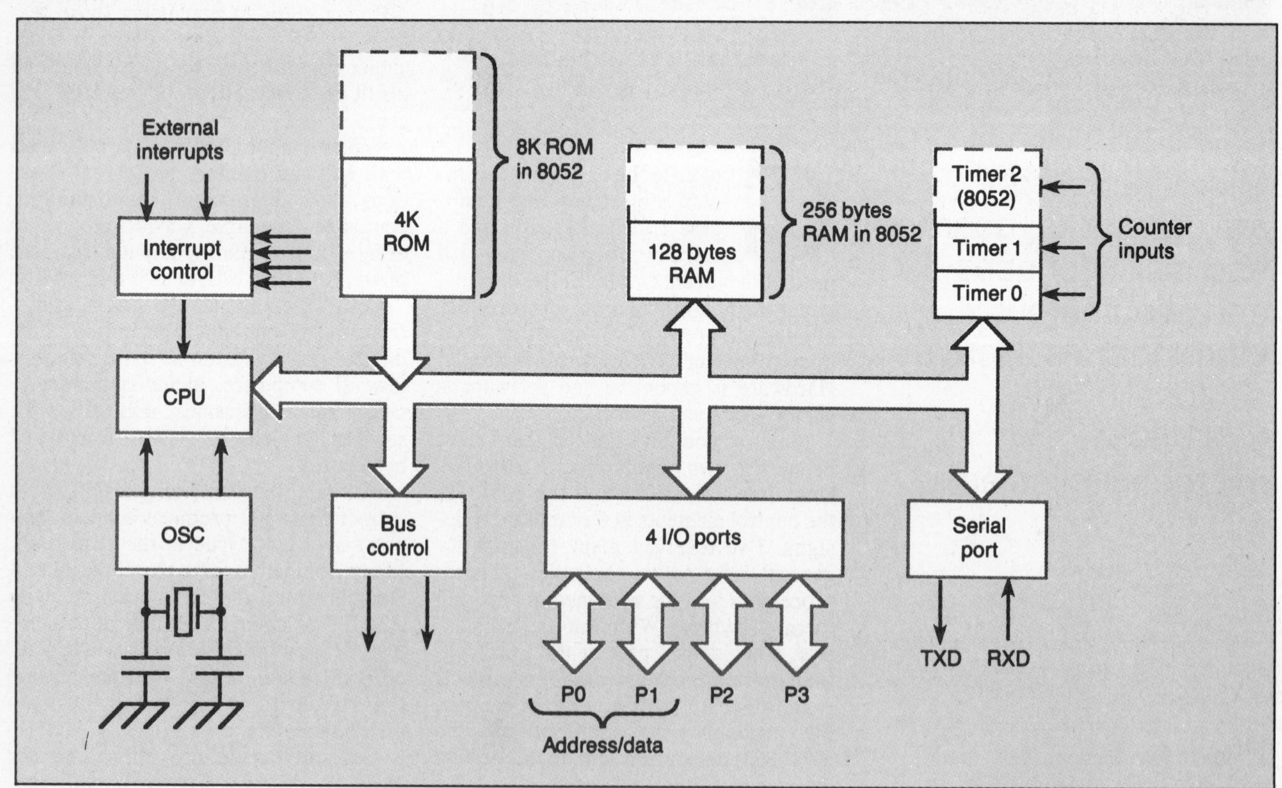

Figure 1: *Inside the 8051.*

microprocessor and peripheral chips. I'll go into the possible expansion techniques later.

Intel and other companies sell variations of the 8051 family with more internal memory, more I/O, lower power, and so forth. An 80C31 is a CMOS low-power version of the 8031, for example. The 8052, which I have also used in projects, is the same as the 8051, except that it has another counter/timer and additional RAM and ROM.

Using an 8051 is as simple as hooking up the power supply and clock crystal. Unlike the 8031, however, you have to supply Intel with a program for the 8051 so it can create a chip mask defining the internal ROM. (If your program is not quite right, it's time for another mask.)

As you might expect, both mask making and chip building take time and money. (Since I did this for a recent project, I thought you might be interested in some manufacturing information. If you plan on making an 8051 microcontroller and need to mask your own chip, the one-time masking charge is about $3000, with a minimum order of 1000 chips.) Mask programming makes sense for an application that uses thousands of identical 8051s a year, but it is not practical for low-volume systems and prototypes.

The 8751 has all the features of the 8051, except that an on-chip EPROM replaces the ROM program storage. Making a program change is as simple as erasing the EPROM with ultraviolet light and burning another program. Many developers use 8751s until the code works, then commit to a large 8051 order with the program in ROM. If the product volume is low enough, it's often worthwhile to use 8751s in the final product. An 8051 costs about $4 to $5 apiece in thousands; an 8751 is about $25 to $40.

The 8031 has no on-chip program storage at all. The system must include an external EPROM and an address latch. Considering the falling prices of EPROMs and the heavy costs of using either 8051s or 8751s in low volumes, the 8031 is a viable alternative despite the additional chips. For many small systems, the 8031/EPROM combination is far more cost-effective than an 8051. (This is the type of system I'll describe in my examples.)

As I mentioned before, all members of the 8051 family have the same core hardware and therefore use the same core instruction set. While some members have one or two additional instructions for features unique to the particular chip, I'll use the term "8051" to describe the "8051 chip family," unless I'm talking about a specific version with unique requirements.

Making It Real

With all that in mind, let's look at configuring a usable "computer/controller" using an 8031 microcontroller chip as part of the system. Remember that in a single-chip microcontroller, internal hardware replaces all the digital logic you'd normally add for control, timing, and so forth. You need add only the keyboard, display, relays, switches, and user-specific I/O that actually makes up the final product.

Figure 2a shows the bare-bones 8031 microcontroller system: the 8031, a 2764 EPROM to hold 8K bytes of program, and a 74LS373 latch to demultiplex the address/data bus. The system has 128 bytes of RAM on the 8031, a bidirectional parallel I/O port, a bidirectional serial port, two counter/timers, and two external interrupt inputs. With a 12-MHz crystal (most often, we select 11.0592 MHz for communications rate compatibility), it executes most instructions in one machine cycle—a peak rate of 1 million instructions per second. Not bad for three chips, is it?

A single I/O port can scan a 16-key matrix. With an additional output bit, it can drive a 2-line by 20-character smart liquid crystal display at the same time. The remaining I/O bits can handle triacs or power field-effect transistors for AC or DC control. Burn a program into the EPROM, and you have a real-time power controller. Run the serial port through a MAX232 RS-232C level converter, and you have a standard serial port for remote control or status monitoring at your master computer.

If you don't need the serial port, counter/timers, and external interrupts, the 8031 can use those special bits as a second parallel I/O port, so the minimum system can have up to 16 I/O bits. Each bit can be tested, set, and cleared individually under program control.

If one or two parallel I/O ports aren't enough for your application, figure 2b (an expansion of figure 2a) shows what's needed to get three more: Add a single 8255 programmable peripheral interface. The 8255's port C can be set up for automatic handshaking, so now you have the basis for a serial-to-parallel (and back) format converter or 24 more I/O bits for a bigger controller. Notice that no "glue" chips are needed between the 8031 and the 8255.

Because the 8255 uses the RD\ and WR\ bits, the second I/O port isn't completely free. The 6 remaining bits can still handle either general I/O or their individual special functions, though.

If your application requires more than 128 bytes of RAM, figure 2c (figure 2a expanded to include 2b and 2c) shows how to get 8K bytes of RAM by adding a 6264 static RAM chip. Now you can build a fancy buffering format converter, a data logger, or a serial-programmable power controller. A 62256 RAM would give 32K bytes with no more effort, still with no glue chips!

Finally, for those of you who need lots of RAM and I/O, figure 2d (figure 2a expanded to include the circuitry of 2b, 2c, and 2d) shows how to connect multiple I/O chips. The 74LS138 decoder generates chip select signals from the 8031's output addresses, with each select covering an 8K-byte range. The system shown has 16K bytes of RAM and seven bidirectional I/O ports. Pretty nice for seven chips.

The point of all this is that the "computer" part of your control system need not require elaborate hardware. For a unit of any reasonable size, you'll spend most of your hardware design time on the I/O devices rather than on the 8031 circuits, which is exactly as it should be.

Perhaps now you understand why I have been using the 8031 frequently. The main benefit of a microcontroller is the ease of adding new features to your system, just by changing the program, not changing the circuit-board connections. A new EPROM can give the hardware a completely new personality. Try doing that by rewiring a board of TTL control logic!

The Software Swamp

Every microprocessor has an instruction set exhibiting the conflict between all the instructions that could possibly be useful and the few that fit on the chip. The 8051 has many bit-manipulation instructions and few general instructions, reflecting its design as a controller rather than as a computer.

Most 8051 instructions are 1 or 2 bytes long, with the remainder requiring 3 bytes. All instructions except `MUL` and `DIV` execute in one or two instruction cycles. An instruction cycle is 1 microsecond (μs) at a 12-MHz clock rate. `MUL` and `DIV` lag along at 4 μs.

If you've written assembly language programs for any other microprocessor, you'll find some of the same instructions in the 8051's code. To understand the 8051's instructions, you must be familiar with the three main address spaces defined on the chip: 64K bytes of pro-

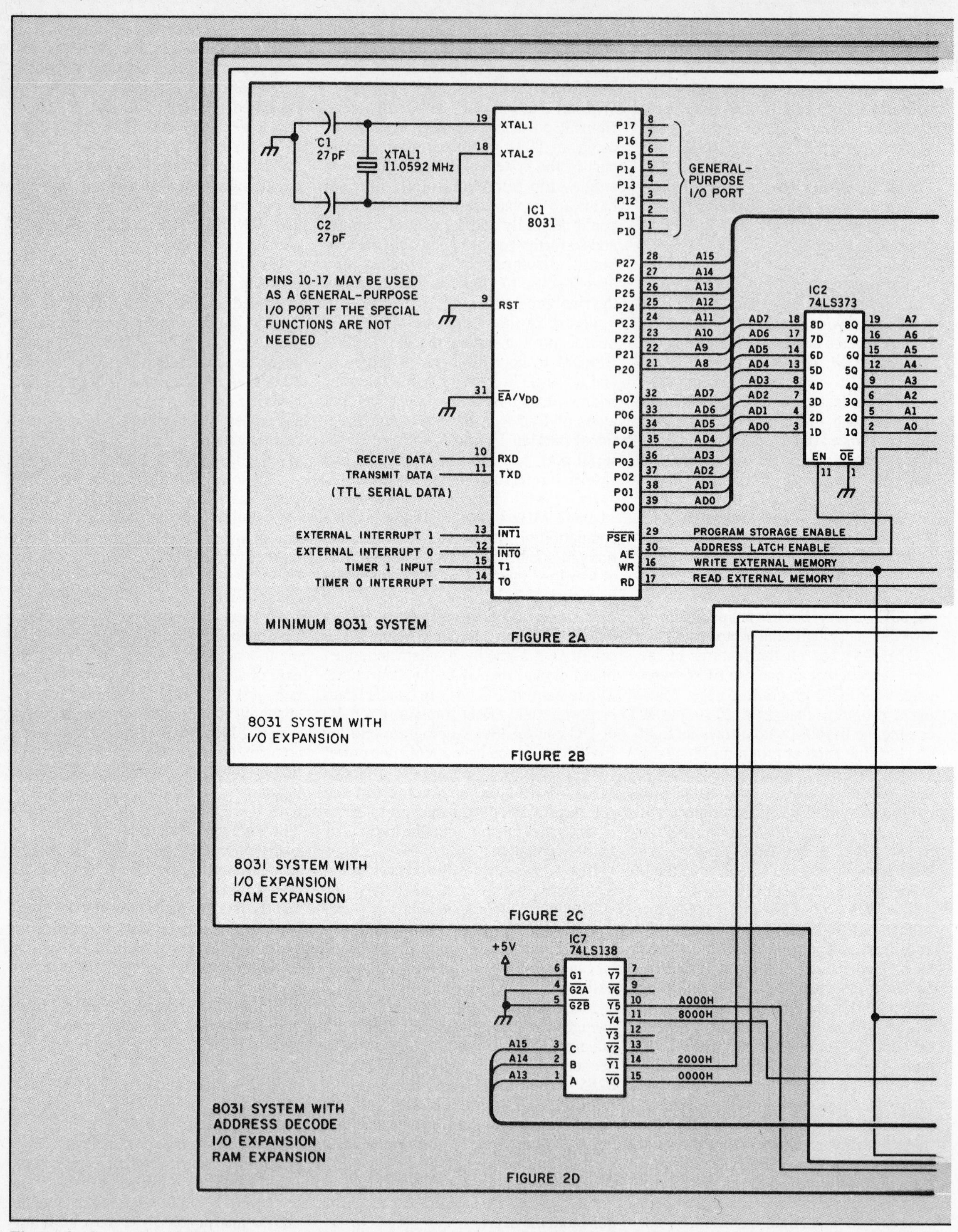

Figure 2: *Growing an 8031 system.* **(a)** *A minimum configuration,* **(b)** *minimum system with I/O expansion,* **(c)** *system with I/O and RAM expansion, and* **(d)** *system with address decode logic and even more RAM and I/O.*

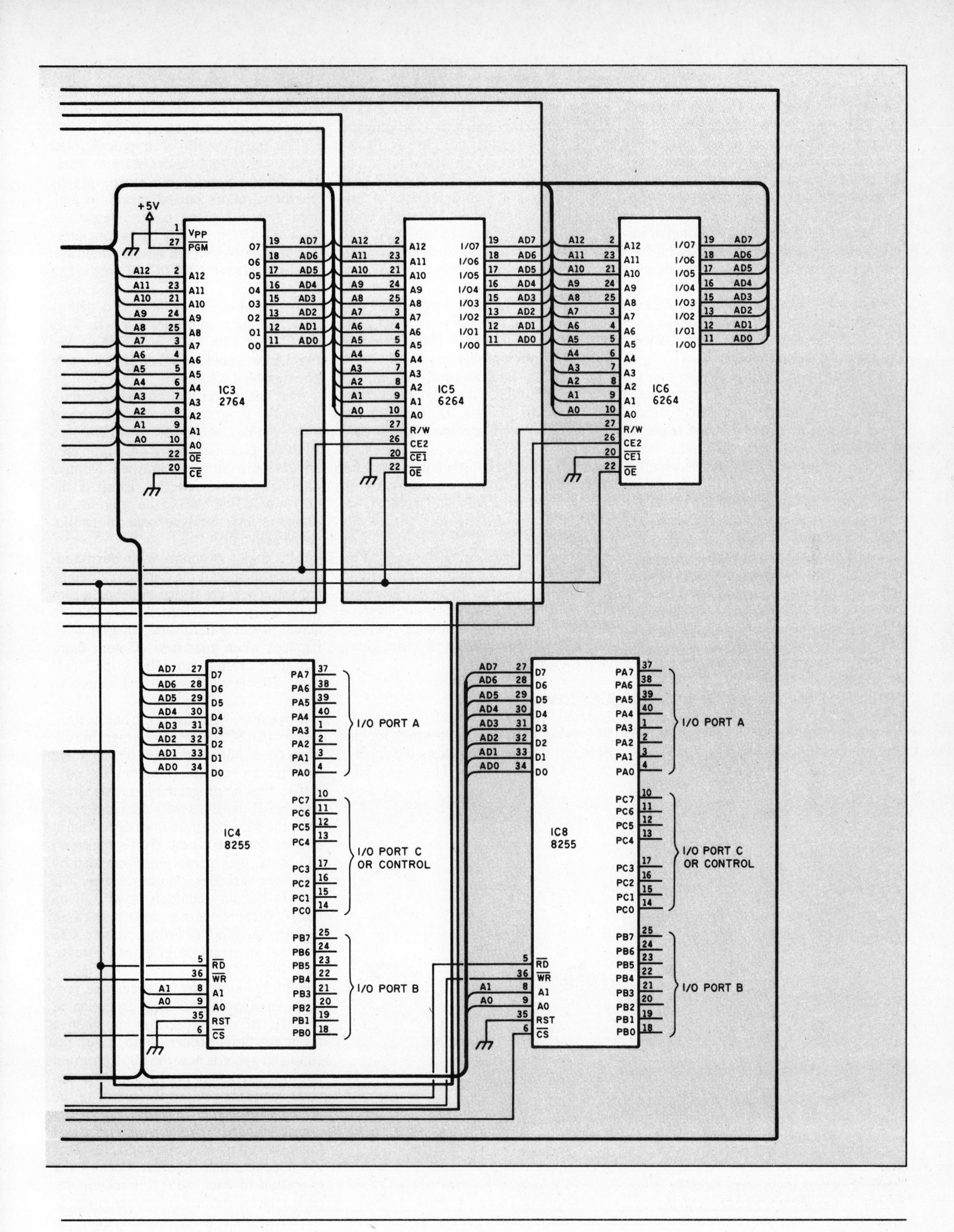
+5V
IC3
2764
IC5
6264
IC6
6264
IC4
8255
IC8
8255
I/O PORT A
I/O PORT C
OR CONTROL
I/O PORT B
I/O PORT A
I/O PORT C
OR CONTROL
I/O PORT B

gram memory addressed by the program counter (PC), 64K bytes of external data memory addressed by the data pointer (DPTR), and the on-chip internal data memory addressed in several different ways. Each instruction implies a particular address space, so you have to know where your data resides to select the right instruction.

There is a sharp distinction between internal and external data addresses. Internal addresses refer to locations on the 8051 chip, which can be accessed in a variety of ways. External addresses are located off the chip, in the 64K bytes of external data memory, and can be accessed only with MOVX instructions.

The four main internal data memory-addressing modes are direct, immediate, register, and register indirect. Direct mode embeds an internal RAM address in the instruction. Immediate mode uses the data value itself. Register and register indirect modes use a register number, with indirect addressing taking the contents of that register as a direct address to access the data.

The MOVX instructions transfer a single byte between the accumulator and external data memory. The DPTR register contains a 16-bit external data memory address, which can be either loaded by a single MOV or incremented. Unfortunately, there aren't any other 16-bit instructions.

The 8051 has a single accumulator, called ACC or A depending on the instruction. Nearly all instructions use the accumulator in one way or another. An auxiliary accumulator (called B, of course) is used by MUL and DIV. Many data-manipulation instructions can move data to or from one of the active banks of eight "working registers" in internal RAM. Four register banks are available.

Because most controller applications require handling at least a few I/O bits, the 8051 has a rich selection of bit-manipulation instructions that are completely separate from the standard byte instructions. A single instruction can set, clear, complement, or copy any bit in internal data memory. The on-chip I/O ports show up in that address space, so there's no need for the "read, mask, set, combine, write" instructions found in most other microprocessors.

Unlike the Intel 8088 or Z80 microprocessor families, the 8051 has no explicit I/O instructions. The on-chip I/O ports are mapped into the internal data memory-address space and accessed with the same MOV instructions used for other transfers. You have to map off-chip I/O into the external data memory-address space and access it with MOVX instructions.

Rather than belaboring the various instructions in detail, I'll introduce them in part 2 of this chapter in short chunks of code that do useful tasks as we build some hardware. With those examples as a base, you should have little trouble designing your own system.

Ugly Reality

The trade-off for not wiring up a board of TTL gates is writing a program for the EPROM. That program tests the inputs, computes the outputs, and handles all the timing to make the system work correctly. Unless you are much better than average, your program won't do the right thing the first time you try it out.

The ugly reality of microcontroller systems is getting the software to work. It's made considerably more ugly by microcontrollers buried inside specialized systems—those never intended to look or act even vaguely like a computer.

For example, which system would you rather debug: an IBM PC AT with a full keyboard, EGA display, hard disk drive, and state-of-the-art editors and debuggers, or a microcontroller in a 3- by 5- by 4-inch box with four push buttons and two LEDs, cabled to a heater in a vat of photographic solution?

The traditional way to debug microcontroller programs is called "burn and crash." You burn the program into EPROM, plug it in, turn it on. . . and then try to figure out why it crashed. Doing a Sherlock Holmes on the listing is the only way to find bugs in the program, although a logic analyzer and an oscilloscope help a lot.

The major problem with burn-and-crash debugging is the damage caused by a crash if you are trying to debug a program when the controller is attached to actual machine hardware. Imagine what happens when your new 10-story hammerhead crane controller goes "full speed counterclockwise" and refuses to reset.

Obviously, burn and crash has its limitations. An 8051 simulator program running on a host computer development system removes most problems and simplifies finding program bugs. The simulator reads the EPROM's data and interprets the 8051 program one instruction at a time. Because all the 8051's registers, I/O ports, and memory are provided by simulator variables, you can display and modify memory contents at will. Even better, because the simulator's software replaces the 8051's hardware, there's no way that an errant program can damage anything.

With the simulator, you can use program breakpoints to stop execution at specific 8051 instructions or when a given condition occurs. Also, since the simulator records how the 8051 program got to a particular instruction, you can undo each step back to the source of the problem at the press of a key.

Unfortunately, while a simulator is a great step up from burning and crashing, it is not a true real-time test. Because the execution of each 8051 instruction requires the execution of many program instructions in the development system, the

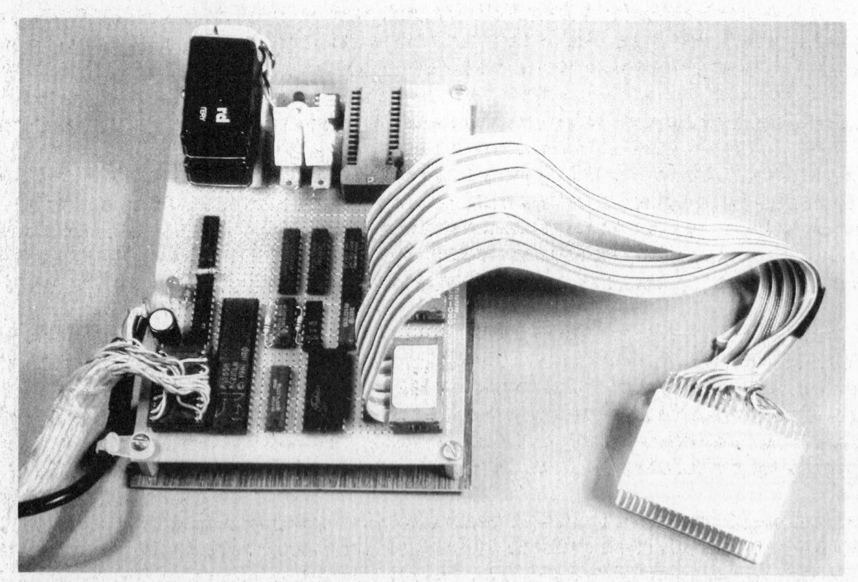

Photo 2: *A prototype of the DDT-51 development system. Note the DIP clip at the end of the cable for attaching onto the target system's processor.*

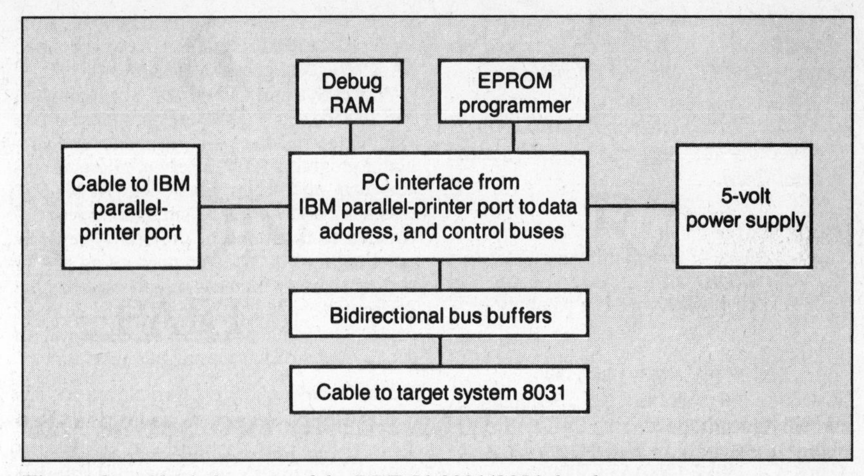

Figure 3: *A block diagram of the DDT-51 8031/8051 development system.*

simulated run time is much slower than the real-time rate on the actual hardware. A further complication is that interrupt timing is not easily duplicated on a simulator.

The ultimate solution is an in-circuit emulator (ICE), which is a special development system peripheral plugged into the 8051's socket in the actual hardware (called the target system). From the target's point of view, the ICE is an 8051 running at full speed. From the user's perspective, the ICE provides many of the features of a simulator, along with the ability to run programs at full speed using the real hardware.

Ideally, an ICE will have no effect on the target system, because all the 8051's features are provided in high-speed hardware. The 8051's internal registers and I/O ports are visible because the ICE uses discrete logic rather than a single chip. The development system directly monitors what's going on, logic comparators control the breakpoints, and there's no interference until the ICE stops at the selected instruction.

All this hardware makes ICE systems rather expensive. If you are developing many 8051 designs, an ICE is the only way to go. As a practical matter, however, an ICE is far beyond the typical user's budget and is generally reserved for the corporate lab.

A More Modest System

There is a middle ground between personal computer–based software simulators and hardware ICE systems. It's often enough to stop at a breakpoint and single-step through instructions while watching the target system's LEDs blink and relays click. By trading off some speed and hardware for time and money, a much simpler program development system can provide many features of an ICE at a fraction of the cost.

The DDT-51 system is an IBM PC–based 8031/8051 development and dynamic debugging tool (see photo 2). It uses a modified parallel printer port and a small circuit board to give the IBM PC complete control over the target system's hardware. The DDT-51 downloads the 8051 program into 8K-byte static RAMs, thus eliminating the need to burn an EPROM for each program change. An on-board 2K-byte static RAM holds the small amount of 8051 code required to support single-stepping and breakpoints. A disassembler shows the current 8051 instruction on the IBM PC's display while single-stepping, as well as the current 8051 registers and internal data memory values.

This system connects between the IBM PC and the target system (see figure 3). It has only about a dozen chips, including the world's simplest (and slowest) 2764 EPROM programmer.

The DDT-51 won't handle all possible 8051 target systems, but it will give you a good start. With that in hand, we can continue on with other interesting Circuit Cellar project designs.

Experimenters

As is the custom with Circuit Cellar projects, the software for the DDT-51 development system is available for downloading from my multiline bulletin board free of charge. The number is (203) 871-1988. Of course, this downloaded software is limited to noncommercial personal use unless licensed otherwise.

In Conclusion

The hardware and software specifics of DDT–51 will be presented in Chapter 8, part 2.

Special thanks for the technical contributions provided on this article from Jeff Bachiochi and Ed Nisley.

Parts Sources

One advantage of using an industry standard like the 8031 is that it's easy to obtain. The following companies are a few possibilities:

Digikey Corp., P.O. Box 677, Thief River Falls, MN 56701, (800) 344-4539
Jameco, 1355 Shoreway Rd., Belmont, CA 94002, (415) 592-8121
JDR Microdevices, 110 Knowles Dr., Los Gatos, CA 95030, (800) 538-5000
Krueger Technology, Inc., 2219 South 48th St., Tempe, AZ 85282, (800) 245-2235

Literature

The 8051 family doesn't have the wealth of publications you're used to seeing for the IBM PC. Fortunately, Intel prints a few useful and surprisingly readable manuals covering the tricky hardware and software details. These manuals are available from your local Intel office for a reasonable fee, or contact Intel Corp., Literature Dept., 3065 Bowers Ave., Santa Clara, CA 95051.

The order numbers include the latest revision after the dash, so you may find that there's a new edition out:

Embedded Controller Handbook, number 210918-005: The straight dope on all Intel's microcontrollers, from the 8096 to the 8048, including the 8051 family. It includes many hardware details, along with quite a bit on 8051 software and applications.
Microsystem Components Handbook, number 230843-004: This two-volume set gives assorted Intel microprocessor and peripheral chip data sheets. These are the definitive words on how the chips are supposed to work: heavy-duty specs, not for beginners.

Hardware

Although the DDT-51 development system includes a simple 2764 EPROM programmer, you'll eventually need a "real" EPROM burner. You need look no further than the Circuit Cellar serial EPROM programmer available from CCI, 4 Park St., Suite 12, Vernon, CT 06066, (800) 635-3355.

Software

The single most essential program is a cross assembler that runs on your IBM PC, eats assembly language files, and spits out hexadecimal files for the EPROM burner. You can spend as much or as little on these as you see fit, but doing assemblies by hand is not much fun. I speak from experience.

The shareware Pseudo-SAM 8031 cross assembler is available on the Circuit Cellar BBS under the name PS51A123.ARC. It has most of the useful features, few frills, and is quite serviceable. A newer version is available from PseudoCode, P.O. Box 1423, Newport News, VA 23601, BBS phone (804) 898-7493.

Avocet Systems (120 Union St., Rockport, ME 04856, (207) 236-9055) produces excellent "industrial-strength" assembler and simulator programs for a variety of microprocessors and microcontrollers. The AVMAC51 assembler and AVSIM51 simulator have contributed to many recent projects. They give excellent results. I recommend using one or both of these as the best way to optimize the performance and cost benefits of the DDT-51 development system and an excellent way to get a system up and running in a short time.

Editor's Note: Steve often refers to previous Circuit Cellar articles. Most of these past articles are available in book form from BYTE Books, McGraw-Hill Publishing Company, P.O. Box 400, Hightstown, NJ 08250, (1–800–2–MCGRAW).

Ciarcia's Circuit Cellar, Volume I covers articles in BYTE from September 1977 through November 1978. *Volume II* covers December 1978 through June 1980. *Volume III* covers July 1980 through December 1981. *Volume IV* covers January 1982 through June 1983. *Volume V* covers July 1983 through December 1984. *Volume VI* covers January 1985 through June 1986.

Circuit Cellar Ink

It's virtually impossible to provide all the pertinent details of a project or cover all the designs I'd like to in the pages of BYTE. For that reason, I have started a bimonthly supplemental publication called Circuit Cellar Ink, which presents additional information on projects published in BYTE, new projects, and supplemental applications-oriented materials. For a one-year subscription (6 issues), send $14.95 to Circuit Cellar Ink. Credit card orders can call (203) 875-2199.

There is a multiline Circuit Cellar bulletin board system (running TBBS 2.0M) that supports past and present projects in BYTE and Ink. You are invited to call and exchange ideas and comments with other Circuit Cellar supporters. The 300-/1200-/2400-bps BBS is online 24 hours a day at (203) 871-1988.

The DDT-51 is now available in kit form. For information call CCI at (203) 875–2751.

8

WHY MICROCONTROLLERS?

PART 2

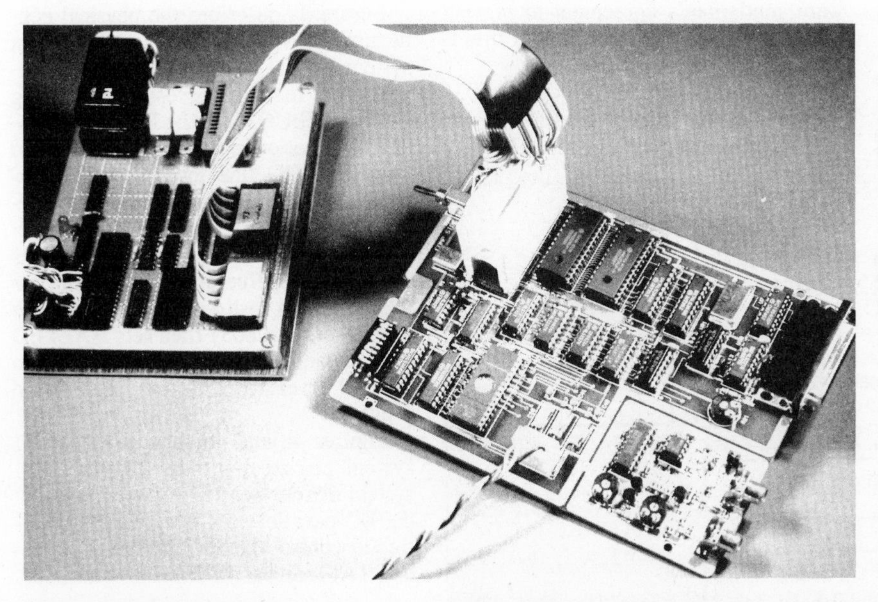

Photo 1: *Prototype of the DDT-51 development system.*

The DDT-51 is a low-cost IBM PC-based Intel 8031/8051 development system

Having explained in part 1 of Chapter 8 why we want to design projects around a single-chip microcontroller, it's time to talk about getting the software into the microcontroller. As I previously suggested, your options are either burn and crash (burn the program into an EPROM and figure out why it crashed) or use a development system.

The DDT–51 8031 development system I'll describe in part 2 of this chapter offers many of the features of more expensive software development systems, at a fraction of the price. Unlike those systems, you can build this one yourself, or modify the features to suit your tastes, and wind up with a system tailor-made for your application.

I'll start with the DDT-51 system hardware, then describe the software interface that lets an IBM PC control the target 8031 system.

The DDT-51 Development Tool

Nothing ever works the first time. When you are debugging microcontroller-based hardware, you need to be able to load and run a program, stop at breakpoints, single-step through critical parts of the code, disassemble instructions, examine and change registers, and continue where you left off. An ideal development system helps you do all that without using any 8031 system resources: no pins, no ICs, no program RAM—nothing at all. In-circuit emulators (ICEs) come closest to that ideal, because they replace the microcontroller with a hard-wired equivalent of the microcontroller chip—an equivalent that lets you directly observe the chip's (normally hidden) inner functions.

The DDT-51 system isn't quite that ideal, but it comes close. It needs one interrupt input, one output pin, and a 2K-byte RAM chip to hold debugging code and data. A 40-pin DIP clip attaches directly to the 8031 IC, and there are no other connectors to the target board. Photo 1 shows the prototype DDT-51 development system board, and figure 1 shows a block diagram of the hardware.

The software that runs on the host IBM PC is called DEBUG31. DEBUG31 shows all the 8031's internal RAM and the main registers when the microcontroller is stopped, as illustrated in figure 2. You can load new values into the registers, set a breakpoint to trap the program, and single-step through the code. You use function keys on the host to control all these operations.

Like an ICE, the DDT-51 supplies the data, address, and control lines needed to simulate an 8031's operations, although at a much slower rate. When the connections are disabled, the 8031 will run as though the DIP clip wasn't there.

Under the Hood

As described in part 1 of this chapter, an 8031 doesn't have an internal program ROM, so it needs an external EPROM (usually a 2764). If you replace that EPROM with a pin-compatible 6264 8K-byte RAM, the DDT–51/PC link can load and change the 8031's program. Once you've got the program loaded, the

8031 can execute it normally, just as though it were in EPROM.

The debug RAM is a key component of the DDT-51 system, but some 8031 systems may not need any RAM. Because the DIP clip connects to the 8031 bus control lines, it's easy enough to put the debug RAM on the DDT-51 board. I used a 2K-byte RAM simply because it is the smallest one that's readily available. An 8K-byte RAM would work just as well.

Unfortunately, the 8031 wasn't designed to share its bus control lines with other hardware. All the bus lines float when the RESET input is high (they use tristate drivers), but the program halts until RESET goes low again. The program restarts from address 0000 hexadecimal after a RESET, rather than continuing where it left off, which is exactly what a RESET should do. What's needed is a dual-ported RAM that can be accessed by either system independently, with separate control, data, and address lines. The two systems can then store and examine data in the common RAM without interaction.

Although the PC and the 8031 can both access the debug RAM, they can't do so simultaneously. To keep the hardware simple, there are no interlocks to prevent collisions (both systems attempting to access the RAM at the same time). If a collision does occur, one processor will always get bad data because the proper buffers are not active. The PC and 8031 programs must adhere to a software gentleman's agreement to prevent RAM conflicts. It turns out that this is not at all difficult to accomplish.

How do they do that? Once the 8031 program starts running, the DDT-51 can pull INT1 low to interrupt normal execution. A few debugging instructions added to the 8031 code handle the interrupt and examine a location in the debug RAM to determine what function was requested. The 8031 copies the results into the debug RAM and returns to normal operation, giving DEBUG31 (on the PC) free access to the RAM.

The gentleman's agreement is controlled by two wires: the IRQ line, triggering an 8031 interrupt, and the 8031 HALTED line, which is set while the interrupt handler is in control. Remember that the 8031 isn't halted after the interrupt, but is in the middle of the interrupt handler.

Figure 3a describes the physical connection between the PC and the target system; figure 3b shows the timing diagram that manages access to the debug RAM. DEBUG31 on the PC writes control information into the debug RAM and pulls the IRQ line low. The 8031 responds to the interrupt, examines the debug RAM, and executes whatever functions are needed. When it's done, it sets the HALTED line active and waits for DEBUG31 to restore the IRQ line to a high level. The 8031 then sets HALTED low, reloads the registers, and returns to normal operation in the interrupted program.

Figures 4a and 4b show the DDT-51 system circuit diagram. The DDT-51 system divides neatly into four sections: the PC parallel-port interface, bus buffers to control the various lines, the debug RAM, and an EPROM programmer.

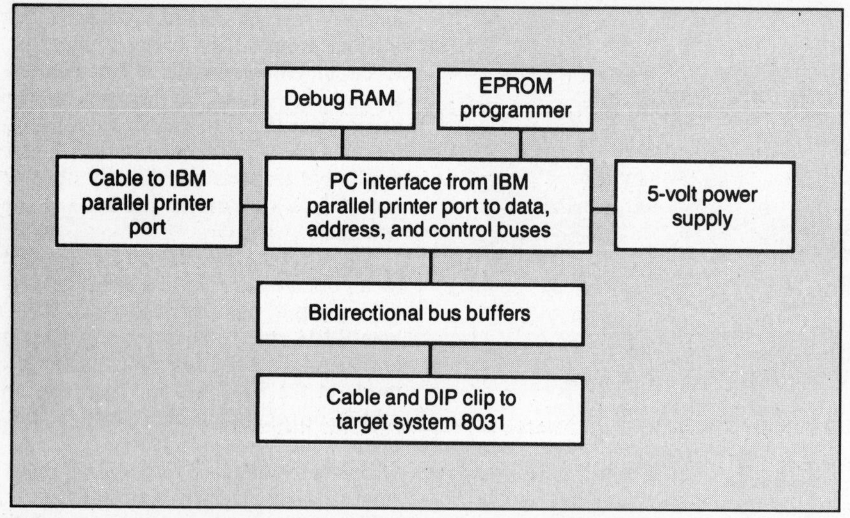

Figure 1: *A block diagram of the DDT-51 development system.*

The Port Swings Both Ways

The DDT-51 system trades hardware for software. The DDT-51's hardware pro-

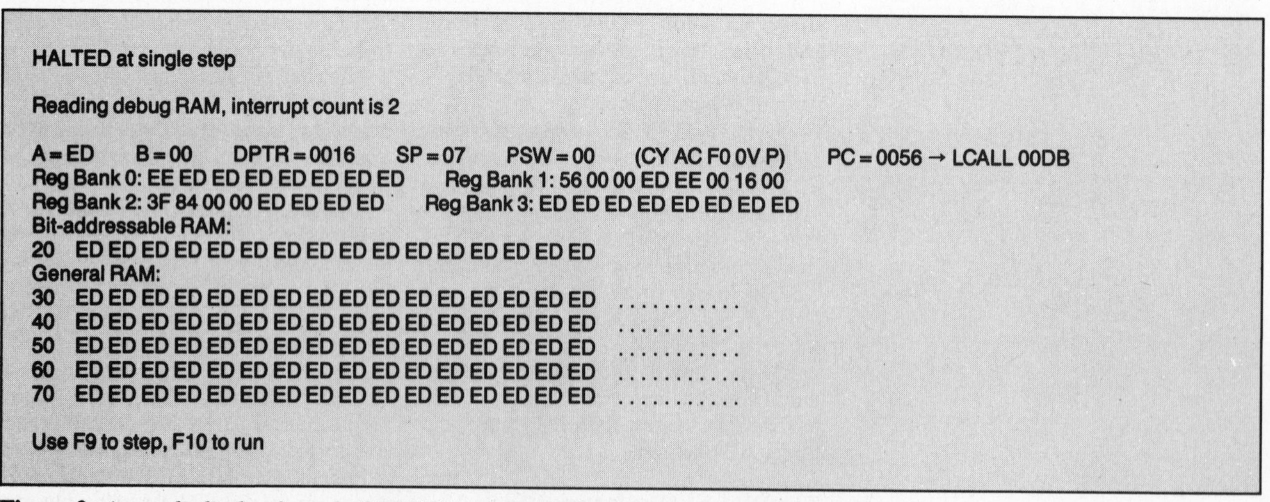

Figure 2: *A sample display from the DEBUG31 program, shown here in single-step mode.*

vides only essential functions; the software handles most of the logic. DEBUG31 merely sets and reads bits in the DDT-51 hardware to find out what's going on in the target 8031. Sounds simple enough.

Because speed of interaction between the host PC and the DDT-51 is important, the DDT-51 connects to the PC via a bidirectional parallel port (using the serial port would greatly increase the hardware complexity of the DDT-51). Unfortunately, an irritating problem with the design of the PC is that its standard parallel printer port is output only.

All the hardware needed to read or write 8 bits of data is already in place, but it lacks a connection to enable that function. While you can buy a custom parallel I/O board, I think changing a single trace on the existing board might be worth it to some readers. Figure 5 shows the single cut and addition to convert a standard parallel printer port to bidirectional operation.

Of course, all the clone boards (even the IBM boards) seem to use different IC numbers and assign the bits to different pins on the ICs, so this modification can be a real mystery. If you are unwilling to chop up your printer port card, buy a $50 clone printer port card. They are usually bidirectional. Some clone cards also omit the input connection between the data bus and pin 14 of the LS174. I suppose they figure that because the output is unused, the input is irrelevant. It's easy to find the two unused pins, however (a modified port will still work correctly with all your other software, simply because the code doesn't know about the change and won't take advantage of it).

DEBUG31 examines each of the three possible printer ports, working down from LPT3, to find which one has the modification installed. This provides a convenient way to check your work. DEBUG31 will tell you which card it's using.

Creating Five Ports from One

The DDT-51 system needs more than one I/O port to control all its hardware. Rather than burden the PC with more ports, I used an 8255 parallel peripheral interface (PPI) chip and a pair of LS374s to get five more ports. DEBUG31 controls these ports using the four standard printer control lines.

IC9, the control register, holds bits that must be active all the time, like the 8031 RESET and IRQ lines. It also supplies the 8255's register addresses and several bits for the DDT-51's bus logic.

IC12, the system register, holds the 8031 system bus control bits. There are 4 unused bits in this register that can be used for additional functions. The outputs are disabled whenever the 8031 drives the DDT-51's address and data buses, so that there is no conflict.

IC8, the 8255 bus interface, drives the high byte of the external address bus through port B and writes and reads the external data bus using port A. Port C samples the 8031 HALTED line. Other port C bits monitor the outputs of the DDT-51 logic circuits when running diagnostics from the PC.

Every 8031 system has a latch that demultiplexes the low-order byte of the address from the combined address/data bus. You might think that, because the outputs of that latch don't connect to the 8031, the DDT-51 system would need another DIP clip to get access to that part of the address bus. Instead, it is simpler to duplicate the latch, using IC10.

The DDT-51's debug RAM presents a similar problem, since the RAM requires a chip select line and the target 8031 system may not have an address decoder. I added IC11 to provide chip selects on 8K-byte boundaries throughout the 64K-byte address space. The EPROM programming socket, which is empty in normal use, also gets a chip select line from this decoder.

Bus Buffering

Both the 8031 and the DDT-51 system have control lines to read and write the program RAM (normally on the 8031 board) and the debug RAM (on the DDT-51 board). There are more combinations of data, address, and control line directions than might seem possible at first, which is why the logic isn't as simple as you'd expect.

IC9 supplies two control lines, called

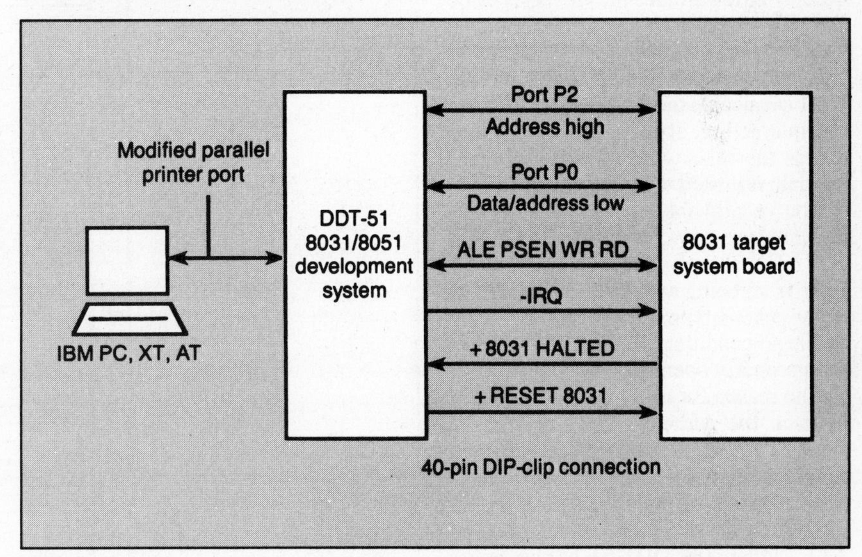

Figure 3a: *Physical connections between the DDT-51 and its host and target systems.*

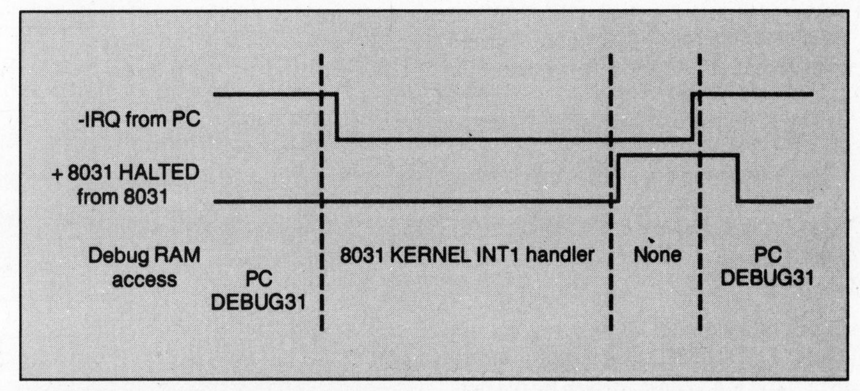

Figure 3b: *A timing diagram showing the system's "gentleman's agreement" for controlling access to the debug RAM.*

CTLS TO 8031 and CTLS FROM 8031, to select one of three states. The 8031 controls the DDT-51 logic when CTLS FROM 8031 is active, DDT-51 logic controls the 8031 system when CTLS TO 8031 is active, and the two systems are isolated from each other when both lines are inactive.

Setting the direction of the data and address lines isn't a simple task. For example, when the target 8031 system writes to the debug RAM, both the address and data come from the 8031. When it's reading from the debug RAM, the address comes and the data goes. A similar situation, in reverse, occurs when the DDT-51 reads the 8031's program RAM.

IC14 sets the DATA TO 8031 and DATA FROM 8031 lines by deciding which way the data must flow based on the current address, read or write activity, and signal source. This must be handled by hardware because it varies cycle by cycle as the 8031 runs its program.

When the two systems are isolated, the PC can access the debug RAM and the 8031 can access the program RAM without interacting. If either system tries to access the other while they're isolated, the data returned will be indefinite.

You've probably noticed that there's no hardware preventing both the CTLS TO 8031 and the CTLS FROM 8031 lines from being active at once. Further, as the schematic shows, this is definitely an illegal condition; it should never occur with properly operating PC code.

The reason for using an LS374 (IC10) to latch the address should now make sense. The bus buffers can't be activated until the control and address lines indicate in which direction the data should go. When the 8031 system supplies the ALE signal, the DDT-51 hardware accepts the 8031 data bus and tries to latch it in IC10. An LS373 is a *transparent* latch, meaning that the outputs track the inputs whenever the clock is high. When the 8031 lowers ALE, the bus buffers turn off and the data goes away before an LS373 can reliably latch it.

The LS374 is an *edge-triggered* latch, which means that the outputs change at the upward transition of the clock. Using ALE to clock the latches gives a sharp edge just before the data buffers turn off. This way, the LS374 records the inputs correctly.

Debug RAM

The debug RAM is simply a 6116 2K-byte static RAM chip, driven by the buffered address and data lines. Either sys-

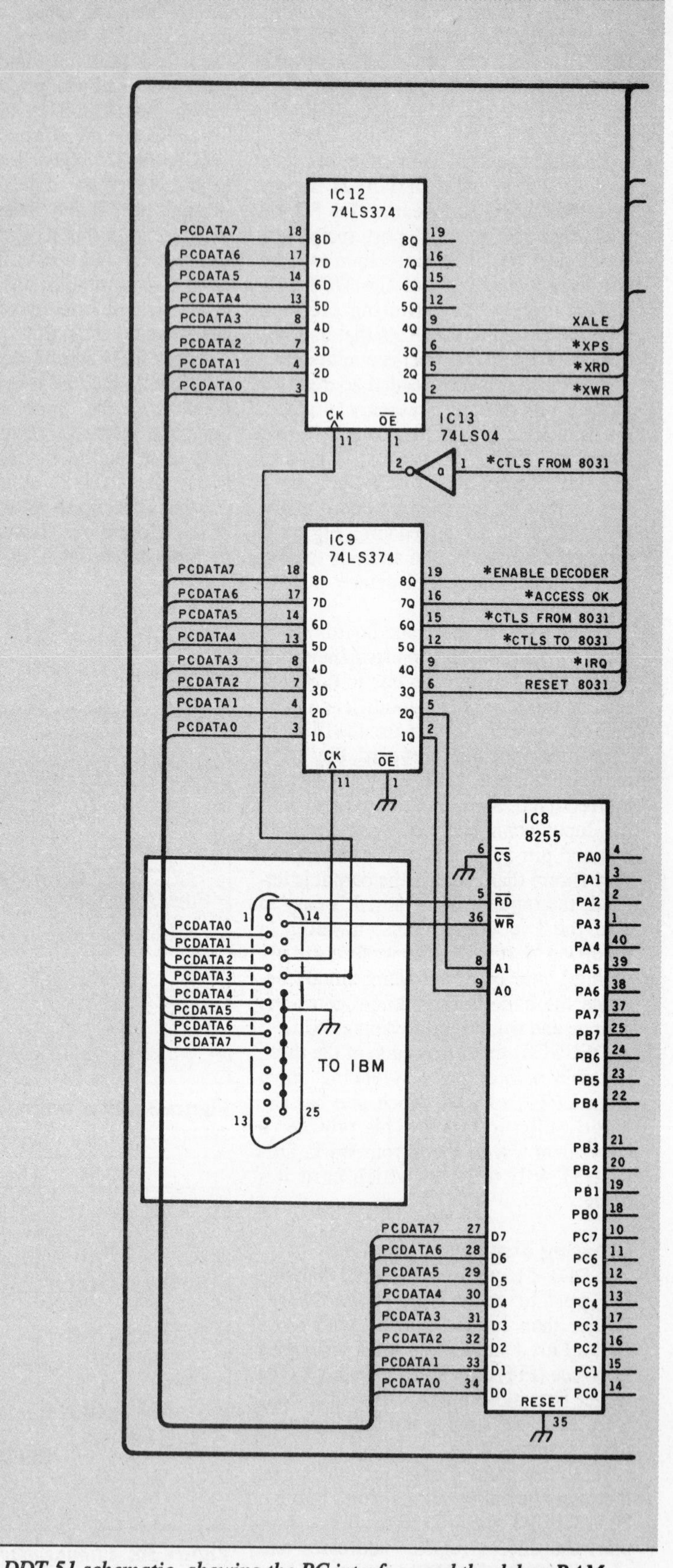

Figure 4a: *The DDT-51 schematic, showing the PC interface and the debug RAM.*

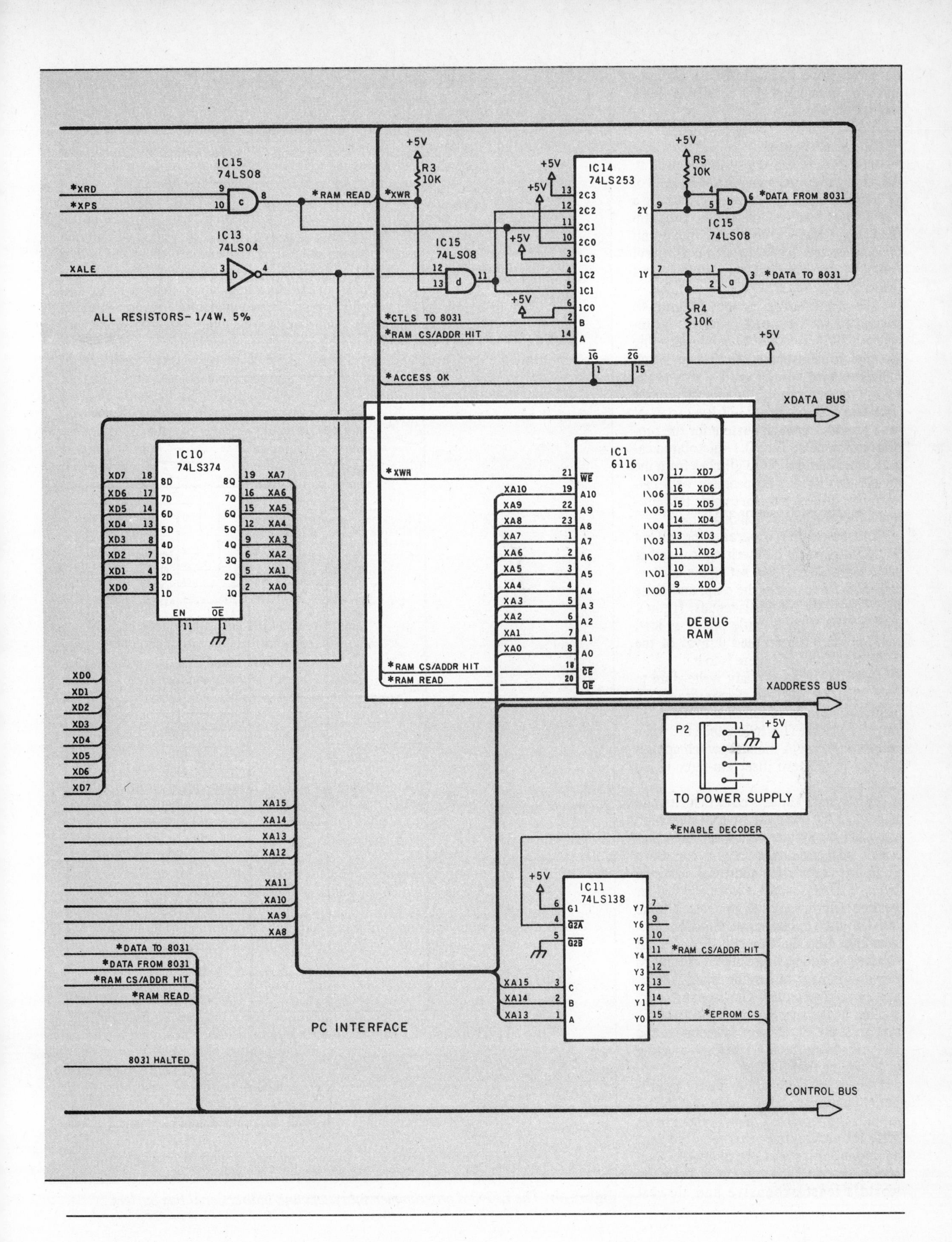
IC15
74LS08
*XRD
*XPS
*RAM READ
*XWR
+5V
R3
10K
IC14
74LS253
R5
10K
*DATA FROM 8031
IC13
74LS04
XALE
IC15
74LS08
*DATA TO 8031
R4
10K
ALL RESISTORS- 1/4W, 5%
*CTLS TO 8031
*RAM CS/ADDR HIT
*ACCESS OK
XDATA BUS
IC10
74LS374
IC1
6116
DEBUG
RAM
XADDRESS BUS
P2
TO POWER SUPPLY
*ENABLE DECODER
IC11
74LS138
*EPROM CS
8031 HALTED
PC INTERFACE
CONTROL BUS

tem can access it. DEBUG31 assumes that the debug RAM is located at 8000 hexadecimal.

EPROM Burning

Although the DDT-51 system is used to develop programs using RAM instead of EPROM, there (presumably) comes a time when you've finished the program and you're ready to ship it. At that point, you need an EPROM programmer. DDT-51 incorporates the world's least expensive 2764 EPROM programmer.

The programmer section of figure 4b shows all the hardware required to program a 2764 EPROM. The programming supply voltage comes from two 9-volt batteries hard-wired to add +18 V to the +5-V power supply. A simple emitter follower reduces this 23-V level to 21 V and provides enough current for the programming pulse. The 0.1-microfarad capacitors filter out noise glitches from the logic. You'll have no problem with battery life unless you start using this programmer for production quantities.

It turns out that programming a 2764 EPROM is much like writing to a rather slow RAM. First, you set up the address and data lines, select the chip, and pulse pin 27 low for 50 milliseconds. If pin 1 has +21.0 V on it while pin 27 is low, you've just programmed 1 byte of the EPROM. That's all there is to it.

DEBUG31 can already write data to any 8031 address, so programming an EPROM simply requires an "EPROM burn" software flag to lengthen the write pulse to 50 ms. A manual switch applies 21.0 V to the zero-insertion-force (ZIF) socket.

This system handles code that fits into a single 2764 EPROM. A DIP switch on the DDT-51 system board sets the chip-select address range so you can burn EPROMs at different addresses. And 8K bytes of EPROM holds a lot of 8031 code, particularly for smaller projects. Unlike the Circuit Cellar serial EPROM programmer (see the October 1986 Circuit Cellar), however, there are no circuits to prevent damage occurring when you insert or remove an EPROM from a "live" socket. Make sure you turn the DDT-51 system's power off and disconnect the DIP clip from the 8031 before inserting or removing the EPROM.

Of course, 2764s allow faster "intelligent" programming methods, but I haven't implemented them (I thought an EPROM programmer that required zero additional chips was pretty good). As a result, the DDT-51 system is both the world's least expensive and slowest

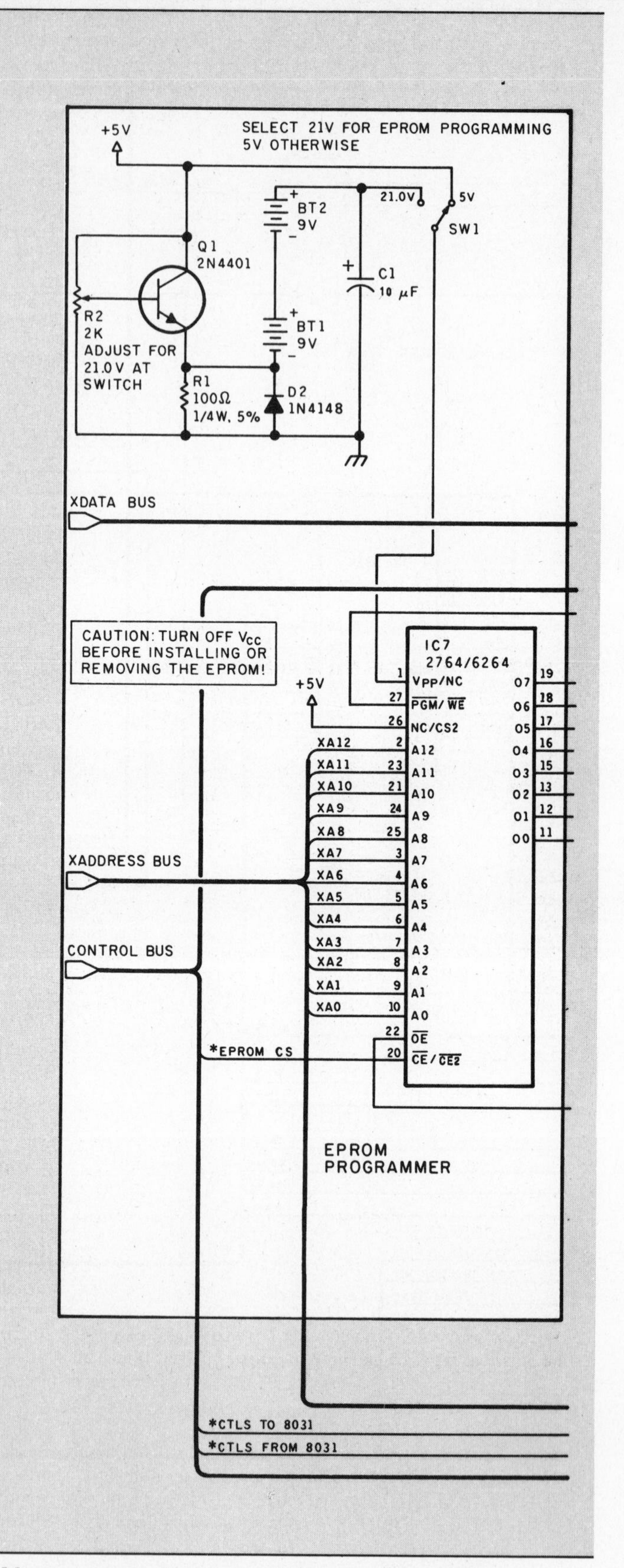

Figure 4b: *The EPROM programmer subsystem and bidirectional bus buffers.*

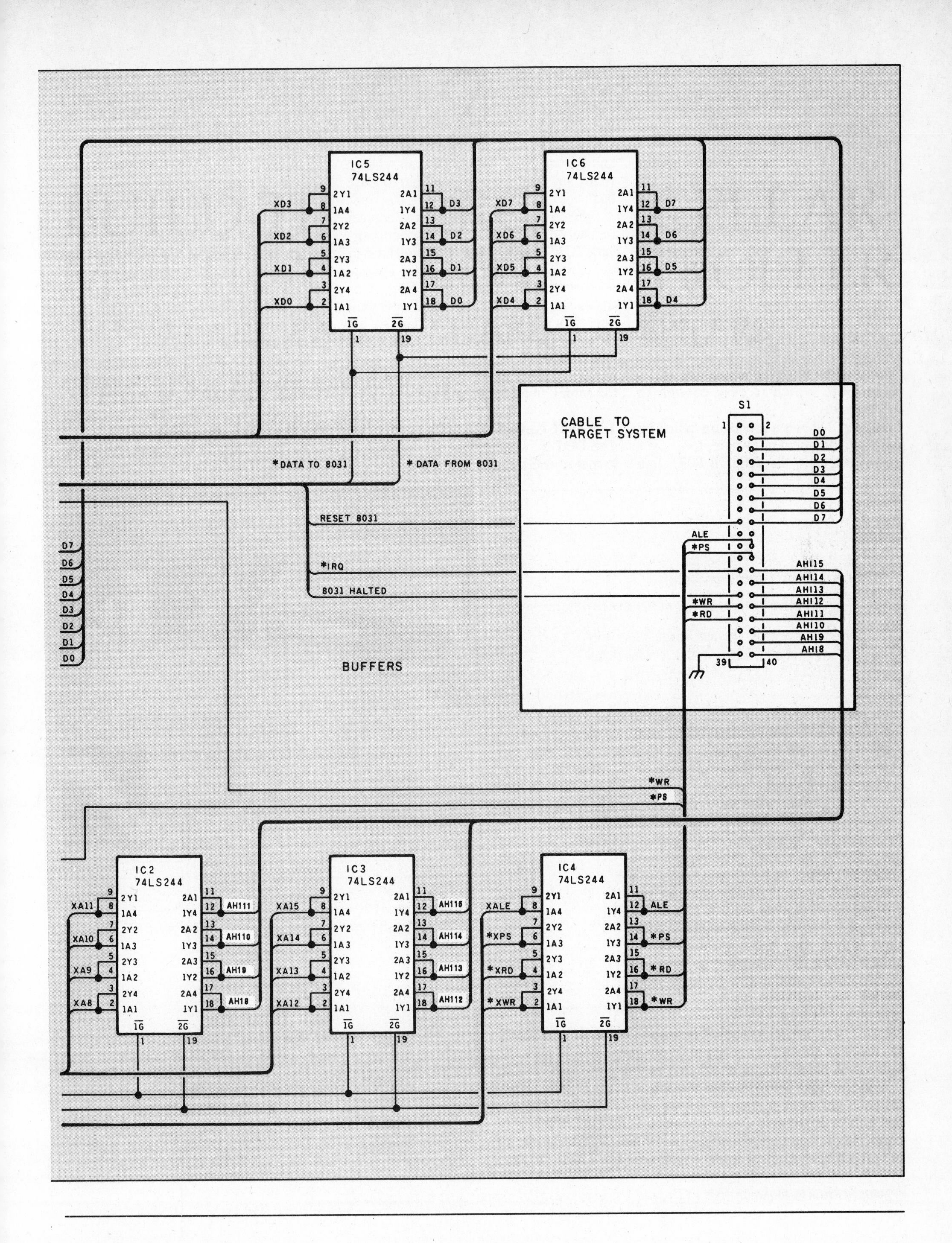
IC5
74LS244
IC6
74LS244
XD3
XD2
XD1
XD0
XD7
XD6
XD5
XD4
D3
D2
D1
D0
D7
D6
D5
D4
*DATA TO 8031
*DATA FROM 8031
RESET 8031
*IRQ
8031 HALTED
BUFFERS
CABLE TO
TARGET SYSTEM
S1
ALE
*PS
*WR
*RD
AH115
AH114
AH113
AH112
AH111
AH110
AH19
AH18
IC2
74LS244
IC3
74LS244
IC4
74LS244
XA11
XA10
XA9
XA8
XA15
XA14
XA13
XA12
XALE
*XPS
*XRD
*XWR

EPROM programmer. If you start writing long programs, you may want to step up to my serial EPROM programmer.

DEBUG31 Essentials

You might think that DEBUG31 on the PC is written in assembly language to handle all the bit manipulation needed to control the hardware. It turns out that it's all done in Turbo Pascal, which is a whole lot easier to read and understand. All the tricky stuff is handled by a few low-level routines.

At the highest level, `DEBUG31.PAS` is a simple loop that waits for function keys to be pressed. Each function key triggers a separate procedure to handle whatever is requested. Listing 1 shows the entire main loop, which is easy enough to understand.

Listing 2 shows a procedure called by the main loop (and elsewhere, too). `SysReset` activates the RESET 8031 line from IC9 and sets up DDT-51 hardware so that the PC has access to the 8031 through the bus buffers. When the RESET line is active, the 8031 has stopped operating and floated all its bus and control lines, so the PC can read or write any hardware on the shared data bus.

`SysReset` calls `SetCR` (set the control register IC9), `SetSR` (set the system register IC12), and `Load8255` (interface to the 8255), which are some of the lowest-level routines in the system. Each controls a single IC connected to the printer-port data lines. The routine handles all the operations required by other code and provides a convenient set of mnemonics that are easier to remember than the actual bit locations and values.

The 8031 Kernel

The DDT-51 system depends on a small kernel of 8031 code to handle interrupts and perform functions on behalf of the PC program. This code runs only during INT1 interrupts and must adhere closely to the gentleman's agreement interface to avoid hardware collisions.

Unlike `DEBUG31.PAS`, `KERNEL.ASM` must be written in 8031 assembly language and incorporated into any 8031 program that uses the debugging features of DEBUG31, but most of it is located in the 2K-byte debug RAM located on the DDT-51 board. The normal program memory must hold only about two dozen bytes of KERNEL's code, and you can move most of that chunk to any convenient location.

Because the 8031 registers and internal RAM are not directly accessible through the DDT-51 interface, KERNEL must copy those values into the debug RAM before DEBUG31 can see them. Once they've been copied, the PC can read, display, and change them as

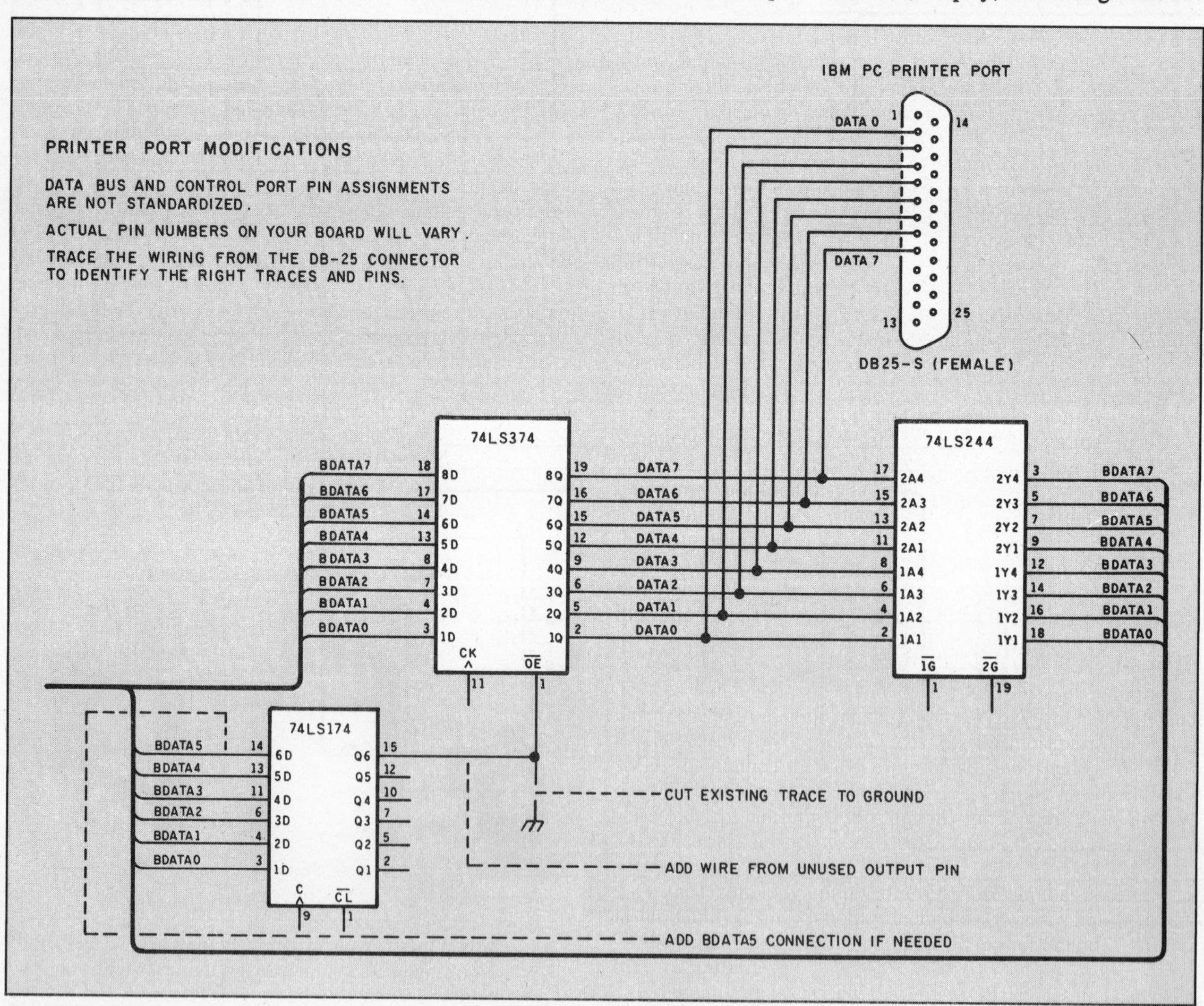

Figure 5: *How to make an IBM PC's unidirectional parallel printer port bidirectional.*

needed. KERNEL then copies the new values back into the internal locations and returns to the interrupted code.

The software handles breakpoints in a similar manner. DEBUG31 stores the breakpoint address in debug RAM and triggers an interrupt. KERNEL decodes the operation and writes an `LJMP` instruction at the specified address. The KERNEL breakpoint handler gets control when the 8031 executes the `LJMP` and copies the internal values into debug RAM.

At the heart of the interrupt-handler routine, the first order of business, as with any interrupt code, is to save all the CPU registers. Later on, these saved values will be copied into debug RAM so that DEBUG31 can reconstruct the actual state of the machine when the interrupt occurred.

If a breakpoint was active, the `UndoBP` routine removes the `LJMP` and restores the program RAM so that the next instruction is ready to go. Because an `LJMP` requires 3 bytes, it may have overwritten one or two instructions following the breakpoint. Regrettably, the 8031 does not have a single-byte software interrupt instruction like the 8088's special `INT 3` breakpoint.

The `DumpRAM` routine copies the internal RAM values into debug RAM as described before, as well as extracting the registers saved on the stack. Remember that the internal RAM reflects the current registers rather than the ones used by the main program. Some internal RAM locations can't be read without introducing side effects: Reading the serial buffer will reset a pending interrupt, which would cause a malfunction in the main code.

Both KERNEL and DEBUG31 increment separate IRQ counters; KERNEL's counter is stored in debug RAM. If DEBUG31's counter differs from KERNEL's version, at least one IRQ got lost, which probably means that the 8031 program is out of control.

Because most of KERNEL is in the debug RAM, DEBUG31 can't read that RAM while the 8031 executes from it. The `StepSpin` routine is located in 8031 program RAM, which frees up the debug RAM for the PC. The `StepSpin` code sets 8031 HALTED to indicate that the debug RAM is available, then waits for the PC to raise INT1 when it's done changing the RAM. At that point, it's safe to begin executing from debug RAM again, so the code returns to `SpinRet` to continue.

The code following `SpinRet` reads the

Listing 1: *The* `DEBUG31.PAS` *main loop.*

```
(*** setup code omitted here ***)

REPEAT
 { Get next key }
 GetKey(key1,key2);
 { Is it an extended key? }
 IF key1 = Chr(ESC)
  THEN BEGIN
   { Which function key? }
   CASE Ord(key2) OF
    59 : ShowHelp;
    60 : SysReset;
    61 : LoadKernel;
    62 : Download;
    63 : SetBurn;
    64 : Writeln('F6 not used');
    65 : SetRegs;
    66 : SetBP;
    67 : SysStep;
    68 : BEGIN
          { Normal, no IRQ }
          SysRun(running,FALSE);
          Writeln('Use F9 to stop/step');
         END;
    ELSE;
   END;
  END;

UNTIL (key1 = Chr(ESC)) AND (key2 = Chr(0));

(*** some cleanup code omitted ***)
```

Listing 2: *The system reset procedure.*

```
{ Force reset and take control of bus }

PROCEDURE SysReset;

BEGIN

 { Ensure a reset }
 SetCR(Reset8031,ON);

 { Get control of bus }
 SetCR(IRQ,OFF);
 SetCR(CtlTo8031,ON);
 SetCR(CtlFrom8031,OFF);
 SetCR(AccessOK,ON);
 SetCR(EnableDec,ON);

 { Set up controls }
 SetSR(XRD,OFF);
 SetSR(XPSEN,OFF);
 SetSR(XALE,OFF);
 SetSR(XWR,OFF);

 { Our data and address }
 Load8255(I55Ctls,I55AoBoCi);

 { Reset IRQ counter }
 IRQctr := 0;
 PutRAMbyte(IRQctrB,$00);

 Writeln('8031 hardware is reset');

 { Indicate reset }
 state := reset;

END;
```

new run-mode byte from debug RAM and handles the request. This code also installs the new breakpoint LJMP instruction.

Finally, the StuffRAM routine copies the (possibly changed) values from debug RAM back into internal RAM. It inserts the new values for the registers into the right stack locations, so that the main program's registers can be altered by the PC. It does not restore registers with side effects (e.g., the serial transmitter buffer) for the same reason that DumpRAM didn't read from them.

After the registers are restored, the 8031 executes a RETI (return from interrupt) and the interrupted main code resumes execution, perhaps with new register values. (The complete listings for DEBUG31.PAS and KERNEL.ASM are available for downloading from the Circuit Cellar bulletin board system.)

A Middle Ground

Does working on the DDT-51 development system convert me into a software jockey? Certainly not. To maintain a high level of performance in Circuit Cellar designs at a time when the rest of the industry is moving to application-specific ICs (ASICs) and other development-intensive reduced-component custom designs, I have chosen a middle ground of using programmable microcontroller chips.

In the same way that a large board of IC logic chips is reduced functionally to an ASIC, I have lumped the controlling elements of the TTL circuit I usually use into a configurable application-specific controller that can simulate these functions. Using simple three- or four-chip microcontrollers as standard computer engines in my projects, I am now better able to concentrate on presenting applications of technology rather than providing redundant explanations of low-level logic.

The DDT-51 development tool will hopefully help other BYTE readers consider the benefits of this middle ground.

The DDT-51 development system won't handle all possible 8051 target systems, but it will give you a good start. With that in hand, we can continue with other interesting Circuit Cellar project designs.

Experimenters

As is the custom with all Circuit Cellar projects, the software for the DDT-51 development system is available for downloading from my multiline bulletin board free of charge. Call (203) 871-1988. Of course, this downloaded software is limited to noncommercial personal use unless licensed otherwise.

Special thanks for the technical contributions provided for this article by Jeff Bachiochi and Ed Nisley.

Editor's Note: Steve often refers to previous Circuit Cellar articles. Most of these past articles are available in book form from BYTE Books, McGraw-Hill Publishing Company, P.O. Box 400, Hightstown, NJ 08250, (1–800–2–MCGRAW).

Ciarcia's Circuit Cellar, Volume I covers articles in BYTE from September 1977 through November 1978. *Volume II* covers December 1978 through June 1980. *Volume III* covers July 1980 through December 1981. *Volume IV* covers January 1982 through June 1983. *Volume V* covers July 1983 through December 1984. *Volume VI* covers January 1985 through June 1986.

Circuit Cellar Ink

It's virtually impossible to provide all the pertinent details of a project or cover all the designs I'd like to in the pages of BYTE. For that reason, I have started a bimonthly supplemental publication called Circuit Cellar Ink, which presents additional information on projects published in BYTE, new projects, and supplemental applications-oriented materials. For a one-year subscription (6 issues), send $14.95 to Circuit Cellar Ink. Credit card orders can call (203) 875-2199.

There is a multiline Circuit Cellar bulletin board system (running TBBS 2.0M) that supports past and present projects in BYTE and Ink. You are invited to call and exchange ideas and comments with other Circuit Cellar supporters. The 300-/1200-/2400-bps BBS is online 24 hours a day at (203) 871-1988.

The DDT-51 is now available in kit form. For information call CCI at (203) 875–2751.

9

THE SMARTSPOOLER

PART 1: THE SPOOLER HARDWARE

Steve clears your printer's throughput bottleneck and comes up with more than a buffer

The discrepancy in speed between fast CPUs and slow printers has existed since the early days of computing, and the magnitude of the speed difference is becoming ever larger. When you upgrade and add performance to a computer system, you generally add processing power rather than printing power. The net result is that many 20-MHz 80386 "million-instructions-per-second-class" machines out there are waiting for 100-character-per-second printers.

Admittedly, many computer users have light printing requirements, and they don't mind waiting 5 to 10 minutes for an occasional printout. Some, I'm sure, welcome this downtime as a mandatory coffee break. However, those who rely on their computers as a tool for making a living know that time is money.

Spooling to the Rescue

This problem is not new, and one solution has always been spooling. *Spool* stands for "simultaneous peripheral operations on-line." Pioneered on mainframes and minicomputers, spooling solves the CPU/printer speed gap by temporarily buffering printer output on intermediate storage (typically a disk drive) at high speed. Then, you move the stored data from storage to the printer at the latter's low speed. In the meantime, the CPU also runs foreground applications.

Traditional spooling has been tried on personal computers with limited success. One problem is that spooling requires an operating system with either multitasking capabilities or special software patches that effect the appearance of simultaneous operation. Even with a multitasking operating system or a patch, however, spooling often taxes the computer's processing power and affects system performance.

Spooling also creates problems in disk priority and data conflicts. If not properly coordinated through the operating system, a computer's disk can get "trashed" if it has to handle print data and application data at the same time. Given the declining cost of processing power and memory, it finally became obvious that a separate computer should handle the printing task, which led to the dedicated printer buffer.

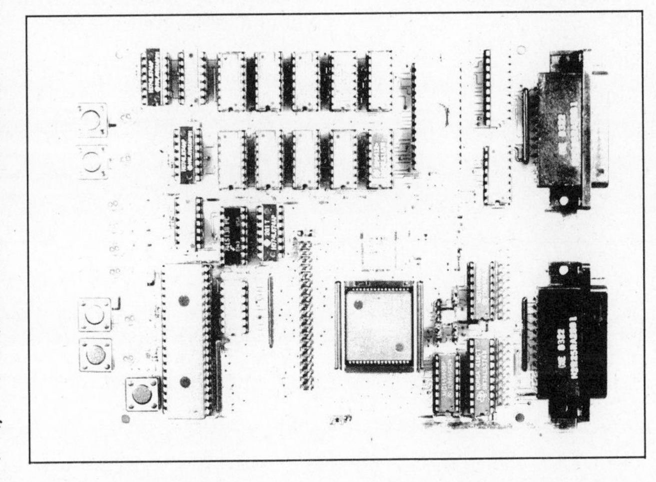

I have received many letters over the years asking when I would present a printer buffer as a project. Generally, I would avoid answering the question or jokingly say "never." Of course, it really wasn't a joking matter. It troubled me greatly that I couldn't do justice to such an important subject and had to live with such an embarrassing answer until I could invest the time and effort to present a Circuit Cellar–quality design. Let me explain.

Most printer buffer projects are cumbersome and simplistic. The ones I remember seeing offer only rudimentary functions and use 40 or 50 chips for a 64K-byte buffer.

Commercial units, on the other hand, have narrowly defined capabilities enhanced by software-implemented system functions and tricky hardware designs intended to reduce manufacturing cost and complexity.

I recall seeing one rather unique printer buffer design that used an 8051 single-chip microcomputer connected to dynamic RAMs (DRAMs) that operated *serially*. The 8051 would constantly cycle through the DRAM addresses at a rate that precluded any need for DRAM refresh. The system stored text serially in the individual 64K- by 1-bit DRAMs rather than in a parallel bank of them, as you'd usually expect. An 8K-byte text buffer therefore consisted of a single 64K-bit DRAM chip, while a 32K-byte buffer used four, and so on.

While I am not totally opposed to off-the-wall hardware designs, fancy technique has its price. Most computer-based technical designs are a trade-off between hardware and software. Anytime a company expects to be manufacturing something in a high-volume quantity, it tries to minimize recurring hardware costs by doing as much as possible in the one-time nonrecurring software development phase.

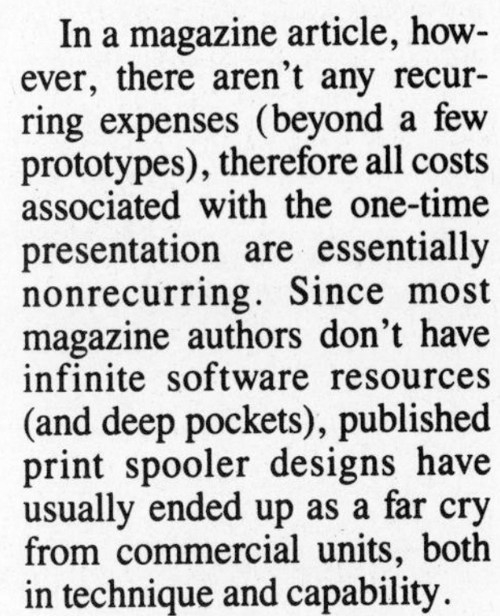

In a magazine article, however, there aren't any recurring expenses (beyond a few prototypes), therefore all costs associated with the one-time presentation are essentially nonrecurring. Since most magazine authors don't have infinite software resources (and deep pockets), published print spooler designs have usually ended up as a far cry from commercial units, both in technique and capability.

I could hardly criticize mediocre designs if I presented a piece of hardware that merely

While I've presented SmartSpooler primarily as a printer buffer, it can function as a complete remote data-processing computer.

met the objective with little regard for the hardware costs and complexity. At the same time, it would be insane to custom-mask an 8051 and present a three-chip printer buffer project that demonstrates nothing to readers who want to learn something about the architecture of printer buffers.

Fast Computers, Slow Printers

With the Circuit Cellar SmartSpooler, I have achieved my goal. This is an efficient integration of commercial and educational objectives. I've complemented its hardware design with interrupt-driven software that does not compromise performance to maintain cost-effectiveness. Designated as a spooler rather than as a simple printer buffer, SmartSpooler vastly improves computing throughput with slow peripherals.

SmartSpooler has 256K bytes of memory, serial and parallel I/O ports, and features that improve versatility and ease of use. These include "switchbox" capability for routing serial or parallel computer input to serial or parallel printer output, the ability to print multiple copies, single-sheet feeding mode, and buffer capacity indicators. Also, you can daisy-chain multiple SmartSpoolers to control a whole network of peripherals.

More important, SmartSpooler is intelligent. While I've presented it primarily as a printer buffer, a host computer can completely control its operation. You can even download executable code to SmartSpooler. It can function as a complete remote data-processing computer that analyzes and interprets the data flowing through it.

After a summary of SmartSpooler's features, I'll cover some printer buffer basics and then describe SmartSpooler's hardware and software specifics.

SmartSpooler has the following features:

- 256K-byte buffer capacity = 100 pages (typically)
- Switchable serial (RS-232C) or parallel (Centronics) inputs and outputs
- Serial port data rates: 300, 600, 1200, 2400, 4800, 9600, 19,200, and 38,400 bits per second (bps)
- Serial port handshaking: hardware ($\overline{RTS}$, $\overline{CTS}$, and so on) and software (XON-XOFF)
- Multiple copy capability: 0 to 4 copies plus original
- Pause mode for single-sheet feeding
- Real-time clock (power required)
- Front-panel buffer capacity gauge
- Host programmed configuration and operation mode
- Built-in interactive diagnostic monitor program
- Low power (5 volts [V] at 0.5 ampere [A], ±12 V at 0.1 A typical)

Generic printer buffers come with a variety of functions. SmartSpooler combines the best features of these other printer buffers and adds a few of its own. SmartSpooler sits between a computer printer port and the printer. Since it provides both parallel and serial I/O ports, it adapts between computers and printers of each type.

When the computer prints, SmartSpooler accepts the data at high speed, buffers it, and outputs it to the low-speed printer. From the computer's point of view, printing a document takes only seconds. During the printout, SmartSpooler can make copies, temporarily suspend printing (pause button), and handle embedded formfeeds for single-sheet printing.

Hardware—Keep It Simple

The hardware requirements for SmartSpooler are simple: a CPU, buffer memory, and host/printer ports (serial and parallel). We also need some parallel I/O ports to handle switches and LEDs for SmartSpooler's front-panel operation.

I used the complementary-metal-oxide-semiconductor HD64180 CPU because it is ideally suited to the task and significantly reduces the hardware complexity. First, it includes two RS-232C serial ports on-chip, eliminating an external dual-UART (universal asynchronous receiver/transmitter) chip.

Second, the need for a large, low-cost buffer demands dynamic memory chips. The HD64180, with its built-in refresh, easily accommodates DRAMs and offers a performance level that can meet today's high-speed communications requirements. (Though the task seems simple, many computers choke when running terminal emulator programs at 19,200 or 38,400 bps.)

The SmartSpooler block diagram in figure 1 shows the three basic components: CPU (i.e., HD64180), memory (i.e., DRAM and SmartSpooler EPROM), and I/O (i.e., host, printer, switches, and LEDs) interface. The SmartSpooler also provides an SB180 XBUS I/O expansion connector for additional I/O (more on that later). See figure 2 for the complete SmartSpooler schematic (less the section covering the DIP-switch settings and LEDs, which I will cover in Chapter 9, part 2).

The EPROM socket (IC7) accepts 8K-byte (2764) to 64K-byte (27512) EPROMs, though the SmartSpooler code presently requires only 16K bytes (27128). Because of the on-chip DRAM refresh controller, connecting 256K bytes of DRAM is relatively easy. All the HD64180 needs are three 74LS157 row/column address multiplexers (IC8 through IC10) and a CAS\

(text continues on page 136)

Figure 1: (a) *The SmartSpooler combines the function of a 256K-byte printer buffer and a serial/parallel converter switch.* **(b)** *Its HD64180 CPU—with built-in serial ports, extensive interrupt handling, address range beyond 64K bytes using direct memory access, and easy DRAM interface—is ideal for this application.*

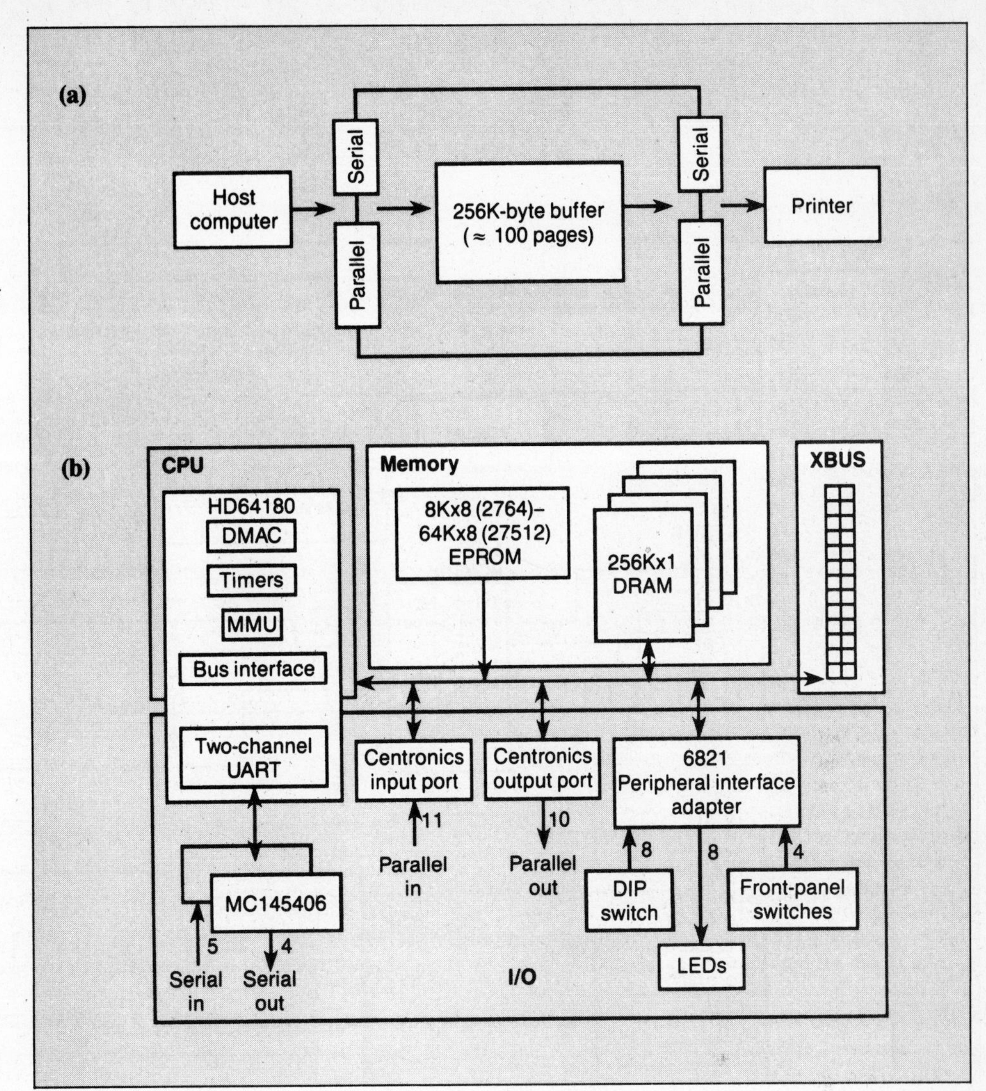

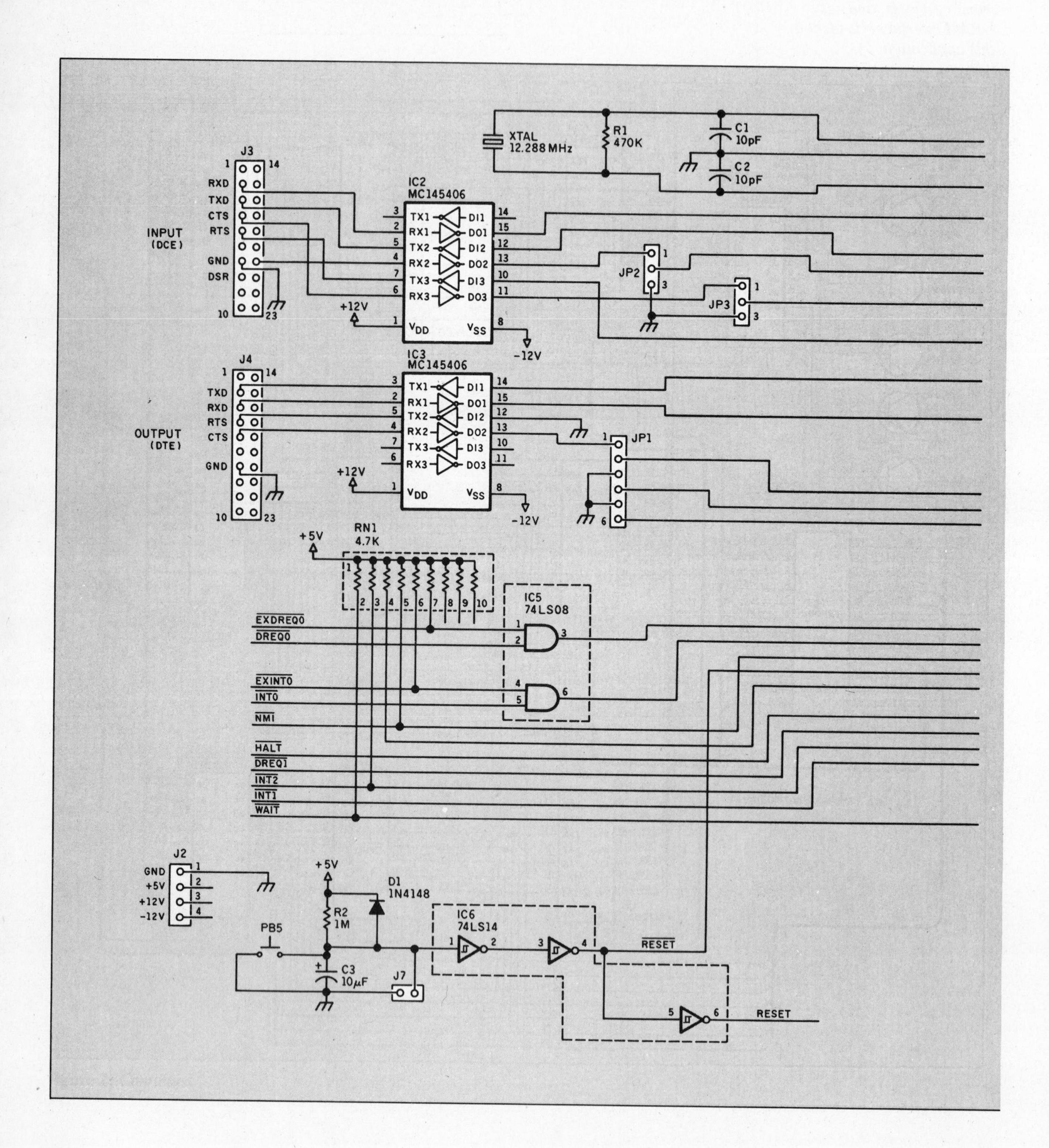
XTAL
12.288 MHz
R1
470K
C1
10pF
C2
10pF
J3
INPUT
(DCE)
RXD
TXD
CTS
RTS
GND
DSR
IC2
MC145406
TX1
RX1
TX2
RX2
TX3
RX3
DI1
DO1
DI2
DO2
DI3
DO3
VDD
VSS
+12V
-12V
JP2
JP3
J4
OUTPUT
(DTE)
IC3
MC145406
JP1
RN1
4.7K
+5V
IC5
74LS08
EXDREQ0
DREQ0
EXINT0
INT0
NMI
HALT
DREQ1
INT2
INT1
WAIT
J2
GND
+5V
+12V
-12V
D1
1N4148
R2
1M
PB5
C3
10μF
J7
IC6
74LS14
RESET
RESET

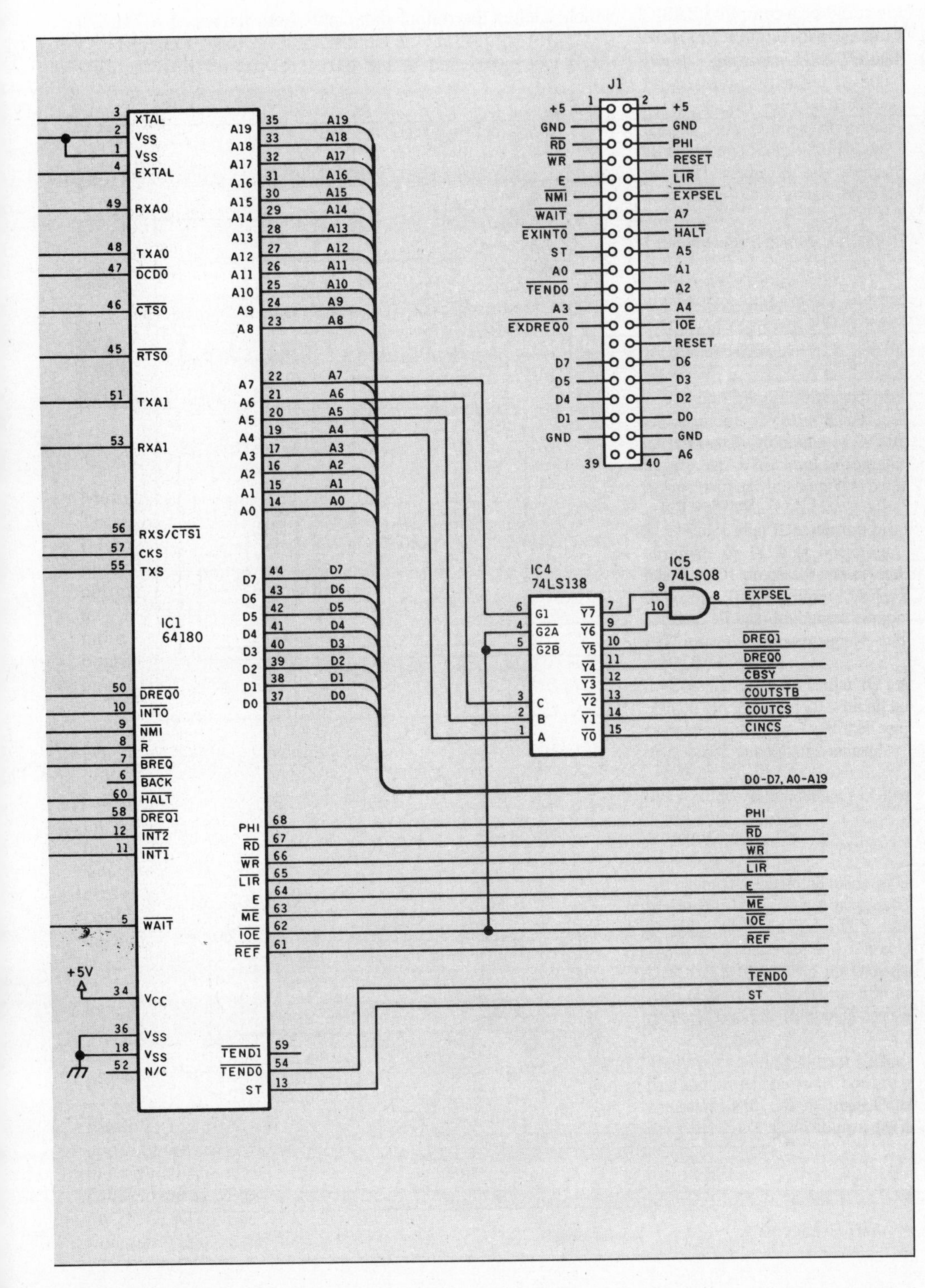

Figure 2a: *SmartSpooler's CPU and serial I/O section. (Not shown here are the 6821 peripheral interface adapter and other circuitry associated with the switches, push buttons, and LEDs; that portion will appear in Chapter 9, part 2).*

generator consisting of a flip-flop and a couple of gates (IC5, IC6, and IC19). The plastic-leaded-chip-carrier version of the HD64180 can address 1 megabyte, but I feel that 256K bytes is more than adequate for typical printing applications, and it keeps power consumption low.

If you examine the SmartSpooler's architecture closely, you will note some similarities with the SB180 single-board computer (see the September 1985 Circuit Cellar). Considering the performance and popularity of the SB180, any similarities are purely intentional. The SmartSpooler's core architecture lets it be more or less software-compatible with the BCC180 multitasking controller (presented in Chapter 5) and the SB180. Remember that you can download executable code to the SmartSpooler using its host programmed mode. This mode could be a compiled BASIC program generated on a BCC180 just as easily as any HD64180 assembly code produced on an IBM PC cross assembler or the SB180.

SmartSpooler's I/O

SmartSpooler has two serial ports and two parallel printer ports. Since the HD64180 has two on-chip serial ports, the only extra hardware needed for the RS-232C ports are a pair of MC145406 RS-232C level shifters (IC2 and IC3). The rest of the I/O hardware consists of the Centronics I/O ports and the front-panel switch/LED interface.

A 74LS138 (IC4) I/O address decoder generates eight chip selects. Two outputs are reserved for XBUS addressing: I/O addresses E0 to FF hexadecimal. Two chip selects are connected to the HD64180 direct-memory-access request inputs, and the remaining four are associated with the Centronics I/O ports.

The Centronics output port consists of a 74LS374 (IC22), which latches the output data, plus both halves of a 74LS74 (IC23) for controlling the handshaking. To send a character to the device connected to the port, the control software first

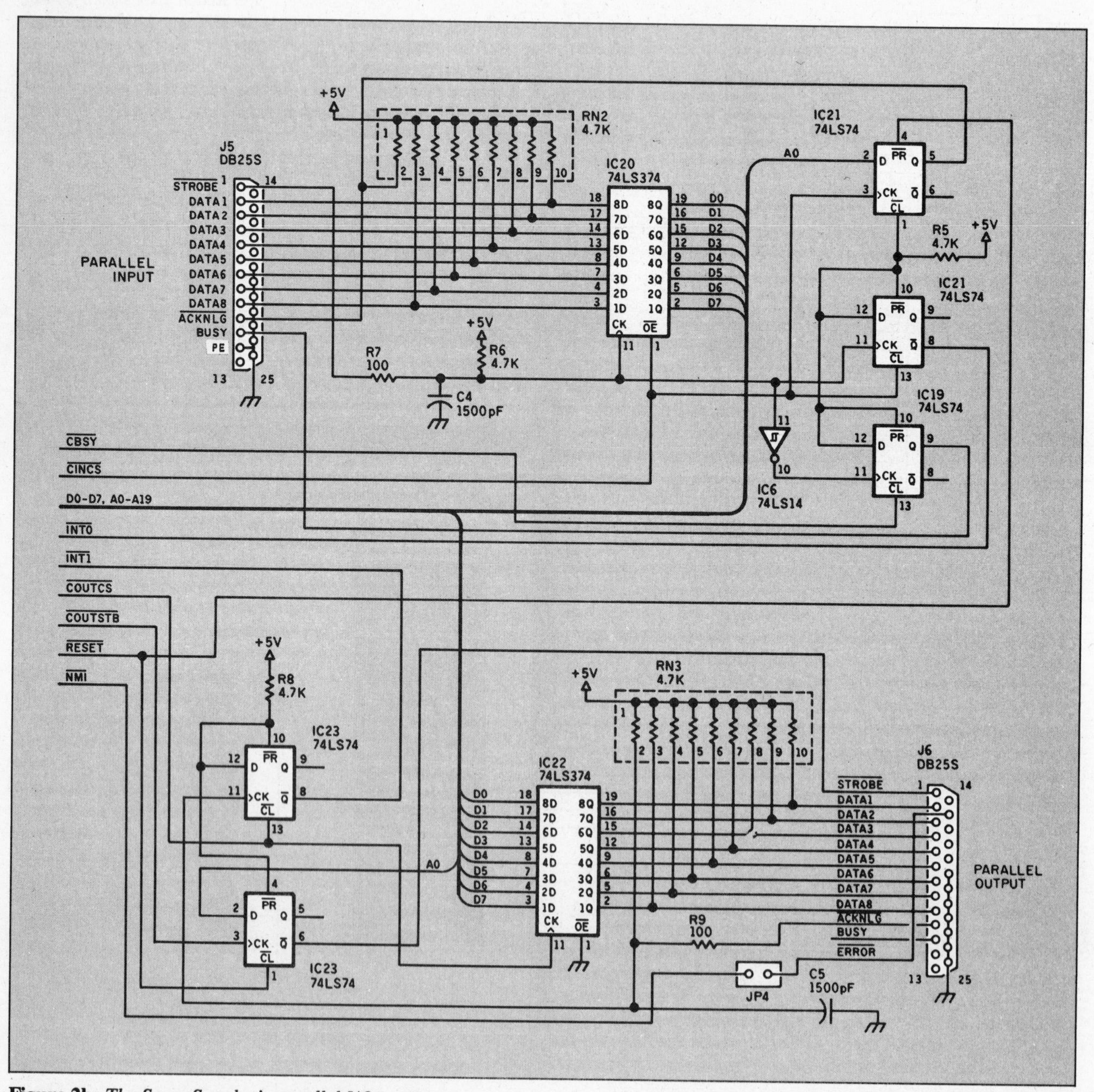

Figure 2b: *The SmartSpooler's parallel I/O section.*

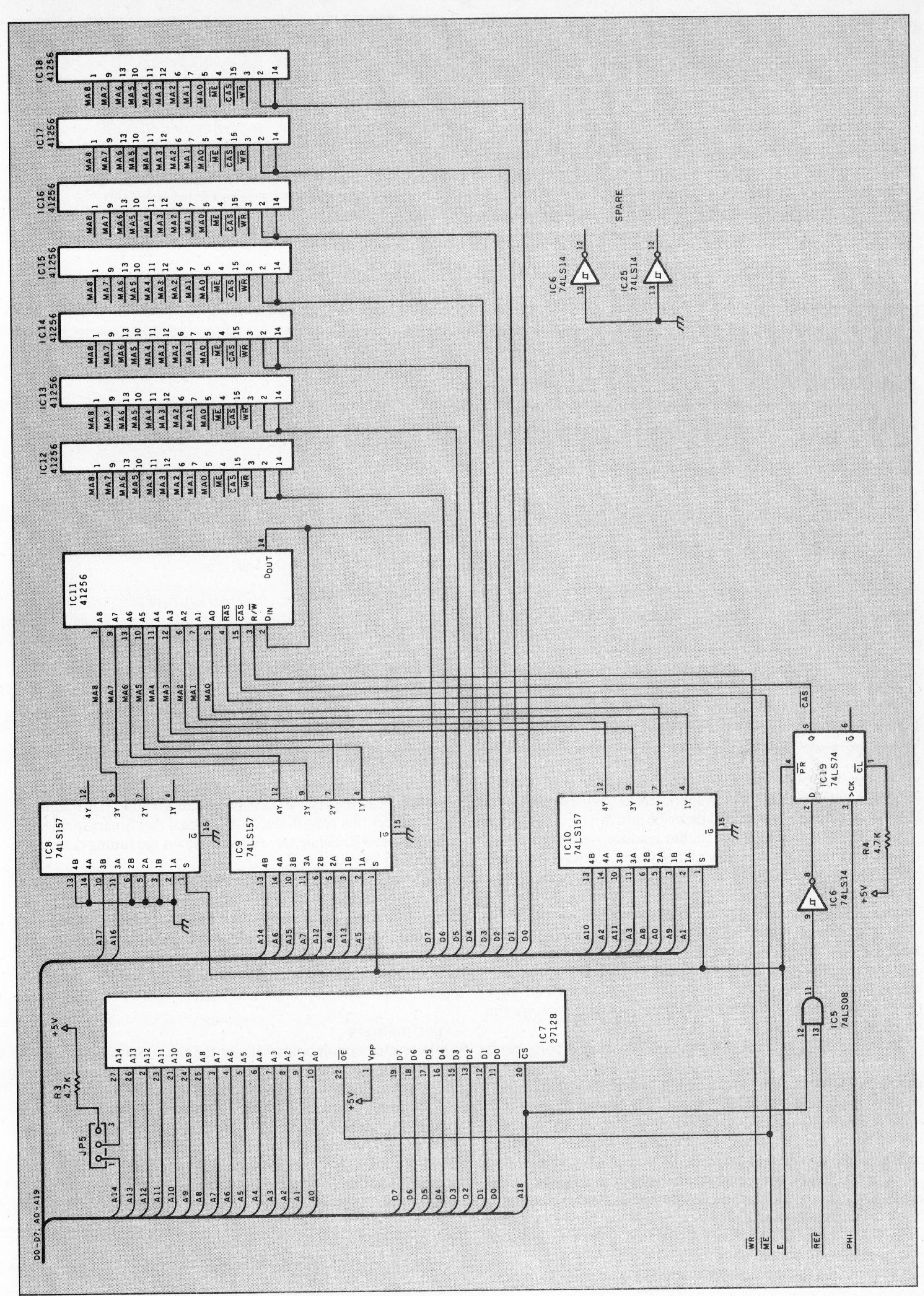

Figure 2c: *The EPROM and DRAM circuitry for the SmartSpooler.*

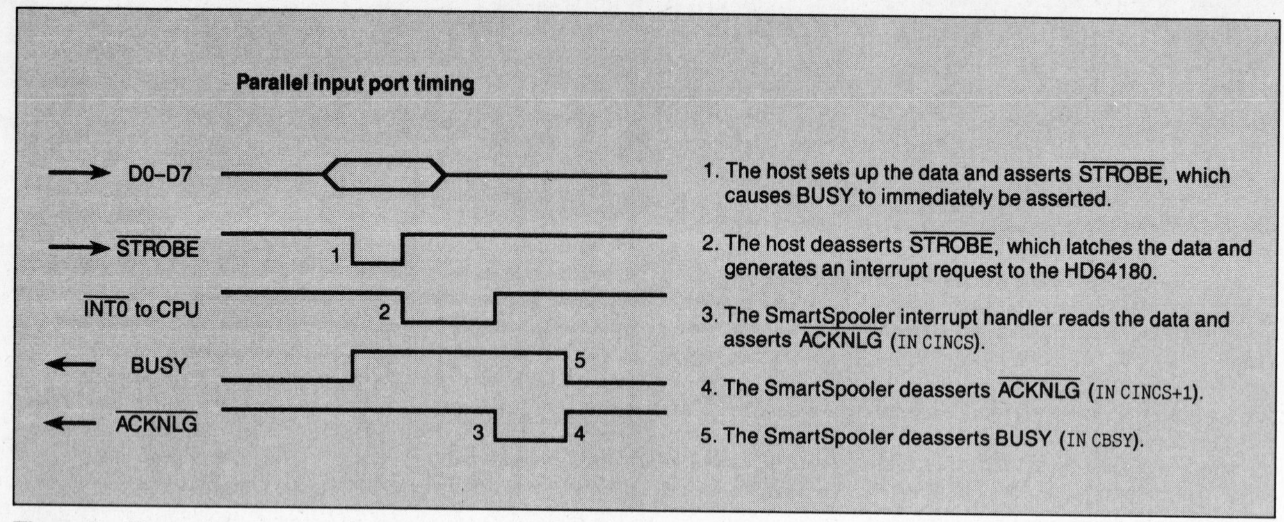

Figure 3a: *Timing diagram for SmartSpooler's parallel input port.*

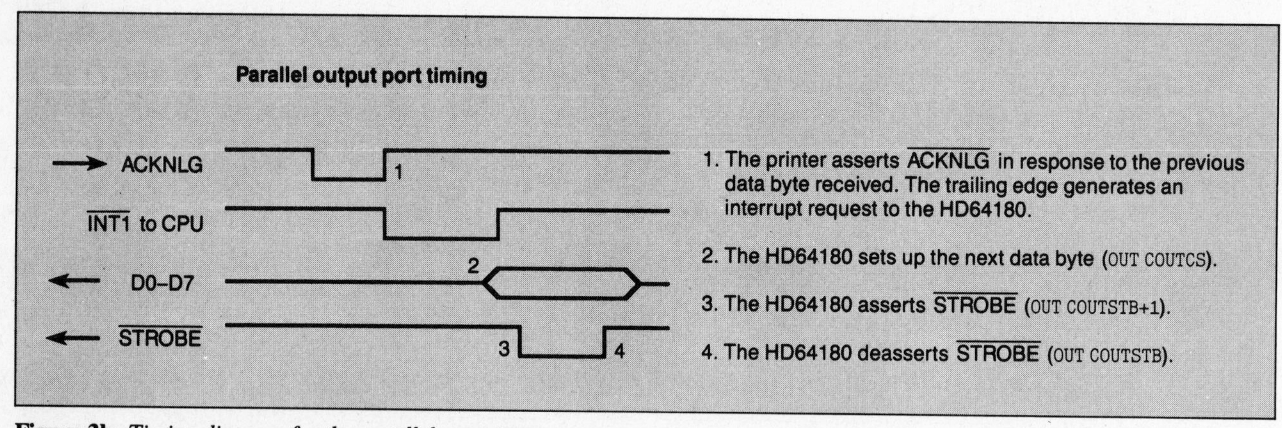

Figure 3b: *Timing diagram for the parallel output port.*

writes the byte out to port 90h. This latches the byte into the 74LS374 and also clears INT1\ high. (INT1\ is used during the acknowledge sequence that I'll describe next.)

Next, the program accesses (either reads or writes) port A1h to assert STROBE\ on J6 low. Finally, the software accesses port A0h to cause STROBE\ to go high, signaling the target device that a byte is ready.

Once the target printer device has retrieved the byte and wants to tell the spooler that it is ready for another byte, the device asserts the ACKNLG\ line on J6 low, then sets it high again. The rising edge clocks the 74LS74 and forces INT1\ low, generating an interrupt. Writing the next byte out to the Centronics port (usually part of the interrupt service routine) clears the interrupt.

The Centronics input port is similar to the output port, but it works in reverse. When the device connected to J5 sends the spooler a character, the device simultaneously generates a low-to-high signal on the STROBE\ input. This latches the data into a 74LS374 (IC20) and sets INT0\ low, generating an interrupt to the HD64180 processor. It also clocks another flip-flop, sending the BUSY line to the input device into a high state.

As part of the interrupt service routine, the processor first reads port 80h to get the character being sent to it by the input device. The port access also clears the interrupt and causes the ACKNLG\ output to go low as the first part of the acknowledge sequence. Next, an access to port 81h sets ACKNLG\ high again, signaling the input device that the character has been read and the spooler is ready for another. Finally, it accesses I/O port B0h, causing the BUSY output to go low. This signals those devices that rely on the BUSY signal that SmartSpooler is ready for another character. Figure 3 shows the timing details of the parallel I/O ports.

Purists may argue that SmartSpooler does not support the full Centronics interface, which, technically, includes additional signals like Paper Error, Error, and Init. However, these signals (and other vendor-unique signals) are inconsistent for different computers, printers, and software. Just in case, SmartSpooler's output port does provide an option for handling a Printer Error input signal. You can jumper it to NMI\.

Experimenters

While printed circuit board kits for the SmartSpooler are available, I encourage you to build your own. If you don't mind doing a little work, I will support your efforts as usual. You can download a hexadecimal file of the executable code for SmartSpooler's system EPROM (27128) from my bulletin board at (203) 871-1988. Alternatively, you can send me a preformatted IBM PC disk with return postage, and I'll put all the files on it for you (add $6 for the SmartSpooler User's Manual). Of course, this free software is for noncommercial personal use.

In Conclusion

In part 2 of this chapter, I'll finish the hardware by explaining the switch and LED configuration and describe SmartSpooler's software.

I'd like to personally thank Tom Cantrell for his extensive work on this project. Without his software expertise, I'd be hopelessly mired in an ocean of bits forever.

Editor's Note: Steve often refers to previous Circuit Cellar articles. Most of these past articles are available in book form from BYTE Books, McGraw-Hill Publishing Company, P.O. Box 400, Hightstown, NJ 08250, (1-800-2-MCGRAW).

Ciarcia's Circuit Cellar, Volume I covers articles in BYTE from September 1977 through November 1978. *Volume II* covers December 1978 through June 1980. *Volume III* covers July 1980 through December 1981. *Volume IV* covers January 1982 through June 1983. *Volume V* covers July 1983 through December 1984. *Volume VI* covers January 1985 through June 1986.

It's virtually impossible to provide all the pertinent details of a project or cover all the designs I'd like to in the pages of BYTE. For that reason, I have started a bimonthly supplemental publication called Circuit Cellar Ink, which presents additional information on projects published in BYTE, new projects, and supplemental applications-oriented materials. For a one-year subscription (6 issues), send $14.95 to Circuit Cellar Ink. Credit card orders can call (203) 875-2199.

The following items are available from

Circuit Cellar, Inc.
4 Park St., Suite 12
Vernon, CT 06066
(203) 875-2751

1. SmartSpooler RAMless full printed circuit board kit for parallel-to-parallel operation. Comes complete with printed circuit board and all printed-circuit-board-mounted components, including HD64180, IC sockets, SmartSpooler software on EPROM, and User's Manual. Less DRAM chips. SS01-KIT $119

Optional serial port ribbon cables are available. Call for pricing.

The SmartSpooler design is available for licensing to qualified manufacturers. Call for information.

All payments should be made in U.S. dollars by check, money order, MasterCard, or Visa. Surface delivery (U.S. and Canada only): add $5 for U.S., $8 for Canada. For delivery to Europe via U.S. airmail, add $14. Three-day air freight delivery: add $10 for U.S. (UPS Blue), $25 for Canada (Purolator overnight), $45 for Europe (Federal Express), or $60 for Asia and elsewhere in the world (Federal Express).

There is a multiline Circuit Cellar bulletin board system (running TBBS 2.0M) that supports past and present projects in BYTE and Ink. You are invited to call and exchange ideas and comments with other Circuit Cellar supporters. The 300/1200/2400-bps BBS is on-line 24 hours a day at (203) 871-1988.

9

THE SMART SPOOLER

PART 2: SOFTWARE AND OPERATION

SmartSpooler can function as a complete remote data-processing computer to analyze data

In part 1 of Chapter 9, I introduced my version of the ultimate printer buffer: SmartSpooler. While SmartSpooler has 256K bytes of memory and supports both serial and parallel printers, it also has features that improve its versatility and ease of use. These features include a "switchbox" capability (i.e., for routing serial or parallel computer input to a serial or parallel printer output), multiple-copy printing, a single-sheet-feeding mode, and buffer capacity indicators. Also, you can daisy chain multiple SmartSpoolers to control a whole network of peripherals.

Most important, SmartSpooler is intelligent. While I've presented it primarily as a printer buffer, SmartSpooler lets a host computer completely control its operation. You can even download executable code to SmartSpooler. Rather than being merely a simple printer buffer, SmartSpooler can function as a complete remote data-processing computer that analyzes and interprets the data flowing through it.

In part 2 of this chapter, I'll finish the hardware discussion by explaining the user interface and then briefly describe SmartSpooler's operation.

Push Buttons and LEDs

A 6821 peripheral interface adapter (PIA) neatly connects four front-panel push buttons, eight front-panel LEDs, and an eight-position DIP switch. The PIA, which is IC24 in figure 1, provides two 8-bit ports (port A and port B) and four multipurpose handshaking lines (CA1, CA2, CB1, and CB2). The DIP switches connect to port A, which is programmed as an 8-bit input port. The switches specify options like data transfer rates and operating modes. Figure 2 contains switch settings and functions.

The four front-panel push buttons set configuration, pause the output, enter copy requests, and clear present settings. The combination of a simple resistor/capacitor circuit and a Schmitt-trigger inverter debounces each push-button input. The conditioned inputs are connected to the handshaking lines, which are programmed so that any switch closure will generate a CPU interrupt.

The eight LEDs with current-limiting resistors connect to port B. SmartSpooler sets this port up as an 8-bit output port. (Be aware that port B has the extra 10-milliampere current needed to drive the LEDs.) Four of the LEDs signify SmartSpooler's operating mode: Config, Copy, Pause, and Clear.

The remaining four LEDs indicate which ports are enabled: Parin (parallel in), Serin (serial in), Parout (parallel out), and Serout (serial out). During initial setup (Clear), these LEDs display the I/O port configuration (serial/parallel). When SmartSpooler is making copies (Copy), these LEDs display the number of copies requested (1 to 4) and then the number of copies remaining to be printed. During normal operation, the LEDs indicate how full the SmartSpooler buffer is: 0 percent, 25 percent, 50 percent, 75 percent, or 100 percent.

The two DB-25S IBM PC–compatible parallel printer input and output connectors are mounted on the printed circuit board. I've mounted two 20-pin headers behind these connectors for the serial ports. These headers accommodate a pair of optional ribbon cables with DB-25P serial connectors on the end. You can operate SmartSpooler with either or both pairs of connectors installed. If you need only parallel-to-parallel operation, you use only the DB-25S connectors.

Buffer-Manager Software

The basic algorithm at the core of SmartSpooler is a FIFO buffer manager. A FIFO, whether a single chip or a box like SmartSpooler, consists of an input port and an output port, connected by a buffer memory (perhaps 256 bytes for a FIFO chip and 256K bytes for SmartSpooler). The buffer memory decouples the input and output data rates: fast dump from the computer, slow dump to the printer.

A good analogy for a FIFO is a water tank with fill (input) and drain (output) pipes, each pipe having a pump (see figure 3). The input can pump faster than the output, so the rate difference is absorbed as the tank fills. As is true with the water tank, the FIFO has to handle two special cases: full and empty. When the tank is full or empty, the respective pump (input or output) should be turned off.

We can immediately dismiss the intuitive software algorithm for implementing a FIFO (i.e., actually moving the data). While suitable for very small FIFOs, such an algorithm would choke on a full 256K-byte buffer. Instead, we use a scheme called a *ring buffer*, which manipulates input and output pointers, instead of actually moving the data (see figure 4).

Interrupts and Direct Memory Access

You can divide the implementation of the FIFO into three components: determining when an I/O port is ready for transfer, performing the transfer, and updating the pointers.

You could use a pure software approach in which you poll the I/O ports for readiness, transfer data with `IN` and `OUT` instructions, and have the program update buffer pointers. However, this scheme has some problems.

First, polling is extremely inefficient. Consider the typical case of simultaneous high-speed input and low-speed output. For each input character, you have to check whether the output is ready, even though it normally won't be. Actually, it's much worse, since you also have to check for handshaking, front-panel switch closure, and a number of other mundane events. The overhead adds up quickly, limiting performance.

Second, software to access the buffer and maintain the buffer pointers is difficult to write for buffer sizes larger than 64K bytes. Eight-bit CPUs like the HD64180—and even 16-bit CPUs like the 8086/80286—must monitor each access and calculate memory-management-unit (HD64180) or segment-register (80286) reload values to manage a large buffer.

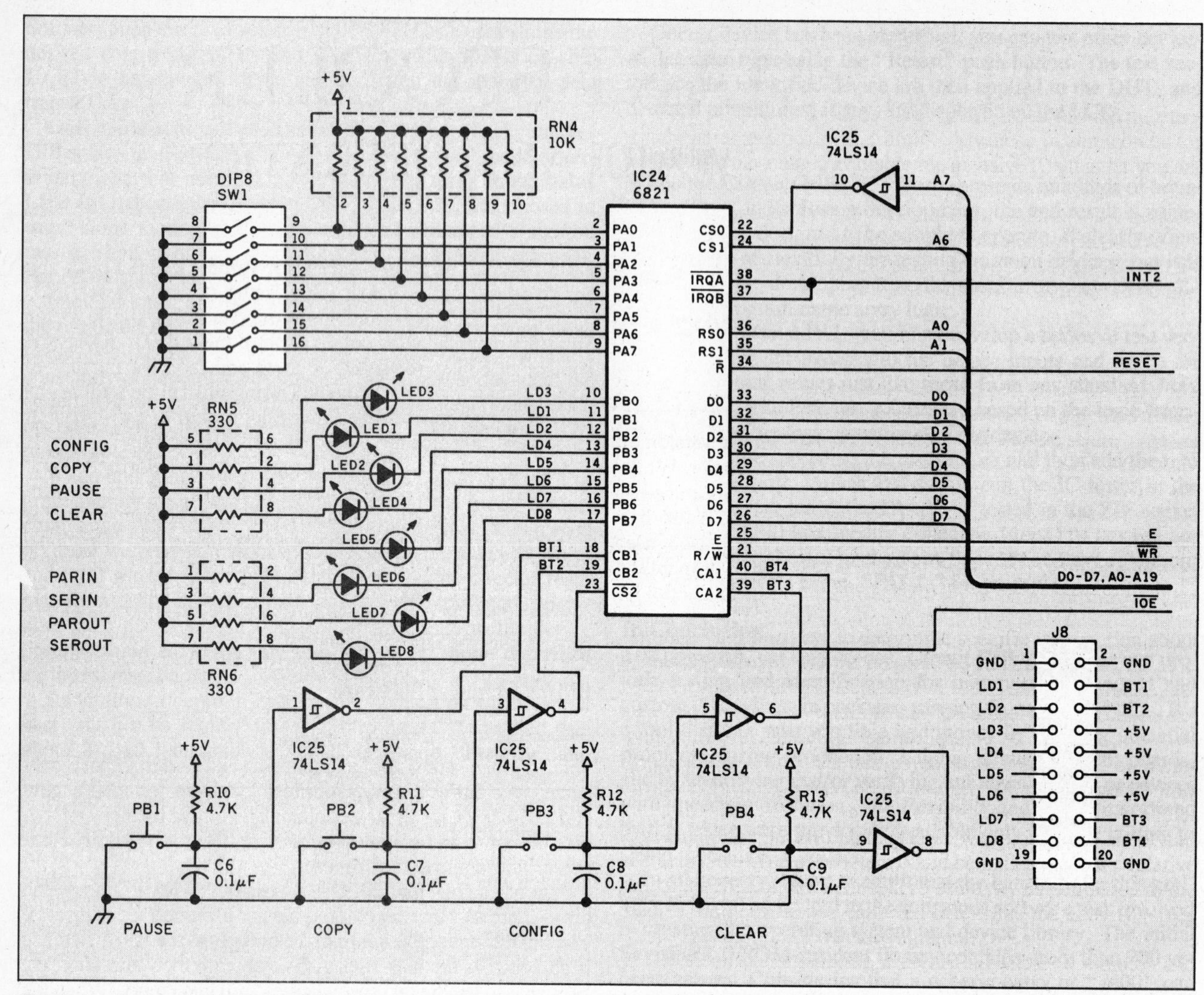

Figure 1: *Schematic for SmartSpooler's DIP-switch and LED I/O circuitry. A single 6821 peripheral-interface-adapter chip handles all the switches and lights.*

Switch	Off/On	Local	Host (Mode)	Test
(MSB) 1		IBR2	On	On
2		IBR1	On	On
3		IBR0	On	On
4		OBR2	Note	On
5		OBR1	Note	On
6		OBR0	Note	On
7		IPORT	Note	On
(LSB) 8		OPORT	Off	On

(Shipping position, all off: Parallel → parallel mode)

Local mode

	38,400	19,200	9600	4800	2400	1200	300	N/A
I/O BR2	Off	Off	Off	Off	On	On	On	On
I/O BR1	Off	Off	On	On	Off	Off	On	On
I/O BR0	Off	On	Off	On	Off	On	Off	On

I/O port Off = parallel, On = serial

Host mode

Note: Positions 4 to 7 off = last node in chain

Test mode

Push any button → memory test

Terminal on serial output port (<cr> . . <cr>) → monitor

Figure 2: *DIP-switch settings and their effects on SmartSpooler's various operating modes.*

Figure 3: *A FIFO buffer is like a water tank, absorbing the difference between input and output rates. In a typical printer buffer task, the input is faster than the output. A basic function of the FIFO/tank is to control the pumps for the special cases of tank full and tank empty.*

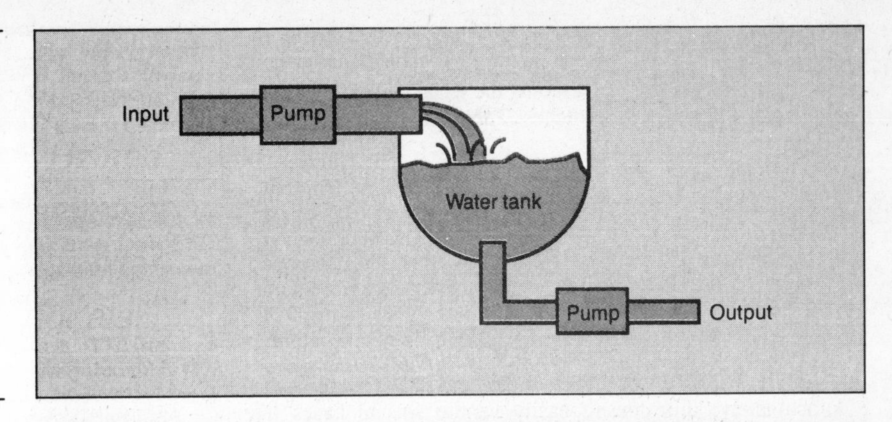

Figure 4: (a) *SmartSpooler's FIFO buffer operates using a ring-buffer algorithm that manipulates input and output data pointers rather than actually moving the data. The flowcharts shown here illustrate two key functions:* **(b)** *handling full and empty conditions and* **(c)** *checking for buffer wraparound.*

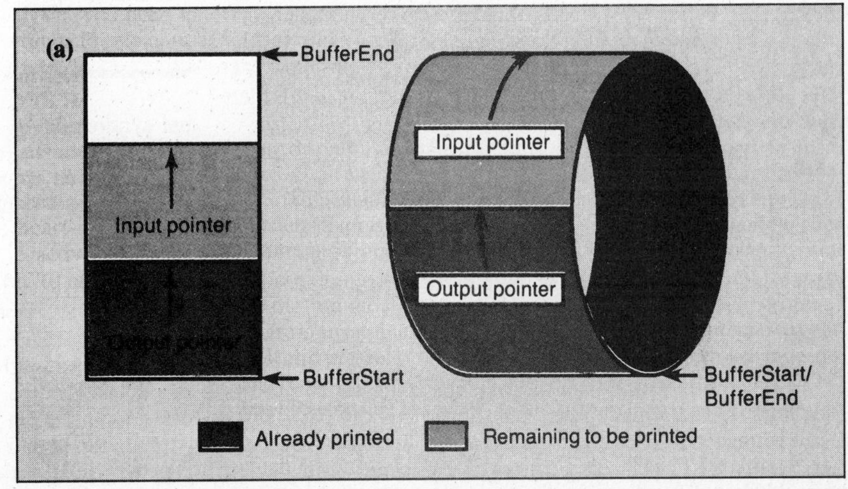

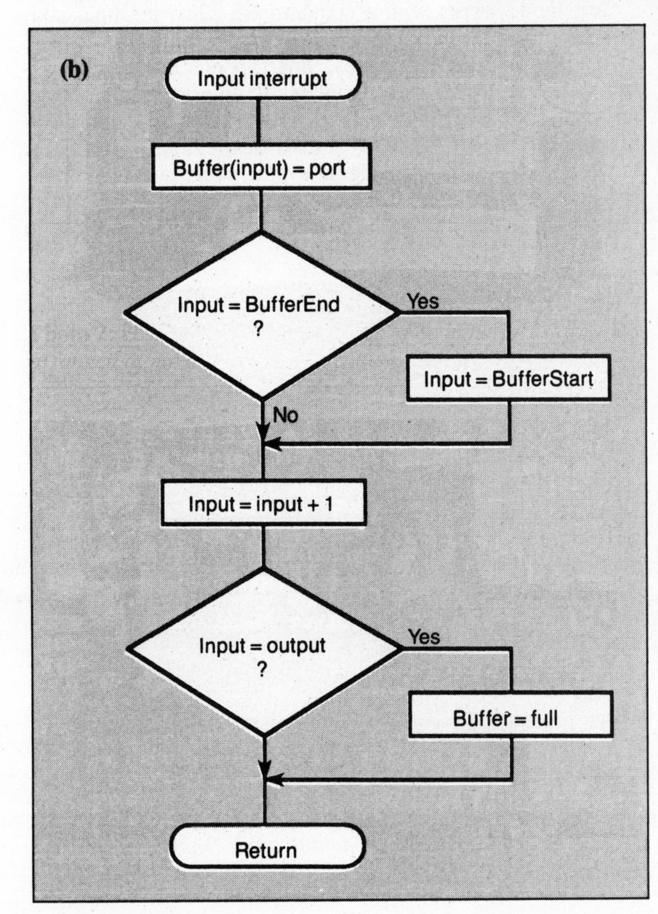

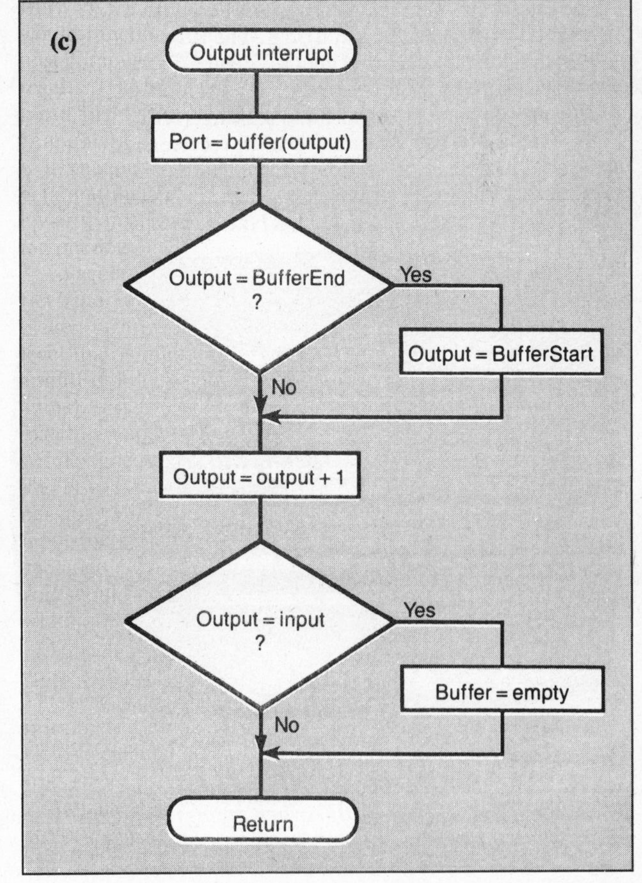

The solution to the problem of inefficiency is to exploit the HD64180's excellent interrupt capabilities. SmartSpooler's I/O is totally interrupt-driven, including the serial and parallel ports as well as the front-panel switches. Table 1 lists these interrupt assignments.

To solve the problem of large buffer maintenance, SmartSpooler uses the HD64180's direct-memory-access controller (DMAC), which has direct access to the entire physical address space. Besides performing the actual `IN` and `OUT` operations, the DMAC maintains the buffer pointers (using built-in DMA address registers with auto-increment). Channel 0 is the input channel, configured to perform I/O-to-memory DMA. Channel 1, as output, is configured for memory-to-I/O DMA.

Usually, I/O DMA occurs by request of the I/O device itself. Unfortunately, this doesn't easily handle special cases like buffer full, buffer empty, Pause button, copies, and handshaking. To get more flexibility, SmartSpooler uses a "soft" DMA technique. The HD64180 DREQ inputs are connected to CPU I/O ports instead of directly to the I/O peripheral ports. This lets the software initiate DMA.

Table 1: *SmartSpooler takes full advantage of the HD64180's interrupt capabilities. Note that the NMI and INT0 signals are also gated with XBUS inputs, allowing I/O expansion boards to use them.*

NMI: Not used (parallel output port ERROR input is optional)
INT0: Parallel input port
INT1: Parallel output port
INT2: Front-panel switches
PRT(Timer)0: Software delay timer
PRT(Timer)1: Real-time timer
DMA0: Not used
DMA1: Not used
CSI/O: Not used
ASCI(UART)0: Serial input port
ASCI(UART)1: Serial output port

Hands Across the Buffer

Both input and output ports need to provide handshaking. On the input side (host to SmartSpooler), the host must be signaled to stop when the buffer fills, to prevent overflow (remember the water tank example). On the output side (SmartSpooler to printer), SmartSpooler needs to pause when the printer is busy printing or goes off-line. In the formal world of data communications, this is known as flow control.

Flow control requires handshaking, which is a way of conveying start/stop information between the various devices. Hardware handshaking uses extra signal lines dedicated to flow control. Software handshaking conveys flow-control information over the data channel itself.

The Centronics parallel interface uses hardware handshaking signals: STROBE, ACK, and BUSY. The RS-232C ports provide both hardware handshaking (RTS and CTS) and software handshaking (XON/XOFF).

The problem with serial handshaking (RTS, CTS, and XON/XOFF) protocols occurs when the receiver can't shut off the sender in time to prevent overflow. Those of you who have spent time trying to get terminals or computers to run at 19,200 or 38,400 bits per second know what I mean (the beeping terminal syndrome). Also, some sending devices check for handshaking only at the end of each line, rather than for each character. To avoid overflow, SmartSpooler's serial port drivers incorporate a 256-byte "pad," allowing plenty of time for handshaking delays.

Local, Test, and Host Modes

SmartSpooler's operating mode is determined at power-on by DIP-switch settings.

Local mode is the normal mode of operation. In this case, SmartSpooler enters the default port configuration (parallel to parallel) and is ready to spool incoming data. Using the front-panel switches and LEDs, you can enter commands to change the port configuration, pause the output, and request copies.

Test mode works with a standard RS-232C terminal connected to one of the serial ports (see photos 1a and 1b). Instead of entering the spooler routines, SmartSpooler executes a built-in monitor program, which contains routines to test the ports, switches, and LEDs, as well as a complement of traditional monitor commands (`display`, `enter memory`, and so on). Test mode is useful for diagnosing hardware, cable, and host driver software problems.

Host mode lets the host computer download commands to

(a)

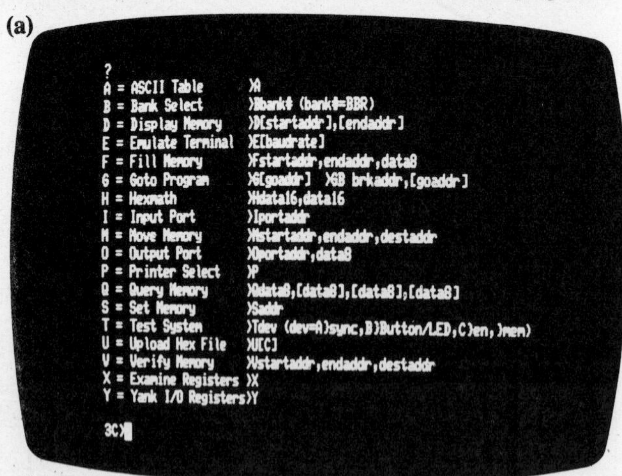

(b)

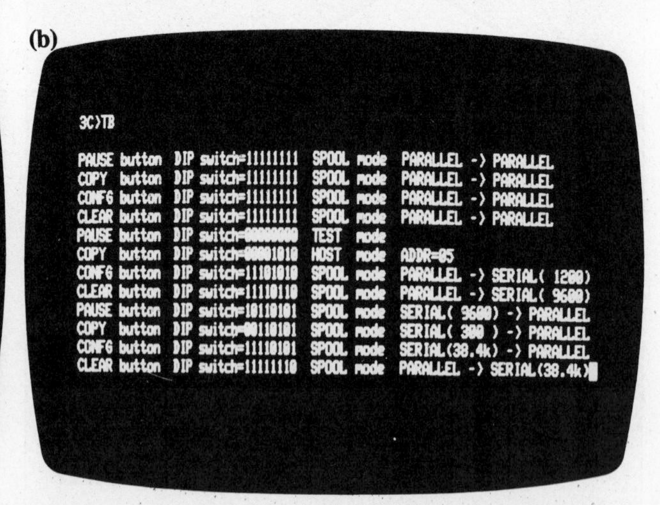

Photo 1: (a) *This screen shows the commands available while SmartSpooler is in test mode.* **(b)** *From this screen, you can individually test all four buttons, all eight LEDs, and all eight DIP switches. The right-hand column shows the SmartSpooler modes for several different DIP-switch settings. Whatever setting is last selected determines the default mode SmartSpooler enters when you reset it or power it up.*

Operation Summary for SmartSpooler

Power-on:
- Clear LED goes on.
- Config LED goes on, and the default port configuration is shown.
- Clear LED goes off, and the LEDs switch to show buffer capacity.
- SmartSpooler is ready for operation.

Changing the port configuration, changing the pause mode, and aborting printout:
- Push the Clear button.
- Push the Config button to change the port configuration and the Pause button to toggle the pause mode on/off.
- Push the Clear button (Clear LED off) to start SmartSpooler operation.
- The LEDs switch from showing the port configuration to monitoring the buffer capacity.

Suspending/resuming printing:
- Push the Pause button to suspend printing.
- Push the Pause button again to resume printing.

Making copies:
- Press Clear prior to sending document to copy.
- Send the document to SmartSpooler.
- Push the Copy button.
- Push the Config button to select the number of copies (0 to 4) desired.
- Push the Copy button again (Config LED off) to finish the copy request.
- The Copy LED will remain on until all copies are printed.

Single-sheet printing:
- Make sure you've selected the pause mode during power-on or Clear setup.
- Make sure your computer transmits a formfeed character to SmartSpooler prior to each new page (including the first).
- When the Pause LED goes on, insert a new page into the printer.
- Push the Pause button to print the next page.

Check buffer capacity, port configuration, and the number of copies remaining:
- During normal operation (Clear LED off), the parallel/serial In/Out LEDs show percent full (0, 25, 50, 75, or 100) the SmartSpooler buffer is, and the Pause LED controls print suspend/resume.

SmartSpooler. One benefit of host mode is that it lets you use software—instead of switches and LEDs—to set the port configuration and serial port format (data transfer rate, start/stop bits, parity, and so on). In fact, you can remove SmartSpooler's switches, LEDs, and corresponding circuits if you never use SmartSpooler for local mode operation.

For the ultimate in versatility, the host can even download a new control program to totally replace SmartSpooler's control program. SmartSpooler's ROM vectors all HD64180 interrupts through a RAM-based vector table, letting the new control program take over interrupt response. Combining SmartSpooler's hardware with optional XBUS expansion boards (e.g., the Circuit Cellar GT180 color graphics display or the COMM180 modem/small-computer-system-interface [SCSI] board) and your own control program opens the door for lots of interesting applications.

Using SmartSpooler

SmartSpooler is easy to use. The following is a summary of specific button functions.

Pushing the Clear button stops any operation (I/O) in progress, initializes SmartSpooler, and lights the Clear LED. Any data in the buffer is lost upon Clear.

You use the Clear button in the following instances:

- to change the port configuration
- to change the pause mode
- before receiving a document that will be copied
- to cancel a printout
- to finish the Clear request (Clear, Config, and Pause LEDs off)

The Pause button has two functions, one after the Clear button is pushed and another during normal operation. After you've pushed the Clear button (Clear LED on), pushing Pause toggles the pause mode on or off. When pause mode is on (Pause LED on), SmartSpooler will suspend output after transmission of a formfeed character to the printer. Use this mode for single-sheet feeding; position the next sheet and push the Pause button (Pause LED off) to resume printing. When pause mode is off (Pause LED off), SmartSpooler will not check formfeed characters. Use this mode when printing continuous (i.e., platen or tractor-fed) forms.

Push the Copy button (Copy LED on) to make copies of everything SmartSpooler has received since the last Clear. Then, increment the number of copies desired by pushing the Config button (Config LED blinks).

The LEDs show how many copies are selected: One LED on means one copy (plus original); four LEDs on means four copies (plus original). After entering the number of copies desired, push the Copy button again (Config LED off) to complete the copy request.

The Config button toggles the I/O (serial or parallel) port configuration when the Clear LED is on. The configuration is reflected on the parallel/serial In/Out LEDs.

Two Functions

SmartSpooler is actually two projects. One of these is a high-performance printer buffer; the second is a configurable intelligent peripheral controller. Most people will assemble it as a printer buffer, but others will find applications ideally suited for the host programmed mode. While SmartSpooler is not a trivial project, you can extend it to perform tasks that separate it from a mere buffer. One possibility is to use SmartSpooler between a host computer and a modem to filter incoming data, initiate and time calls, or encrypt and decrypt data.

As a printer buffer, 256K bytes is more than adequate. However, as a specific-application peripheral controller, SmartSpooler might need additional capability.

As I mentioned previously, it is not inconceivable to add 8 or 16 additional I/O ports, a 20-megabyte SCSI hard disk drive, or the GT180 color graphics display to SmartSpooler through its XBUS expansion connector. The necessary hardware for such peripherals already exists for SmartSpooler from previous Circuit Cellar articles.

Experimenters

While printed circuit board kits for SmartSpooler are available, I encourage you to build your own. If you don't mind doing a little work, I will support your efforts as usual. You can download a hexadecimal file of the executable code for SmartSpooler's system EPROM (27128) from my bulletin board at (203) 871-1988. Alternatively, you can send me a preformatted IBM PC floppy disk with return postage, and I'll put all the files on it for you (add $6 for the SmartSpooler User's Manual). Of course, this free software is for noncommercial personal use only.

I'd like to personally thank Tom Cantrell for his extensive work on this project. Without his software expertise, I'd be hopelessly mired in an ocean of bits forever.

Editor's Note: Steve often refers to previous Circuit Cellar articles. Most of these past articles are available in book form from BYTE Books, McGraw-Hill Publishing Company, P.O. Box 400, Hightstown, NJ 08250, (1–800–2–MCGRAW).

Ciarcia's Circuit Cellar, Volume I covers articles in BYTE from September 1977 through November 1978. *Volume II* covers December 1978 through June 1980. *Volume III* covers July 1980 through December 1981. *Volume IV* covers January 1982 through June 1983. *Volume V* covers July 1983 through December 1984. *Volume VI* covers January 1985 through June 1986.

It's virtually impossible to provide all the pertinent details of a project or cover all the designs I'd like to in the pages of BYTE. For that reason, I have started a bimonthly supplemental publication called Circuit Cellar Ink, which presents additional information on projects published in BYTE, new projects, and supplemental applications-oriented materials. For a one-year subscription (6 issues), send $14.95 to Circuit Cellar Ink. Credit card orders can call (203) 875-2199.

The following items are available from

Circuit Cellar, Inc.
4 Park St., Suite 12
Vernon, CT 06066
(203) 875-2751

1. SmartSpooler RAMless full printed circuit board kit for parallel-to-parallel operation. Comes complete with printed circuit board and all printed-circuit-board-mounted components, including HD64180, IC sockets, SmartSpooler software on EPROM, and User's Manual. Less DRAM chips. SS01-KIT $119

Optional serial port ribbon cables are available. Call for pricing.

The SmartSpooler design is available for licensing to qualified manufacturers. Call for information.

All payments should be made in U.S. dollars by check, money order, MasterCard, or Visa. Surface delivery (U.S. and Canada only): add $5 for U.S., $8 for Canada. For delivery to Europe via U.S. airmail, add $14. Three-day air freight delivery: add $10 for U.S. (UPS Blue), $25 for Canada (Purolator overnight), $45 for Europe (Federal Express), or $60 for Asia and elsewhere in the world (Federal Express).

There is a multiline Circuit Cellar bulletin board system (running TBBS 2.0M) that supports past and present projects in BYTE and Ink. You are invited to call and exchange ideas and comments with other Circuit Cellar supporters. The 300-/1200-/2400-bps BBS is on-line 24 hours a day at (203) 871-1988.

10

BUILD AN INTELLIGENT SERIAL EPROM PROGRAMMER

Steve's new and improved device includes on-board CPU and intelligent firmware

I don't like admitting that I made a mistake, but apparently I did. Well, not actually. You see, I was dragged into . . . Let me start from the beginning.

My February 1985 Circuit Cellar article was a project on how to build a serial EPROM programmer, about which I said: "The latest Circuit Cellar EPROM programmer is a serial-port programmer that has the speed of a turtle, the intelligence of the mightiest computer (that is, it has absolutely no smarts of its own), and is as functional as a doorstop between uses. On the positive side, it's fully documented, universally applicable, and easily expandable to accommodate future EPROM types."

What a mess after it was published! Everybody must have built this programmer. BYTEnet almost shut down the Peterborough phone company as people downloaded the BASIC listings, and my staff developed "postage tongue" replying to the correspondence. Needless to say, the project was well received.

In truth, it was an experimenter's project intended to satisfy a certain core of supporters yet enlighten the larger audience of readers about EPROM programming in general. Because I could not gauge its potential reception, and also because I didn't see it as having any greater performance than low-cost bus-compatible programmer boards, I didn't arrange to have it made into a printed circuit board as are most of my projects. I'm embarrassed to say that even after all these years I underestimated the number of experimenters who wanted to build a serial EPROM programmer.

It's too late to go back now, but I have to make up for past indiscretion and find some way to save face. I know that there are warmed soldering irons all across the country waiting for me to apologize appropriately. I trust you'll accept this improved rendition on an old theme as proper recompense.

As the title indicates, this programmer is still intended for serial-port operation. Thus, it retains computer and bus independence. The primary difference between then and now, however, is the addition of a microprocessor that greatly enhances its functions. The new Circuit Cellar intelligent serial EPROM programmer (CCSP for short) programs more types of EPROMs faster and more reliably. It also functions as a stand-alone programmer for copying or verifying EPROMs. (See photo 1.) The following is a list of CCSP features:

- RS-232–compatible (no handshaking necessary)
- internal V_{pp} power generation
- menu-selectable EPROM types (no programming configuration jumpers)
- default power-up selectable data rates
- automatic power-down of EPROM for installation/removal
- stand-alone or computer system/terminal–connected operation
- menu-driven operation
- single-byte or full-buffer write modes
- 32K-byte on-board memory buffer
- read, copy, or verify EPROM
- Intel hexadecimal file upload/download
- verify after write
- verify EPROM erasure
- screen dump by page or byte
- BASIC driver that can be modified by the user
- program EPROMs in standard 50-millisecond and 1-ms fast algorithm modes
- support V_{pp} settings of 25, 21, and 12.5 volts
- program all 27xxx 5-V single-supply EPROMs, including 2716, 2732, 2732A, 2764, 2764A, 27C64, 27128, 27128A, 27C128, 27256, 27512, and any functional equivalents

Obviously, a list this impressive would take a great deal of effort to put together as a single chapter's project. The potential software development nightmares of assembly language serial drivers, menu displays, and table manipulations hardly made it worth adding a microprocessor to my original BASIC-language-manipulated unit. Besides, how could it be done in one month?

BASIC allowed a significant level of interactive menus and help displays while requiring little software overhead. Unfortunately, using a high-level-language interpreter to simplify software development is of little value when the primary goal of producing a better programmer requires fast data manipulations that are best accomplished in assembly language.

Rather than be thwarted by this apparent dilemma, I decided to design a hybrid system that used both BASIC and assembly language. The obvious choice was the BASIC-52 computer/controller I presented in the August 1985 project. With the help of software guru and friend Bill Curlew, the CCSP was designed, built, and tested in two weeks flat.

The CCSP uses an Intel 8052AH-BASIC microprocessor that contains an 8K-byte ROM-based BASIC interpreter. Besides manipulating strings, tables, and menus, the BASIC contains serial communication drivers and easily links to assembly language routines. It seemed the perfect engine for a quickly designed user-modifiable project.

Photo 1: *Finished printed circuit prototype of the serial EPROM programmer. The digital section and memory buffer are at the bottom center. The analog-voltage-level switching section is at the top center.*

A Hybrid Approach

The CCSP is a stand-alone microcomputer with an application-specific I/O configuration. It supports 40K bytes of operating system and buffer RAM and 16K bytes of program ROM. It uses six parallel I/O ports to drive the programming-pin level-shifter voltage-control circuitry, EPROM address and data lines, and user-interactive buttons and display. The CCSP can be used by itself to copy EPROMs or, when connected to a terminal or computer, as a full-function programmer/verifier. It requires no programming jumpers or personality modules and is completely automatic. It programs/examines/verifies all 5-V EPROMs from 2716s through 27512s in both standard and fast modes (on applicable devices).

In the sections that follow, I'll describe the configuration of the microcomputer and its unique I/O structure. Once you have the hardware in hand, I'll describe the system software and how the different modes operate. First, I'll go over some EPROM basics.

A Review

A personal computer, even in its minimum configuration, always contains some user-programmable memory, or RAM, usually in the form of semiconductor-memory integrated circuits. This memory can contain both programs and data. Any machine-word-level storage element within the memory can be individually read or modified (written) as needed.

Any of several kinds of electronic components can function as bit-storage elements in this kind of memory. TTL-type 7474 flip-flops, bistable relays, or tiny ferrite toroids (memory cores) are suitable, but they all cost too much, are hard to use, and have other disadvantages.

In personal computer and other microprocessor-based applications, the most cost-effective memory is made from MOS integrated circuits. Unfortunately, data stored in these semiconductor RAMs is volatile. When the power is turned off, the data is lost. Many ways of dealing with this problem have been devised, with essential programs and data usually stored in some nonvolatile medium.

In most computer systems, some data or programs are stored in ROM. A semiconductor ROM can be randomly accessed for reading in the same manner as the volatile memory, but the data in the ROM is permanent. The data in a mask-programmed ROM is determined during the manufacturing process. Whenever power is supplied to the ROM, this permanent data (or program) is available. In small computer systems, ROM is chiefly used to contain operating systems and/or BASIC interpreters—programs that don't need to be changed.

Another type of ROM is the PROM, which is delivered from the factory containing no data. The user decides what data to put in it and permanently programs it with a special device. Once programmed, PROMs exhibit the characteristics of mask-programmed ROMs. You might label such PROMs "write-once" memories.

The ultraviolet-light erasable EPROM is a compromise between the "write-once" kind of PROM and volatile memory. You can think of the EPROM as a "read-mostly" memory, used in read-only mode most of the time but occasionally erased and reprogrammed as necessary. The EPROM is erased by exposing the silicon chip to ultraviolet light at a wavelength of 2537 angstroms. Conveniently, most EPROM chips are packaged in an enclosure with a transparent quartz window.

HOW AN EPROM WORKS

EPROMs from several manufacturers store data bits in cells formed from stored-charge FAMOS (floating-gate avalanche-injection metal-oxide semiconductor) transistors. Such transistors are similar to positive-channel silicon-gate field-effect transistors, but with two gates. The lower or floating gate is completely surrounded by an insulating layer of silicon dioxide, and the upper control or select gate is connected to external circuitry.

The amount of electric charge stored on the floating gate determines whether the bit cell contains a 1 or a 0. Charged cells are read as 0s; uncharged cells are read as 1s. When the EPROM chip comes from the factory, all bit locations are cleared of charge and are read as logic 1s; each byte contains hexadecimal FF.

When a given bit cell is to be burned from a 1 to a 0, a current is passed through the transistor's channel from the source to the gate. (The electrons, of course, move the opposite way.) At the same time, a relatively high-voltage potential is placed on the transistor's upper select gate, creating a strong electric field within the layers of semiconductor material. (This is the function of the +12.5-V, +21-V, or +25-V V_{pp} charging potential applied to the EPROM.) In the presence of this strong electric field, some of the electrons passing through the source-drain channel gain enough energy to tunnel through the insulating layer that normally isolates the floating gate. As the tunneling electrons accumulate on the floating gate, it takes on a negative charge, which makes the cell contain a 0.

When data is to be erased from the chip, it is exposed to ultraviolet light, which contains photons of relatively high energy. The incident photons excite the electrons on the floating gate to sufficiently high-energy states that they can tunnel back through the insulating layer, removing the charge from the gate and returning the cell to a state of 1.

When data is to be erased from the chip, it is exposed to ultraviolet light.

The 27xxx EPROMs contain bit-storage cells configured as individually addressable bytes. This organization is often called "2K by 8" for a 2716 or "8K by 8" for a 2764. The completely static operation of the device requires no clock signals. The primary operating modes include read, standby, and program (program-inhibit and program-verify modes are important primarily in high-volume applications).

Control inputs are used to select the chip and configure it for one of these operating modes. In the program mode, particular bit cells are induced to contain 0 values. Both 1s and 0s are in the data word presented on the data lines, but only a 0 causes action to take place. For example, the 27128 is in the programming mode when V_{pp} input is at 21 V and CE and PGM are both at TTL low. The data to be programmed is applied 8 bits in parallel to the data output pins. For regular programming, CE should be kept TTL low at all times while V_{pp} is kept at 21 V. When the address and data are stable, a 50-ms (55 ms maximum) active-low TTL program pulse is applied to the $\overline{PGM}$ input. A program pulse must be applied at each address location to be programmed.

STANDARD VS. FAST

In the old days, all we had to contend with were 50-ms timing pulses (neglecting obsolete 1702 and 2708 EPROMs). Today, the newest EPROMs can use a fast closed-loop programming algorithm that lessens programming time (realize that a 27512 takes about 1 hour to program in 50-ms increments). The CCSP supports fast programming.

The fast algorithm uses closed-loop margin checking. To ensure reliable program margin, the fast algorithm utilizes two different pulse types: initial and overprogram. The duration of the initial PGM pulse(s) is 1 ms, which will then be followed by a longer overprogram pulse of length 4*x* ms; some chip types use 3*x* (*x* is the number of initial 1-ms pulses applied to a particular location before a correct verify occurs). Once it is verified, four times that number of pulses are applied to the same location to permanently burn the data. If 15 (some chip types require 25 pulses) 1-ms pulses are applied to any single-byte location without reaching the margin, the overprogram pulse is applied automatically.

The entire sequence of program pulses and byte verifications is performed at V_{cc} = 6.0 V and V_{pp} = 21.0 V (V_{pp} may be 12.5 V on some EPROMs). When the fast programming cycle has been completed, all bytes should be compared to the original data with $V_{cc} = V_{pp}$ = 5.0 V.

The fast algorithm may be the preferred programming method since it allows certain EPROMs to be programmed in significantly less time than the standard 50-ms-per-byte programming routine. Typical programming times for 27128s, for example, are on the order of 2 minutes, a sixfold reduction in programming time from the standard method.

(a)

PIN #	2716	2732	2764	27128	27256	27512
1	N/C	N/C	VPP 21V 12.5V 5V	VPP 21V 12.5V 5V	VPP 12.5V 5V	A15 5V 0V
28	N/C	N/C	VCC* 6V 5V	VCC* 6V 5V	VCC* 6V 5V	VCC* 6V 5V
27	N/C	N/C	$\overline{\text{PGM}}$ 5V 0V	$\overline{\text{PGM}}$ 5V 0V	A14 5V 0V	A14 5V 0V
26/24	VCC 5V	VCC 5V	N/C	A13 5V 0V	A13 5V 0V	A13 5V 0V
23/21	VPP 25V 5V	A11 5V 0V	A11 5V 0V	A11 5V 0V	A11 5V 0V	A11 5V 0V
22/20	$\overline{\text{OE}}$ 5V 0V	$\overline{\text{OE}}$/VPP 25V 21V 0V	$\overline{\text{OE}}$ 5V 0V	$\overline{\text{OE}}$ 5V 0V	$\overline{\text{OE}}$ 5V 0V	$\overline{\text{OE}}$/VPP 12.5V 0V
20/18	$\overline{\text{CE}}$ 5V 0V	$\overline{\text{CE}}$ 5V 0V	$\overline{\text{CE}}$ 5V 0V	$\overline{\text{CE}}$ 5V 0V	$\overline{\text{CE}}$ 5V 0V	$\overline{\text{CE}}$ 5V 0V

*6.0 VOLTS ONLY ON EPROMS THAT ALLOW FAST PROGRAMMING

(b)

27512	27256	27128	2764	2732A	2716	27XXX		2716	2732A	2764	27128	27256	27512
A15	VPP	VPP	VPP			1	28			VCC	VCC	VCC	VCC
A12	A12	A12	A12			2	27			$\overline{\text{PGM}}$	$\overline{\text{PGM}}$	A14	A14
A7	A7	A7	A7	A7	A7	3	26	VCC	VCC	N.C.	A13	A13	A13
A6	A6	A6	A6	A6	A6	4	25	A8	A8	A8	A8	A8	A8
A5	A5	A5	A5	A5	A5	5	24	A9	A9	A9	A9	A9	A9
A4	A4	A4	A4	A4	A4	6	23	VPP	A11	A11	A11	A11	A11
A3	A3	A3	A3	A3	A3	7	22	$\overline{\text{OE}}$	$\overline{\text{OE}}$/VPP	$\overline{\text{OE}}$	$\overline{\text{OE}}$	$\overline{\text{OE}}$	$\overline{\text{OE}}$/VPP
A2	A2	A2	A2	A2	A2	8	21	A10	A10	A10	A10	A10	A10
A1	A1	A1	A1	A1	A1	9	20	$\overline{\text{CE}}$	$\overline{\text{CE}}$	$\overline{\text{CE}}$	$\overline{\text{CE}}$	$\overline{\text{CE}}$	$\overline{\text{CE}}$
A0	A0	A0	A0	A0	A0	10	19	O7	O7	O7	O7	O7	O7
O0	O0	O0	O0	O0	O0	11	18	O6	O6	O6	O6	O6	O6
O1	O1	O1	O1	O1	O1	12	17	O5	O5	O5	O5	O5	O5
O2	O2	O2	O2	O2	O2	13	16	O4	O4	O4	O4	O4	O4
GND	GND	GND	GND	GND	GND	14	15	O3	O3	O3	O3	O3	O3

Figure 1: *(a)* EPROM *programming-pin functions by* EPROM *type. (b) The great* EPROM *pin-out maze, illustrating the configuration of those* EPROMs *the* CCSP *is designed to handle.*

CONFIGURATION MAZE

The first problem encountered in any EPROM programmer design is to compare the pins of the various EPROMs (see figure 1b). Among the 28 defined pins (four are unused on 24-pin devices), 21 are used for the same functions (address and data). Evidently, semiconductor manufacturers never thought very far ahead or talked to each other, because the remaining seven pins are a complicated switching maze. Among the different EPROMs, the same pin location can supply power, address, or programming pulses. Figure 1a illustrates the differences in detail.

In inexpensive programmers, configuration jumpers are frequently used to select the specific wiring configuration for different EPROM types. Wire jumpers rather than semiconductor switches are used because of the high currents involved. Take pin 26 (pin 24 on 24-pin EPROMs) with either a 2732 or 27128 installed, for example. In both cases, the voltage level is 5 V. On a 27128 it is a TTL A13 address line; on a 2732 it is a 150-milliampere V_{cc} power line. Similarly, pin 22 (all pin numbers are referenced to a 28-pin layout) has to be set at 0 V, 5 V, 12.5 V, 21 V, or 25 V at currents ranging from 400 microamperes to 50 mA, depending upon the EPROM.

Fortunately, only five of the seven configuration pins require elaborate voltage and current control. Rather than use mechanical jumpers, I designed a voltage-control circuit that could be preset to the voltage limits of the desired EPROM type and easily pass high current when required. Figure 2 illustrates this basic circuit that is duplicated for each of the five pins (pins 28, 26, 1, 22, and 23).

The level shifter uses an LM317 voltage regulator as a programmable voltage controller. The basic LM317 output voltage is set by two resistors: R1 between the adjustment pin and ground and R2 between the adjustment pin and the output. As the formula shows, with R1=665 ohms and R2=221 ohms, the output is 5.0 V.

In this configuration, various R1 resistors can be connected from the adjustment pin to ground through open-collector 7407 drivers. These were used since they operate at up to 30 V (don't substitute a 7417). The four drivers from top to bottom set 5 V, 12.5 V, 21 V, and 25 V, respectively (not all sections are required for each EPROM pin). Their inputs are fed by a parallel output port.

Normally, the regulated output of an LM317 is 1.2 V to 32 V. An additional two-transistor control circuit is added to allow the output to go to 0 V on command. Rather than providing a resistance path to ground, however, this is accomplished by applying a negative 1.2 V to the adjustment pin. Because there is no way to know how many of the control circuits will be set to 0 V at any one time or if the 7407 drivers are enabled concurrently, the −1.4-V bias source is itself a regulated supply.

The CCSP level-shifter circuit can simulate a variety of programmable conditions. For example, by setting the 7407 driver that limits the output to 5 V and pulsing the 0-V enable line, we have a TTL-level $\overline{\text{PGM}}$, $\overline{\text{OE}}$, or $\overline{\text{CE}}$ control line. (In the tests I conducted, the circuit easily responded to control input changes of 20 kilohertz with little overshoot on the output. At those speeds, however, the output filter capacitor should be small.) Since the circuit is also capable of supplying 500 mA at 5 V, it is also appropriate to use this same circuit to supply and control V_{cc}.

The heart of the CCSP is found in the analog switching system and the management of the seven control lines in figure 1a. While I haven't explained yet how these level shifters are individually controlled, it still seems appropriate to show how they are ultimately configured. Figure 3a demonstrates how they are connected to the ZIF socket (zero insertion force programming socket), and figure 3b outlines their power source connections.

I designed a circuit that could be preset to the voltage limits of the desired EPROM.

8-BIT MICROCOMPUTER INTELLIGENCE

As I mentioned earlier, the CCSP's intelligence is provided by an Intel 8052AH microcomputer. BASIC-52 is particularly suited for this application.

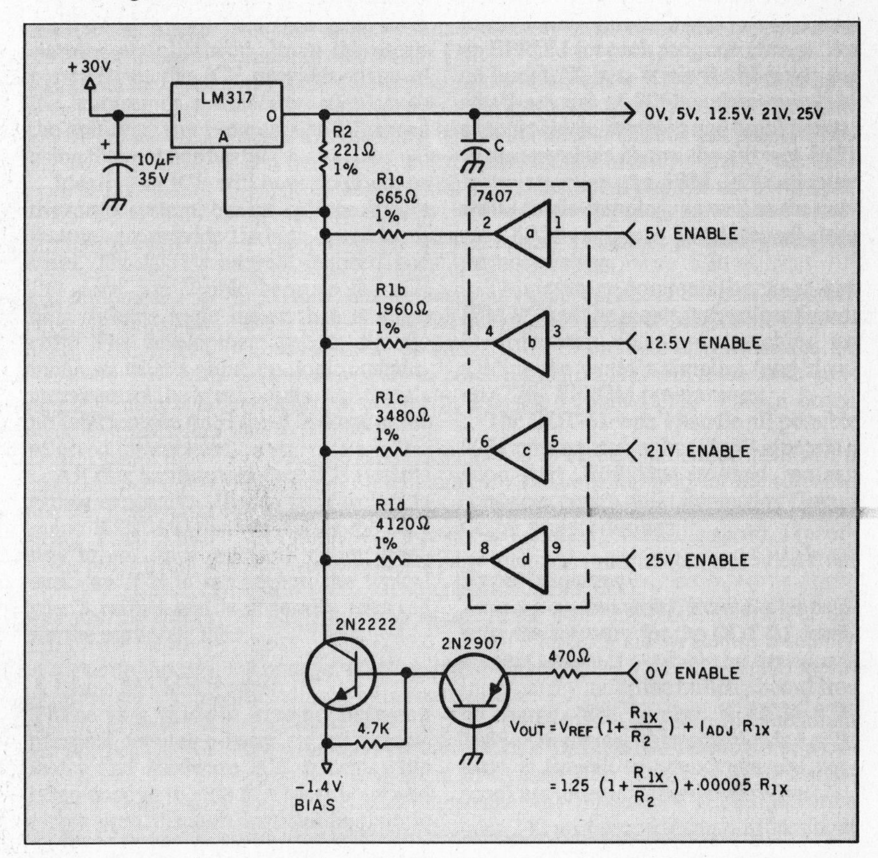

Figure 2: *Typical programmable voltage-control circuit for* EPROM *pins* 28, 26, 1, 22, *and* 23.

Three control lines—$\overline{\text{RD}}$, $\overline{\text{WR}}$, and $\overline{\text{PSEN}}$—are gated to allow 64K bytes of combined program and data memory.

providing IF. . .THEN, FOR. . .NEXT, DO. . .WHILE/UNTIL, ONTIME, and CALL statements as well as a broad repertoire of 8051 assembly language instructions. Calculations can be handled in integer or floating-point math.

The 8052AH contains an 8K-byte BASIC interpreter in ROM, 256 bytes of RAM, three 16-bit counter/timers, six interrupts, and 32 I/O lines that are redefined as a 16-bit address and an 8-bit data bus. A minimum of 1K byte of RAM is required for BASIC-52 to function, and any RAM must be located starting at 0000 hexadecimal. (I won't go into great detail on this computer since it closely resembles the BCC-52 presented in August 1985.) The microcomputer section of the CCSP is outlined in figure 4.

Three control lines—$\overline{\text{RD}}$ (pin 17), $\overline{\text{WR}}$ (pin 16), and $\overline{\text{PSEN}}$ (pin 29)—are gated to allow 64K bytes of combined program and data memory. The three most significant address lines (A13–A15) are connected to a 74LS138 decoder chip, IC4, which separates the addressable range into eight 8K-byte memory segments, each with its own chip select (Y0–Y7). The four most significant chip selects are connected to 8K-byte 6264 static RAMs, ICs 7–10. This area is the RAM buffer for reading or writing EPROMs. IC6, addressed at 0000 hexadecimal, must be another 6264 RAM for BASIC-52 to function. IC11 (2000–3FFF hexadecimal) contains the programmer software and is intended for either a 2764 or 27128.

All together, 56K bytes of memory are defined on the CCSP if you use five 6264 RAMs (as ICs 6–10) and a 27128 EPROM in IC11. To use the programmer, you need only the one RAM chip installed in IC6 (such a limited buffer area will require many passes to write or copy any large

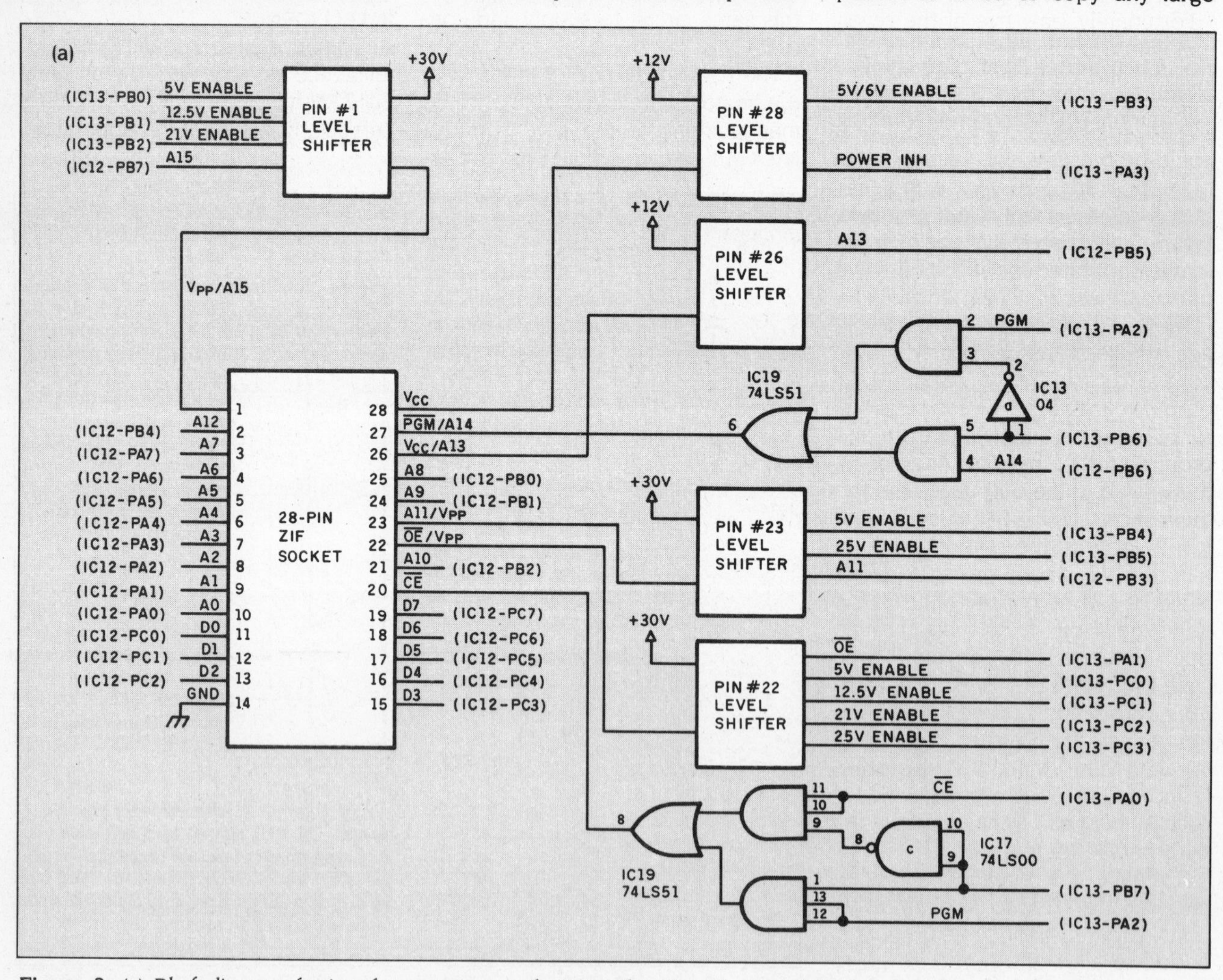

Figure 3: *(a) Block diagram showing the connections to the ZIF socket. Note the level-shifter circuitry connections for those pins that require programming voltages or that differ across EPROM types.*

EPROM). The memory cannot be expanded since the rest of the address space is decoded as I/O.

The address range of 6000-7FFF hexadecimal is divided into two I/O strobes at 6000 and 7000 through IC17. [*Editor's note: For the remainder of the chapter, all addresses will be in hexadecimal.*] Two 8255A-5 peripheral interface adapters providing three 8-bit I/O parallel ports each are controlled by a strobe line. The three I/O ports—labeled A, B, and C—and a write-only mode-configuration port on each 8255 occupy four consecutive addresses at 6000–6003 (IC12) and 7000–7003 (IC13), respectively. The ZIF socket and level-shifting circuitry outlined in figure 2 are connected to 41 of these parallel I/O bits. The lines attached to IC12 (the control PIA) are used primarily for presetting the level shifters and providing the programming pulses. IC13 (the address and data PIA) supplies the address and data bus lines to the EPROM. Figure 5 details the configuration and connection of the level shifters and power distribution.

The CCSP communicates with a terminal or host computer through an on-board serial port. The port's data rate is hard-coded in the program ROM and is preset at 1200 bits per second, but you can reprogram it to any standard value between 300 and 19,200 bps. (The 8052AH chip has the capability for automatic data-rate selection on the console port. Because the CCSP has both a local and a remote operating capability triggered by the GET command, the automatic data rate cannot be used.) MC1488 and MC1489 drivers/receivers (ICs 14 and 15) convert the 8052's serial I/O line TTL logic levels to RS-232.

Power supplies with the required output are readily available, so I excluded an on-board supply to keep costs lower.

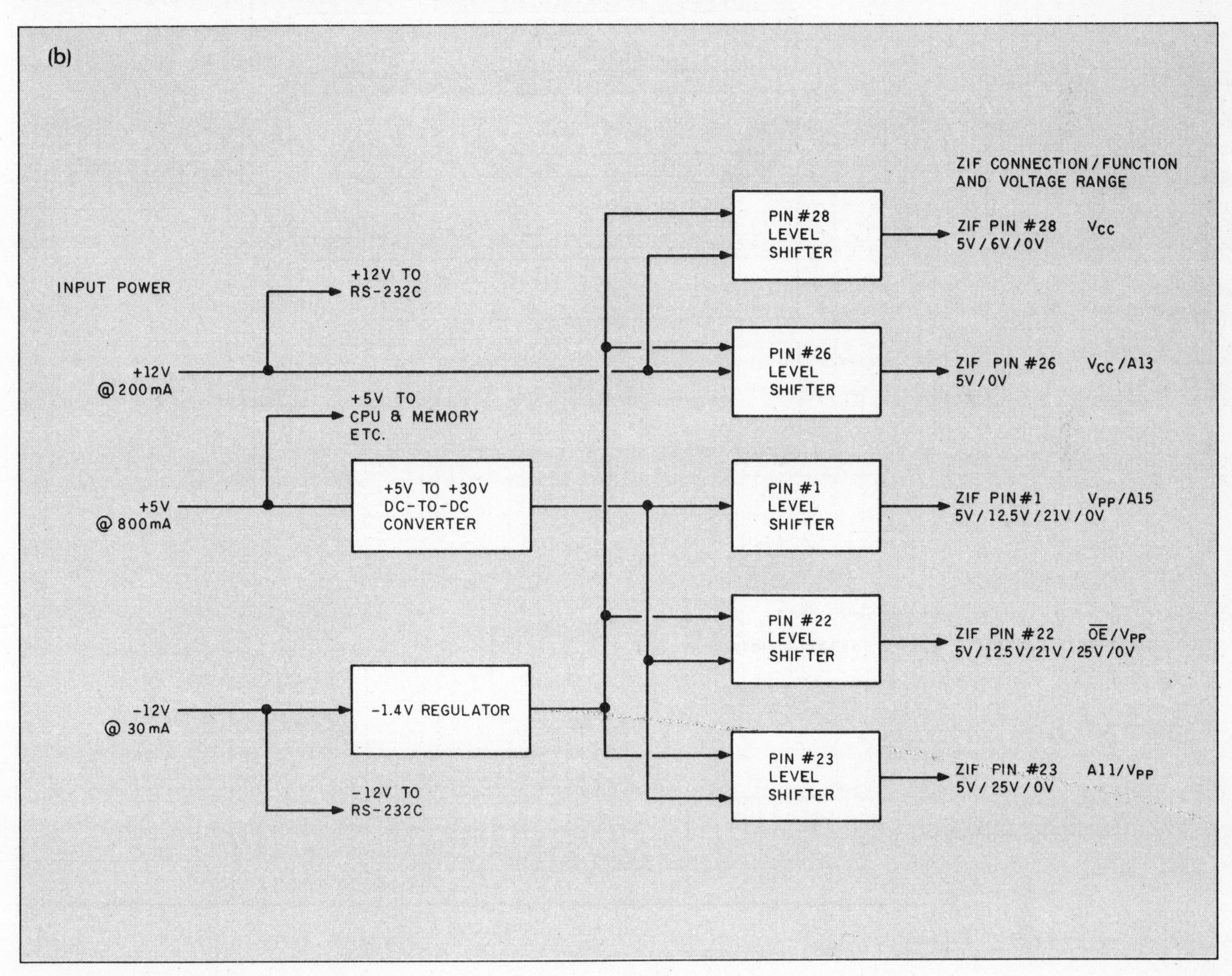

(*b*) CCSP *power-distribution block diagram.*

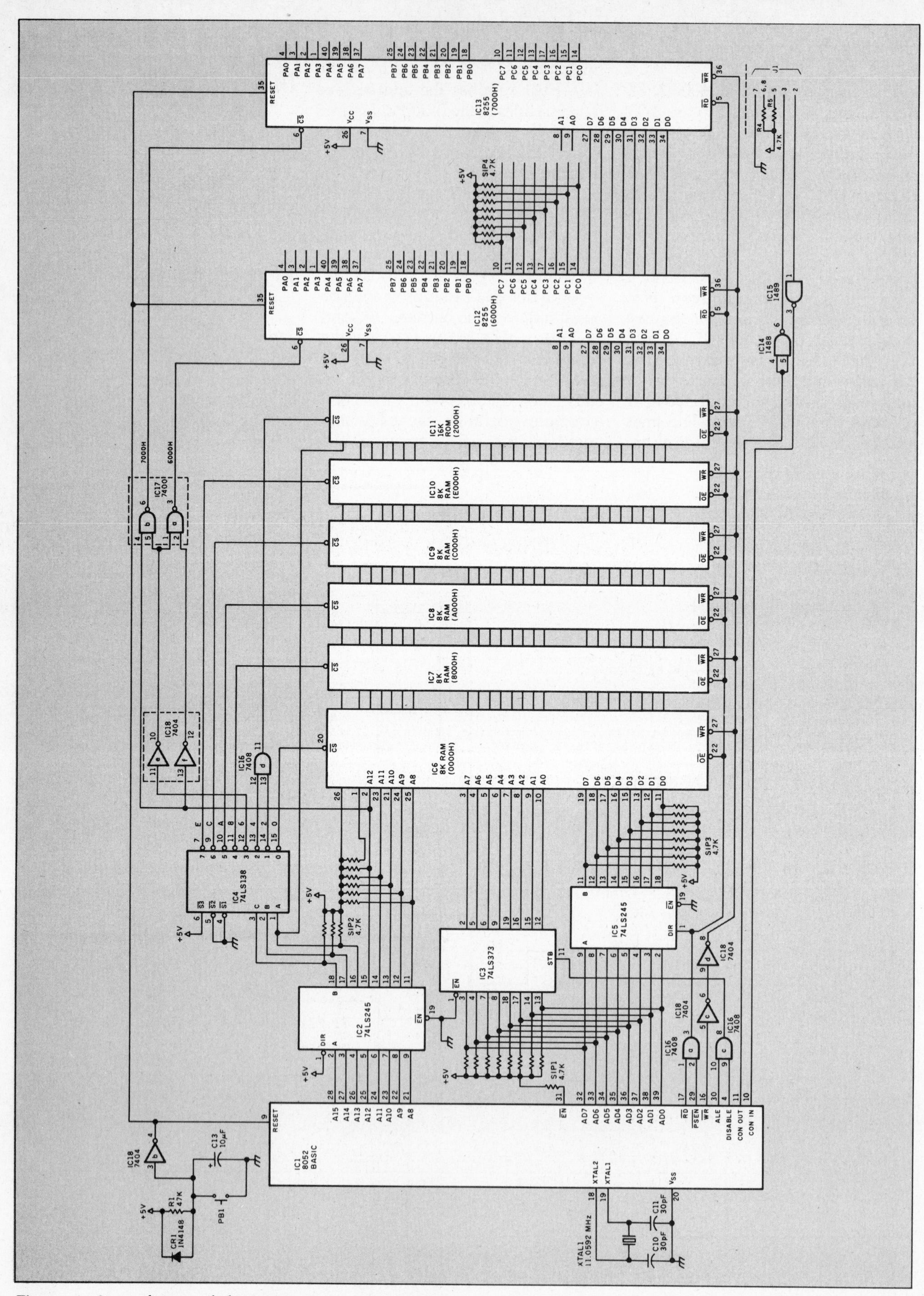

Figure 4: *Circuit diagram of the CCSP's microcomputer section.*

Power for the CCSP is provided by an external supply that must deliver +12 V at 200 mA, +5 V at 800 mA, and −5 V to −12 V at 30 mA. Power supplies with these outputs are readily available on the surplus market, so I excluded an on-board supply to keep costs lower. In fact, a perfect unit is the Coleco computer power supply available from Radio Shack for $4.95 (part #277-1022).

Three V_{pp} voltages must be contended with: 12.5 V, 21 V, and 25 V. All are derived from the +30-V output of the DC-to-DC converter circuit shown in figure 5. IC24 is a 78S40 switching regulator configured as a voltage multiplier. This circuit is capable of producing 30 V at 50 mA from a 5-V input. (For more information on this regulator and this specific circuit, see my November 1981 article, "Switching Power Supplies: An Introduction.")

The user entry/display interface is shown in figure 6. It consists of a two-button entry panel through which you operate the programmer in local mode, a local/remote LED indicator, EPROM power-on indicator, and a seven-segment display through which the computer displays EPROM type and errors. To save I/O bits, I used a somewhat unorthodox display driver rather than the usual parallel port and seven-segment decoder configuration. The seven-segment LED is attached to an 8-bit shift register that has each output connected to drive an individual segment and the decimal point. To display a character, the seven-segment information is extracted from a memory-resident table and quickly shifted into the shift register. Ordinarily, I wouldn't use such a software-intensive approach, but I didn't have to write the software.

PROGRAMMER SOFTWARE

The CCSP is controlled by a program that is a combination of BASIC and 8051 assembly language. The BASIC-52 program provides all initialization and control functions, including local mode support and menu processing in the remote mode. The assembly language routines are used only where speed is critical, as in reading, comparing, verifying erasure, and programming EPROMs. In addition, the Intel hexadecimal file upload and download routines are written in

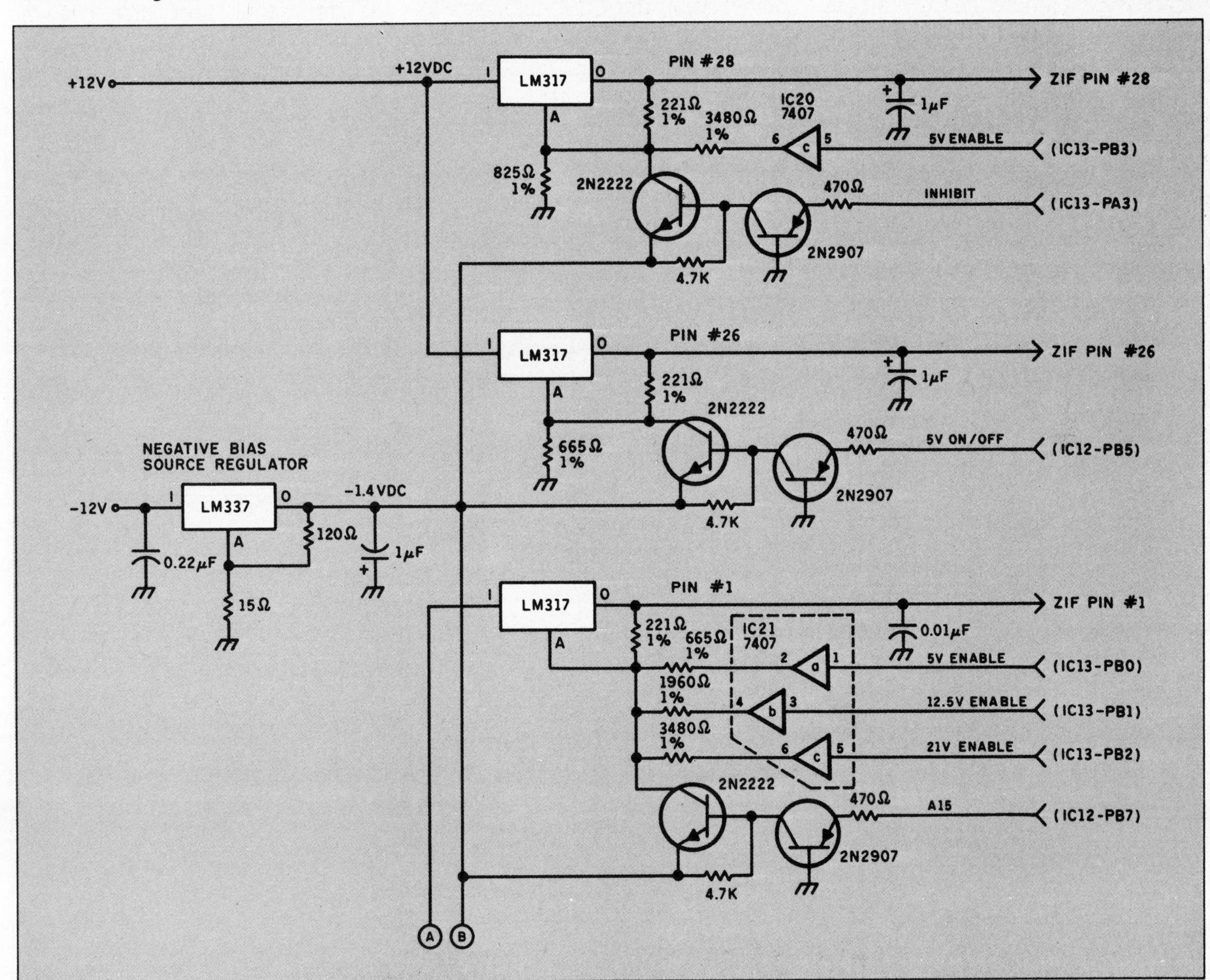

Figure 5: *Detailed schematic of the programmer's level-shifting circuitry.*

When working in local mode, you can copy any 27xxx EPROM by swapping the original and copy EPROMs multiple times.

assembly language to keep up with the attached ASCII terminal device. Figure 7 outlines the CCSP software logic flow.

The software that drives the CCSP is capable of running in two basic modes: local, where the CCSP acts as an EPROM copier controlled by buttons, and remote, where the CCSP acts as a full-featured programming workstation serially connected to the user's terminal. When working in local mode, you can copy any 27xxx EPROM regardless of its size by swapping the original and copy EPROMs multiple times. The larger the RAM buffer is, the fewer times you will have to change the EPROMs.

Power-up and Reset

When the CCSP is first powered up or reset, its software configures itself for a 2716 EPROM, the default type. After setting up the hardware, the software outputs a 0 in the seven-segment LED display to indicate the EPROM type, turns on the local mode LED, and sizes the RAM buffer.

If no memory is located at 8000 (the buffer area), the CCSP allocates 4K bytes of system RAM in IC6 as the buffer area. If it is unable to accomplish this, it will stop and display an alternating error code, E and 0, in the seven-segment LED display. Pressing a button or sending a character to the serial port will force the CCSP to reattempt sizing memory (memory sizing is destructive). If you have RAM chips plugged into locations IC7 through IC10, this will provide additional buffer memory. After memory is sized, the CCSP enters a loop to determine what mode you want the programmer to operate in.

During the mode-setting loop, the CCSP will decide if it is going to run in local or remote mode. The mode selected is determined by which event occurs first: If one of the buttons is pressed first, the CCSP establishes local mode; if a character is detected at the serial port first (via the BASIC-52 GET command), the pro-

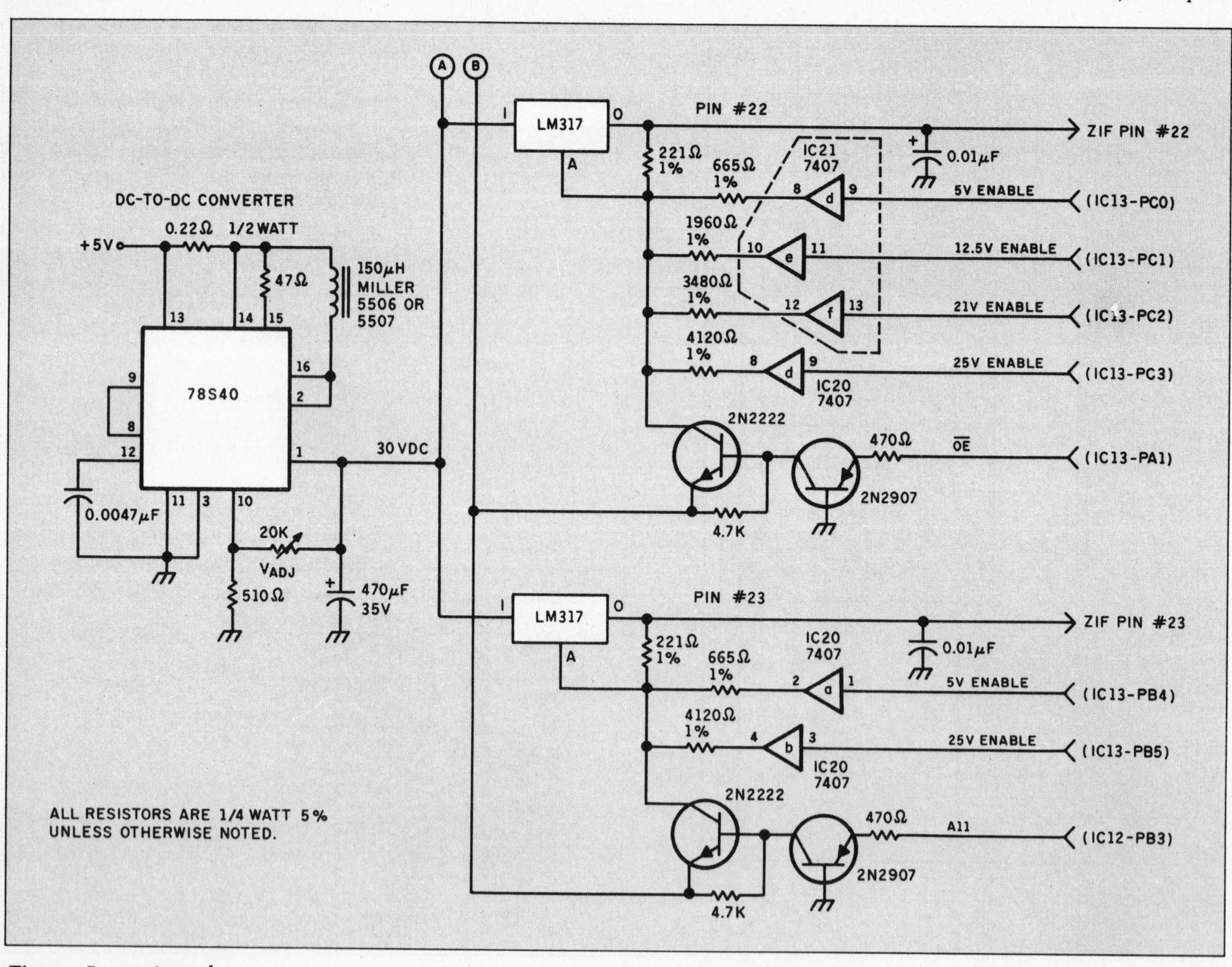

Figure 5 continued.

grammer enters remote mode. Once a mode has been selected, the CCSP must be reset or powered off/on to change modes.

STAND-ALONE LOCAL MODE

In local mode, the CCSP is controlled by two buttons called Type and Start/Next. Displays to the user are made via the seven-segment LED display. When local mode is initially entered, or at any point between completed programming cycles, you can change the designated EPROM type by pressing Type. Each press of the button steps the CCSP to the next EPROM type, and the seven-segment LED display is updated with the number that indicates the currently selected type. The designations are shown in table 1.

After setting the type of EPROM to work with, you begin the copy cycle by pressing Start/Next. At this point the seven-segment LED will display an alternating L and O, indicating that you should insert the original EPROM into the ZIF socket. You then load the original EPROM and press Start/Next again to begin the next step: reading the EPROM.

When the CCSP has read as much of the EPROM data as the memory buffer will allow, it signals you to remove the original EPROM and insert the copy EPROM by displaying an alternating L and C on the seven-segment display. After doing this, you again press Start/Next.

The CCSP will now attempt to program the contents of the RAM buffer into the copy EPROM. After verifying that the target area of the copy EPROM is erased, the letter "P" is displayed on the seven-segment display to indicate that programming is in progress (LED2 will be red, indicating that power is on to the EPROM and it should not be removed). When programming is complete, the contents of the EPROM are compared to the memory buffer contents. During this time, the letter "C" is displayed on the seven-segment display (LED2 will be green, indicating power off).

If the target EPROM is not erased or the programming was not successful (bad compare), the seven-segment LED will display an alternating E and a numeral, either a 1 for an unerased target EPROM or a 2 for a failed comparison. If an error does occur, you will be returned to the "between copies" state at the next press of the button.

Assuming all went well, the CCSP checks to see if the entire EPROM has been copied. If it has, the CCSP returns to the "between copies" state and displays the current EPROM type selected on the seven-segment display.

If the entire EPROM has not yet been copied, the effective starting address of the RAM buffer will be incremented by the size of the RAM buffer, and the CCSP will prompt you to insert the original EPROM again. This time, the programmer reads the EPROM starting at the new address. The amount of data read will be either the RAM buffer size or the remaining bytes to be copied from the EPROM, whichever is less. After reading the original, the CCSP calls for the copy EPROM, and programming continues as described above.

These steps will continue until the entire contents of the original EPROM have been transferred into the copy EPROM. Using this approach allows any size EPROM to be copied, regardless of the amount of memory in the RAM buffer.

REMOTE MODE OPERATION

When used in remote mode, the CCSP turns into a menu-driven programming workstation, controlled by an

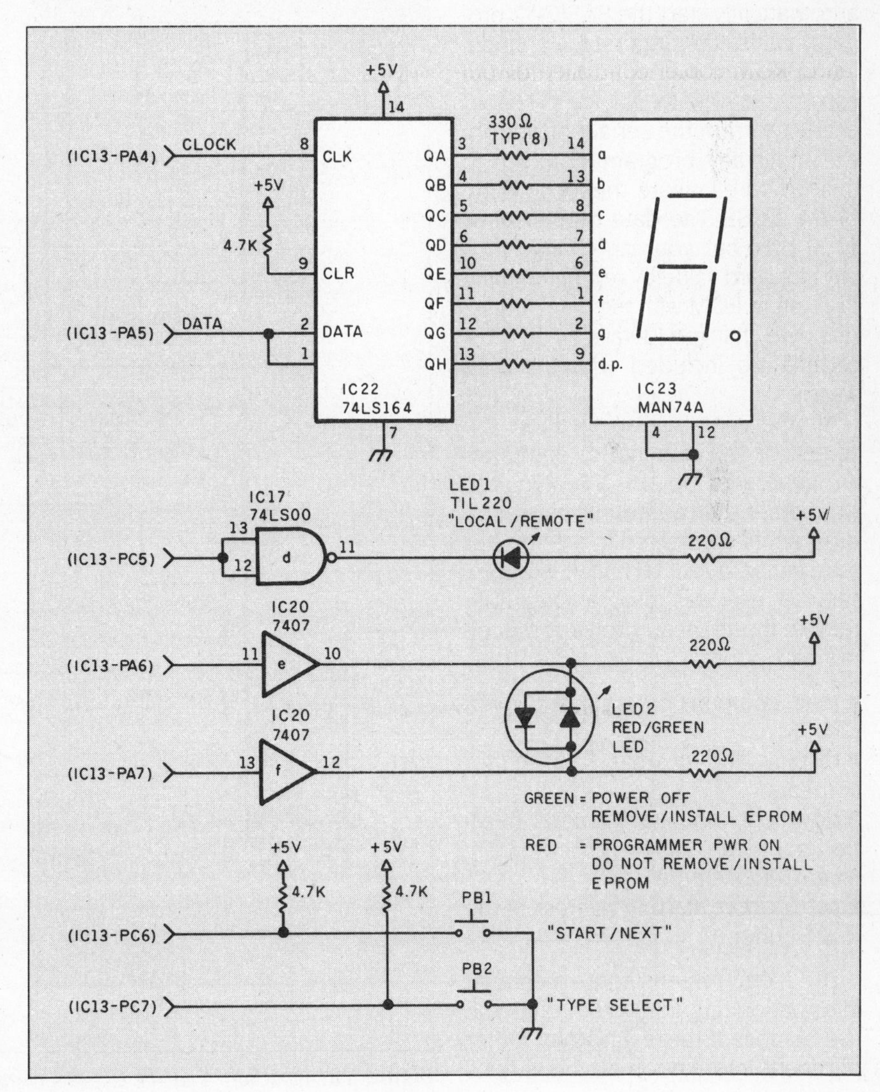

Figure 6: *The CCSP's entry/display circuit.*

All menus displayed on the terminal are generated by the CCSP. A terminal-emulation program is the only software necessary to use this programmer.

ASCII terminal. (See photo 2.) The data rate of the terminal must be hard-coded because the 8052 cannot automatically start the BASIC-52 program unless the data rate is defined. Using the automatic data-rate feature causes the 8052 to wait for a "space" character from the serial port before executing any program stored in it; this would eliminate the local mode of the CCSP. The data rate is set at 1200 bps, but you can change it to any standard value by reprogramming the system ROM with the default data-rate byte changed (details on this procedure are included with the software).

All the menus displayed on the screen of the terminal or computer are generated by the CCSP. No software other than a terminal-emulation program (if connected to a computer rather than a real terminal) is necessary to use this programmer. The remote mode menu provides the following options:

- read, compare, program, and verify EPROM
- display and change RAM buffer contents
- download and upload Intel hexadecimal files
- set EPROM type
- set effective starting address of the RAM buffer

The menu screens contain enough information to guide you through the use of most of these functions. Other pertinent information on the various options is given below.

Read, compare, program, and verify EPROM all depend on the RAM buffer, and they usually use the effective starting address and the length of the RAM buffer to determine the area of the EPROM that is being worked with. Think of the RAM buffer as a window into the contents of the EPROM. If the RAM buffer is not large enough to show you the whole EPROM, you can move it around by changing its effective starting address.

Let's use an example. The EPROM type is a 27512, which is 64K bytes, and the RAM buffer is 16K bytes. It should be pretty obvious that you can't get the whole 27512 into the RAM buffer at the same time. In this case, you would set the starting address of the RAM buffer to 0000 to work with the first quarter of the EPROM and then set it to 4000 to work with the second quarter, 8000 for the third, and C000 for the last. The READ, COMPARE, and PRO-

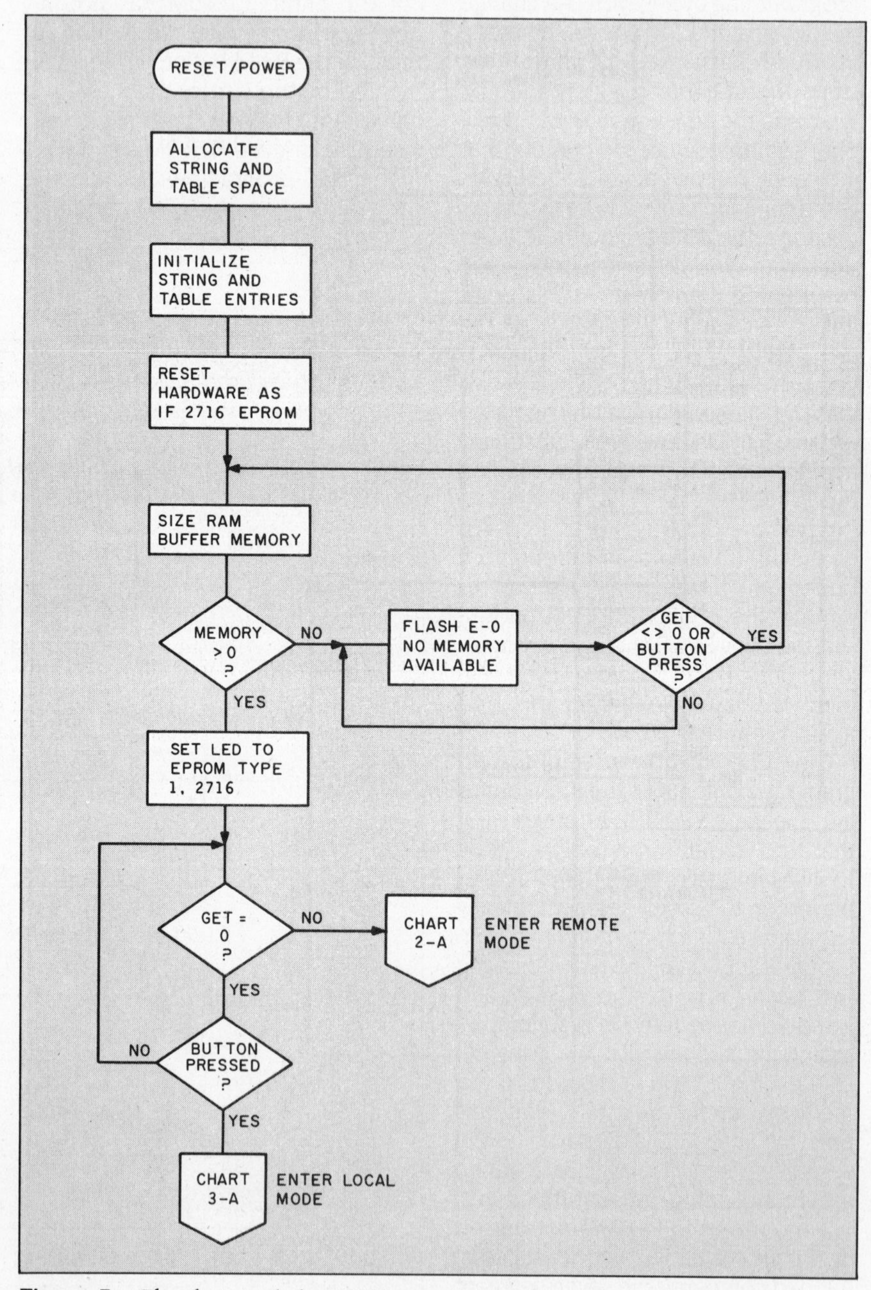

Figure 7a: *Flowchart 1 of the CCSP's overall logic flow, showing the power-up and reset routines.*

GRAM commands would use the starting address of the RAM buffer to see where to read data from or write data to the EPROM. The greatest length of the transferred data would be the size of the RAM buffer or the remaining number of bytes in the EPROM, whichever was smaller.

Even though the VERIFY command does not care about the size of the buffer, its default start and end addresses are controlled as described above. This is because VERIFY generally precedes a programming cycle (you use VERIFY to confirm that the EPROM is properly erased), and the RAM buffer addressing controls programming default start and end addresses.

The following functions—display and change RAM buffer contents and download/upload Intel hexadecimal files—are also tied into the RAM buffer. Since the RAM buffer is supposed to mirror the equivalent area of the EPROM, displays, changes, and uploads/downloads must be addressed to the RAM buffer, just as they would be in the real EPROM. This means that the software will reject addresses outside the range of the current RAM buffer area, which is especially important when doing uploads and downloads. These loads *must* be broken up to fit into the current RAM buffer area

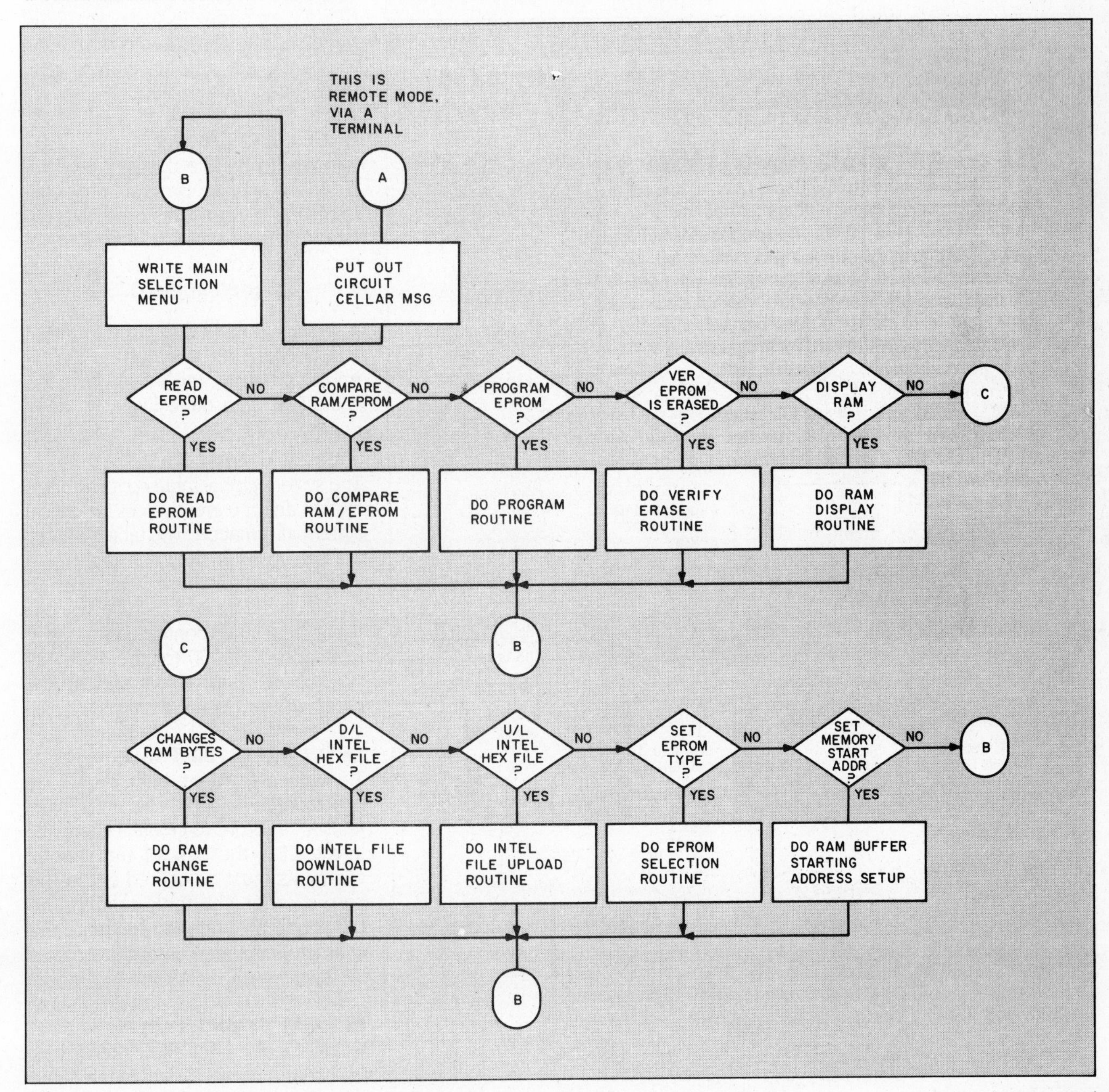

Figure 7b: *Flowchart 2, logic flow for the remote mode.*

Table 1: *The CCSP's EPROM selection number and corresponding EPROM types. Note that this list is of generic EPROM types, and other manufacturer designations should be cross-referenced to it. Also, since CMOS programming cycles are equivalent to those in standard EPROMs, separate 27Cxxx designations are not included.*

Number	EPROM	Type	Number	EPROM	Type
0	2716	25 V	5	27128	21 V
1	2732	25 V	6	27128A	12.5 V
2	2732A	21 V	7	27256	12.5 V
3	2764	21 V	8	27512	12.5 V
4	2764A	12.5 V			

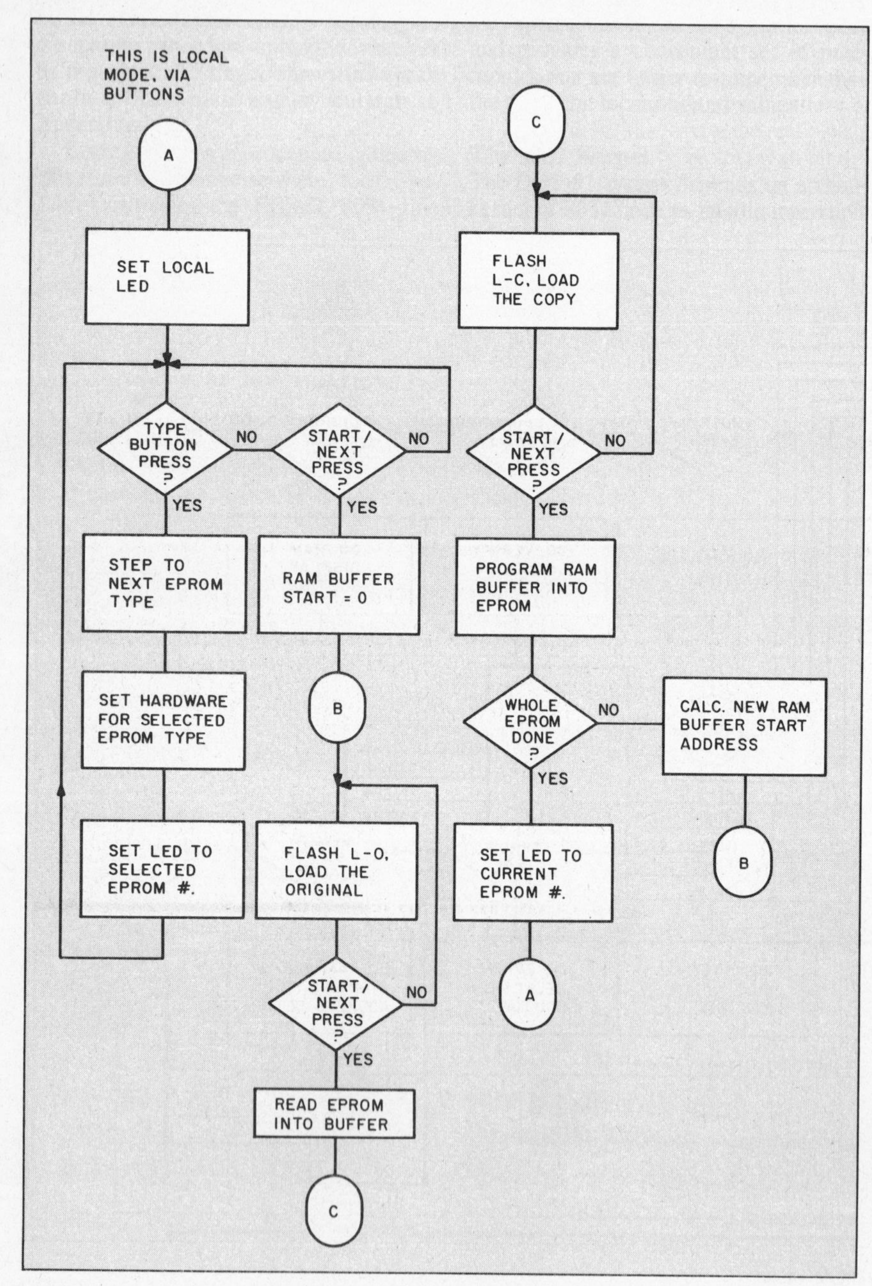

Figure 7c: *Flowchart 3, logic flow for the local mode.*

address range. Trying to go outside the range will abort the display/change/load processes.

The remaining two functions—set EPROM type and set RAM buffer starting address—let you deal with various EPROM types and manipulate the starting address of the RAM buffer. Setting the RAM buffer address lets you change the location of the window into the EPROM. This should be necessary only if the size of the EPROM exceeds the size of the RAM buffer. Otherwise, there is no reason to change the starting address from its default value of 0000.

UNDER THE COVERS

In order to handle the various combinations of sizes, programming voltages, and control lines used with different EPROM types, the software incorporates control tables. Four such tables are used in the CCSP:

- system global table G(*x*)
- LED character table L(*x*)
- EPROM string table $(*x*)
- EPROM data table E(*x*)

SYSTEM GLOBAL TABLE

The system global table contains information about current values for critical system information. The table's entries are set up as is shown in table 2.

The values for these items change based on the type of EPROM you are using, how much contiguous memory is at address 8000 (or the 4K bytes stolen from system RAM), and the last bytes written to the 8255s.

LED CHARACTER TABLE

The seven-segment LED display is controlled by a serial-to-parallel shift register. In order to create a character in the LED, the bits for the various segments must be shifted out in the correct order. This table contains the LED code byte needed to create the characters that can be displayed (see table 3).

EPROM STRING TABLE

BASIC-52 does not allow the mixing of text and numeric data in the same table, so the $(*x*) string table function is used to store this information. This table contains the EPROM designator and the programming voltage used with that type of EPROM. Actually, the programming voltage indicated in the table is only a reminder. You have to set the correct bits in the EPROM data table to ensure that the programmer uses the proper voltage.

EPROM Data Table

The EPROM data table contains all the information the system requires to work with the different EPROM types. The items in each record of the EPROM data table are shown in table 4.

Listing 1 illustrates how this is handled in BASIC. This data is maintained for the use of both the BASIC and assembly language routines. BASIC passes data from the EPROM data table to the assembly language routines via the free registers of the 8052 device.

The Assembly Language Routines

The CCSP software is a hybrid of BASIC and assembly language. Besides reading and verifying, the EPROM programming pulses are accurately timed in assembly language routines. The derivation of the timing's accuracy is given in table 5.

In Conclusion

At first look, the CCSP appears to be considerably more complicated than my programmer of 18 months ago. I think at this point I can change my new description to more accurately state that "this programmer is a serial-port programmer that has the speed of lightning, the intelligence of the mightiest computer (on-board), and is far too functional to be used as a doorstop between uses."

In actuality, only the explanation is more involved. With microcomputer intelligence, the CCSP achieves performance levels approaching kilobuck commercial units yet is flexible enough to be adapted to the next V_{pp} change when it happens.

I'm quite satisfied with my two-week miracle, but I still have to contend with a potential horde of builders. To make amends for my past indiscretion,

Table 2: *Contents of the system global table.*

Index	Use
0	Type number of the current EPROM.
1	Amount of RAM buffer available in 256-byte increments. increments.
2	Current starting address of RAM buffer.
3	Number of items in EPROM table entry.
4	Reserved.
5,6,7	Value of the last data byte written to IC12 address/data PIA (3 bytes, one for each port of the 8255).
8,9,10	Value of the last data byte written to IC13 control PIA.
11	Number of EPROM types in the EPROM string and data tables.

Table 3: *Contents of the* LED *character table.*

Index	Use
0–9	Characters 0–9 (no decimal point)
10–15	Characters A–F (no decimal point)
16–25	Characters 0–9 (decimal point)
26–31	Characters A–F (decimal point)
32	Blanks LED
33	Character H
34	Character L
35	Character P
36	Character U

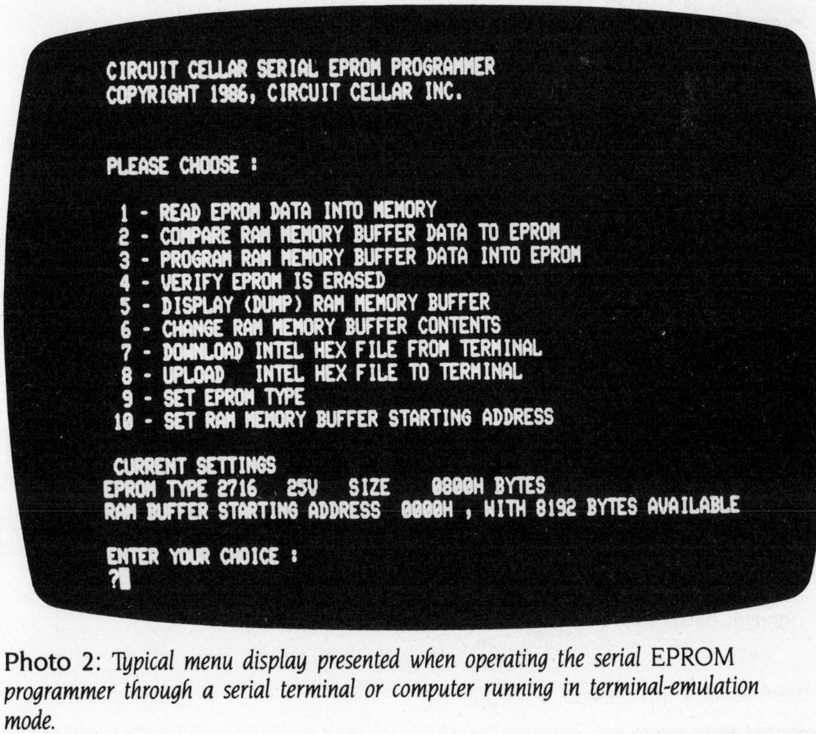

```
CIRCUIT CELLAR SERIAL EPROM PROGRAMMER
COPYRIGHT 1986, CIRCUIT CELLAR INC.

PLEASE CHOOSE :

 1 - READ EPROM DATA INTO MEMORY
 2 - COMPARE RAM MEMORY BUFFER DATA TO EPROM
 3 - PROGRAM RAM MEMORY BUFFER DATA INTO EPROM
 4 - VERIFY EPROM IS ERASED
 5 - DISPLAY (DUMP) RAM MEMORY BUFFER
 6 - CHANGE RAM MEMORY BUFFER CONTENTS
 7 - DOWNLOAD INTEL HEX FILE FROM TERMINAL
 8 - UPLOAD   INTEL HEX FILE TO TERMINAL
 9 - SET EPROM TYPE
10 - SET RAM MEMORY BUFFER STARTING ADDRESS

 CURRENT SETTINGS
EPROM TYPE 2716   25V   SIZE    0800H BYTES
RAM BUFFER STARTING ADDRESS  0000H , WITH 8192 BYTES AVAILABLE

ENTER YOUR CHOICE :
?
```

Photo 2: *Typical menu display presented when operating the serial* EPROM *programmer through a serial terminal or computer running in terminal-emulation mode.*

Table 4: *Contents of the EPROM data table.*

Index	Use
0	Number of the EPROM.
1	Size of the EPROM in 256-byte increments.
2	The pin V_{cc} is applied to, referenced to a 28-pin socket.
3,4,5	The initialization values for IC13 control PIA's ports A, B, and C.
6,7	The programming mode values for IC13's ports A and B.
8	Logic true value of the programming pin (CE and PGM).
9	Normal mode programming pulse width in milliseconds.
10	Fast mode programming pulse in milliseconds (0 if no fast programming allowed).
11	Maximum number of fast programming pulses before forced overprogram pulse occurs.
12	Overprogram pulse multiplication factor.

Table 5: *Derivation of the timing for the CCSP's 1-ms timing routine.*

Label	Instruction	Frequency of Execution	Clock Periods Used
PPLOOP	EQU $	A	0
	PUSH B	C	24
	CALL ONEMS	C	24
	POP B	C	24
	DJNZ B,PPLOOP	C	24
	RET	O	24
ONEMS	EQU $	A	0
	MOV B,#MSDLAY	C	24
MSLOOP	EQU $	A	0
	NOP	L	12
	NOP	L	12
	DJNZ B,MSLOOP	L	24
	RET	C	24
MSDLAY	EQU 227	A	0
	END	A	0

Frequency codes:
A—Assembler only, not executed
C—Executed once per 1-ms count
O—Executed only once per entry to subroutine
L—Executed in each loop of the 1-ms routine

Timing Calculations:

1MS=11059.2 clock periods (clock is 11059200 Hz)

$11059.2 = 24 + (24+24+24+24+24+24) + N(12+12+24)$

$11059.2 = 168 + (N \times 48)$

$(11059.2 - 168)/48 = N$

$N = 226.9$ (round to 227)

Error for 1-ms pulse is $0.1 \times 48 \times (1/11059200) = 0.000443$ ms

Cumulative error for 100-ms pulse is:

$$\frac{1105920-24-14400}{4800} = 227.395 \text{ (use 227)}$$

$0.395 \times 48 \times (1/11059200) = 0.00171$ ms

Listing 1: *The* BASIC-52 *code to initialize the* EPROM *data table.*

```
14170 REM
14180 REM INITIALIZE EPROM DATA TABLE
14190 REM
14200 FOR X=0 TO A:READ E(X):NEXT X
14210 REM  TYPE SIZE PWR 7XPA 7XPB 7XPC 7PCP 7PBP PLV NPL FP FXP FACTR
14220 DATA 2716,008H,026,0BBH,061H,001H,008H,000H,001,050,000,000,000
14230 DATA 2732,010H,026,0BFH,051H,001H,008H,000H,000,050,000,000,000
14240 DATA 2732,010H,026,0BFH,051H,001H,004H,000H,000,050,000,000,000
14250 DATA 2764,020H,028,0BFH,014H,001H,000H,000H,000,050,001,025,003
14255 DATA 2764,020H,028,0BBH,012H,001H,000H,000H,000,050,001,025,003
14260 DATA 27128,040H,028,0BBH,014H,001H,000H,000H,000,050,001,015,004
14265 DATA 27128,040H,028,0BBH,012H,001H,000H,000H,000,050,001,015,004
14270 DATA 27256,080H,028,0BBH,052H,001H,000H,000H,000,050,001,025,003
14280 DATA 27512,0100H,028,0BBH,051H,002H,000H,000H,000,050,001,025,003
14281 REM SET UP EPROM NAME TABLE
14283 $(1)="2716   25V  "
14284 $(2)="2732   25V  "
14285 $(3)="2732A  21V  "
14286 $(4)="2764   21V  "
14287 $(5)="2764A  12.5V"
14288 $(6)="27128  21V  "
14289 $(7)="27128A 12.5V"
14290 $(8)="27256  12.5V"
14291 $(9)="27512  12.5V"
14295 RETURN
```

there is indeed a printed circuit board and kit for this programmer. For those of you with the components at hand and a desire for wire-wrapper's cramp, the finished code for the programmer is available on a 27128 EPROM or is downloadable from the Circuit Cellar BBS and BYTEnet Listings. The phone number for BYTEnet Listings is (617) 861-9764. This file contains the 12K bytes of executable code that should be put into a 27128 EPROM and installed in IC11.

Special thanks to Bill Curlew for his software expertise.

There is an on-line Circuit Cellar bulletin board system that supports past and present projects. You are invited to call and exchange ideas and comments with other Circuit Cellar supporters. The 300/1200/2400-bps BBS is on-line 24 hours a day at (203) 871-1988.

Editor's Note: Steve often refers to previous Circuit Cellar articles. Most of these past articles are available in book form from BYTE Books, McGraw-Hill Publishing Company, P.O. Box 400, Hightstown, NJ 08250, (1-800-2-MCGRAW).

Ciarcia's Circuit Cellar, Volume I covers articles in BYTE from September 1977 through November 1978. *Volume* II covers December 1978 through June 1980. *Volume* III covers July 1980 through December 1981. *Volume* IV covers January 1982 through June 1983. *Volume* V covers July 1983 through December 1984.

The following items are available from

Circuit Cellar, Inc.
4 Park St., Suite 12
Vernon, CT 06066
(203) 875-2751

1. Serial EPROM programmer experimenter's kit. Includes PC board, 8052AH-BASIC chip, 11.05-MHz crystal, operating system software on preprogrammed 27128 EPROM, manual, and detailed parts list $89
2. Complete serial EPROM programmer kit. Includes all board-mounted components, programmed 27128 EPROM, 8052AH microprocessor, and manual. Less case and power supply $199
3. Preprogrammed 27128 EPROM containing serial EPROM programmer system software $22

The serial EPROM programmer is currently available only in kit form. It is available assembled and tested only in volume OEM quantities (telex: 643331). Price and delivery information available on request.

All payments should be made in U.S. dollars by check, money order, MasterCard, or Visa. Surface delivery (U.S. and Canada only): add $5 for U.S., $10 for Canada. For delivery to Europe via U.S. airmail, add $20. Three-day air freight delivery: add $8 for U.S. (UPS Blue), $25 for Canada (Purolator overnight), $45 for Europe (Federal Express), or $60 (Federal Express) for Asia and elsewhere in the world. Shipping costs are the same for one or two units. Connecticut residents **please include 8 percent sales tax.**

Most of the individual components are available from JDR Microdevices, 1224 South Bascom Ave., San Jose, CA 95128, (800) 538-5000.

11

BUILD THE CIRCUIT CELLAR AT COMPUTER

PART 1: AT BASICS

New chip technology lets Steve put an AT on a card

The personal computer industry can best be described as competitive and fast-moving. Hundreds of manufacturers around the world are turning out IBM PC, XT, and AT clones, as well as board products for those who wish to mix and match to build a desktop computer. Heavy competition forces these manufacturers to continually bring out higher performing products that are less expensive than their competitors' products. Talk about a dog-eat-dog business!

Ordinarily, given such market volatility, I would not stick my neck into the "clone wars," but I just couldn't pass up the opportunity to show some new technology that would one-up all these companies. The two-part project in Chapter 11 uses a set of four high-integration ZyMOS ICs that contain most of the peripheral chips needed to build an IBM PC AT. Using this advanced technology, I will present a faster, smaller, and more efficient 100 percent compatible AT CPU board called the CCAT (Circuit Cellar AT). With the addition of Award Software's AT BIOS, the CCAT and your imagination can configure an unbeatable (should I say uncloneable?) 80286 computer system.

Technology to the Rescue

The ZyMOS POACH (which stands for PC on a chip) set is really an ASIC (application-specific IC) set that was originally developed to show just how much could be integrated on one 230-pin chip (it contained 22,000 logic gates). Eventually, it was divided into more cost-effective 84-pin devices (see photo 1).

ZyMOS uses standard-cell CHMOS (complementary high-speed metal-oxide semiconductor) technology for its ASIC products. Some methods for developing ASICs are gate arrays, programmable logic arrays, and standard-cell technology, to name a few. Standard-cell technology produces chips that are highly integrated (they can squeeze 25,000 logic gates on a single device) but also very efficient in their use of silicon area.

Gate arrays are essentially a mass of predefined gates that are interconnected by the designer, but that most often leave a lot of unconnected gates and waste silicon. The standard-cell approach uses just the logic elements the designer requires. The result is higher densities of utilized gates requiring smaller die size. Less silicon means lower cost per device.

Standard cells are predefined logic units that correspond to commercially available devices like inverters, AND gates, flip-flops, and more complex parts (like the 82xxx peripheral chips used in the IBM PC AT).

A chip designer developing an IC first lays out a schematic, just as for any project. Such a schematic typically contains off-the-shelf components like 8254 counter-timers, 7474 flip-flops, 7408 AND gates, and 7432 OR gates. Next, the designer enters this schematic into a computer using specialized graphics soft-

Photo 1: *This section of the Circuit Cellar AT circuit board shows the POACH (PC on a chip) integrated circuits.*

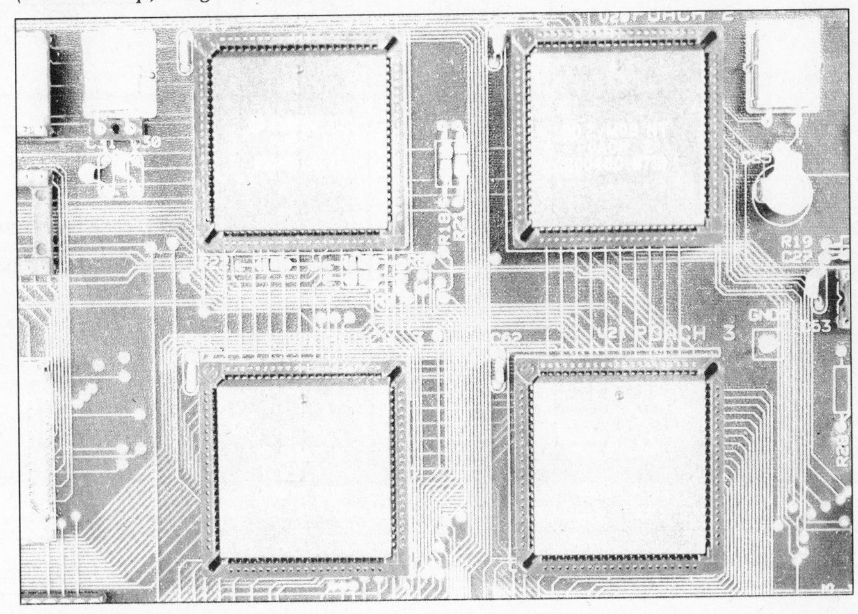

ware that generates a file called a *net list* (Circuit Cellar projects are currently done on Schema).

The net list defines which logic elements from a library of standard cells are needed and how those cells are interconnected. The net list is input to a logic simulator, which the designer uses to debug the design, and then fed to a program called a *router*. The router actually lays out the final chip, transistor by transistor. Sounds easy! It's not—it's just easier and faster with the computerized tools. Even so, it took seven months to develop the four-chip POACH set used in the CCAT.

Understanding the AT's Design

Before we look closely at the POACH chips and the CCAT, we should get some understanding of the IBM PC AT motherboard's design so we can better understand what it is that we are trying to improve. The AT is an Intel 80286 16-bit microprocessor design that can optionally support the 80287 math coprocessor for fast floating-point operations. It also uses an Intel 8742 microcontroller as a keyboard processor.

In addition to the processors, the AT uses 10 VLSI peripherals that work in

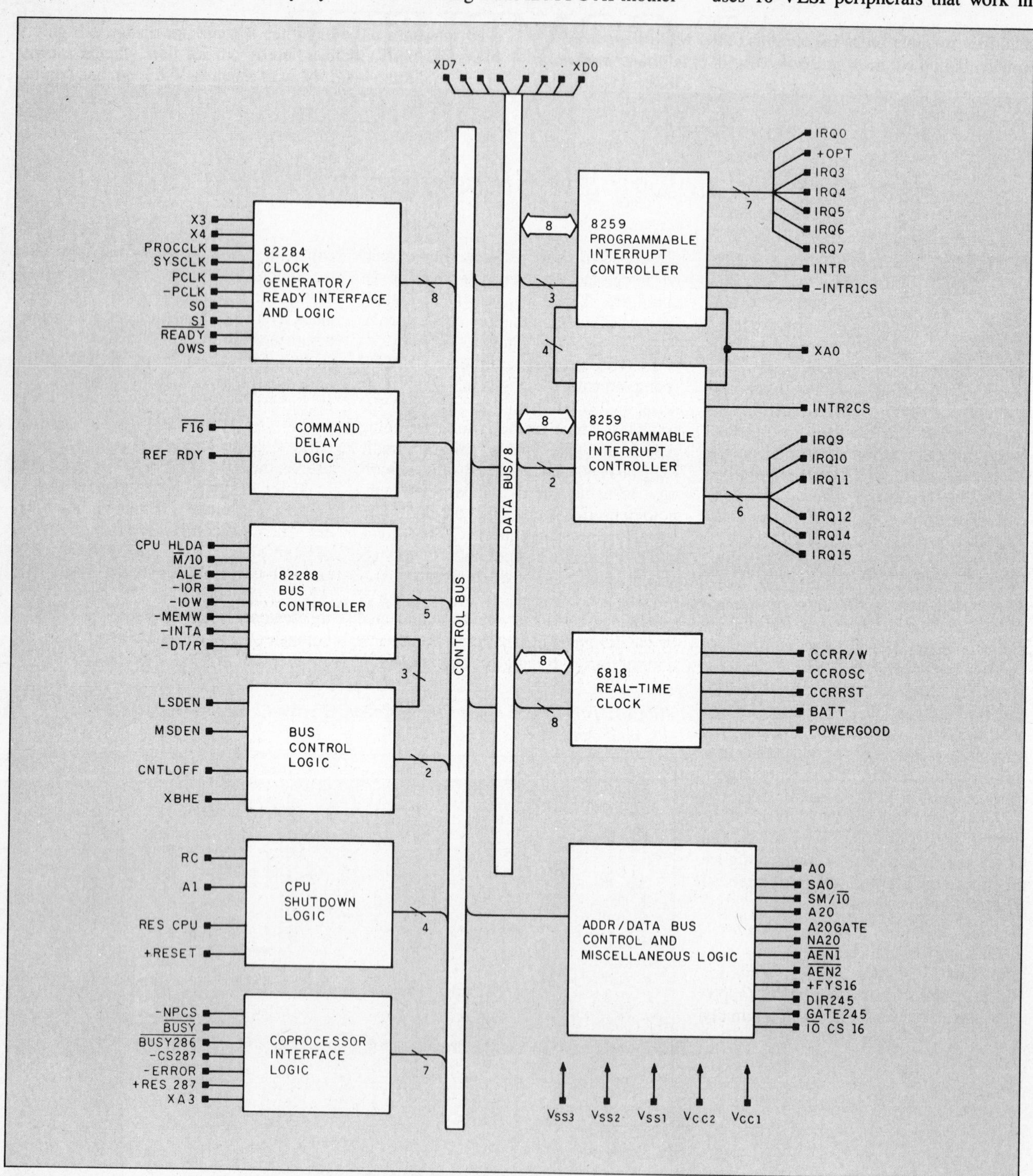

Figure 1: *The block diagram of the internals of POACH1.*

conjunction with the 80286 to perform functions like bus timing, interrupt control, and direct-memory-access operations. These devices integrate much necessary logic that a designer would ordinarily have to build up from primitive logic functions to get a design to perform properly. They are the building blocks that, with integrated microprocessors, have shrunk the computer's physical size and made prices affordable.

The peripherals in the AT include two 8259A programmable interrupt controllers, an 82284 clock generator and ready interface, an 82288 bus controller, a 6818 clock/calendar/RAM, two 8237A DMA controllers, a 74LS612 memory mapper, an 8284 clock generator, and an 8254 programmable interval timer.

The interrupt controllers sort out and prioritize interrupt requests to the microprocessor. Each interrupt controller can handle up to eight interrupts, but Controller-2 (CTLR2) interrupts are directed through CTLR1, which uses up one of CTLR1's interrupt lines. Interrupt requests (IRQ) are mapped as shown in table 1. The 82288 and 82284 perform general system clocking, some decoding of 80286 control signals, and bus-control functions in the AT.

The 6818 contains the clock/calendar and 64 bytes of CMOS RAM. The clock function uses 14 bytes of the RAM to hold time and date data. The rest of the RAM holds the system's configuration information, like the type of floppy and hard disk drives and low- and high-memory bytes. The 6818 is kept alive when the machine is powered down with battery power and a continually running clock frequency.

The two 8237As provide seven DMA channels. DMA CTLR1 supports 8-bit data transfers between 8-bit I/O adapters and 8- or 16-bit system memory. Data transfers can occur throughout the 16-megabyte address space in 64K-byte blocks. DMA CTLR2 supports 16-bit data transfers between 16-bit I/O adapters and 16-bit memory and can perform data transfers in 128K-byte blocks throughout the full 16-megabyte address range. Since the DMA controllers generate only 16-bit addresses, the system uses the LS612 memory mapper to extend the addressing to 16 megabytes.

A 14.318-megahertz crystal drives the 8284 clock generator. The 14.318-MHz clock is routed directly to the expansion slots.

The 8254 programmable interval timer provides 16-bit counters on three independent channels. Channel 0 produces the system timer signal (18 ticks per second), channel 1 generates the dynamic RAM-refresh request, and the system uses channel 2 for the speaker's tone generator.

You should begin to see some of the characteristics of the AT emerging. It is a 16-bit interrupt-driven system with DMA capability for fast memory data transfers. The PC's speaker is still there to prompt you with those annoying beeps and to add some dimension to game playing. And we've added a real-time clock to keep track of time and date.

If we tack on 512K bytes of DRAM and a couple of ROMs to hold the BIOS, the system starts taking form. It would be great if we could stop here, with about 43 chips making up the system. But we have to glue all this together and provide a means for the processor to talk to memory and the outside world—so we have address and data buses.

The AT has a number of address and data buses, with many buffers, latches, and multiplexers separating the individual buses. In fact, it has *five* distinct buses: local, system, X, memory, and L address. (The first four have both address and data components.)

Table 1: *Request mapping for the interrupt controllers on the CCAT.*

CTLR1	CTLR2	
IRQ0		Timer output 0
IRQ1		Keyboard (output buffer full)
IRQ2		Interrupt from CTLR2
	IRQ8	Clock/calendar/RAM
	IRQ9	Software redirected to INT 0AH (IRQ2)
	IRQ10	Reserved
	IRQ11	Reserved
	IRQ12	Reserved
	IRQ13	Coprocessor
	IRQ14	Fixed disk controller
	IRQ15	Reserved
IRQ3		Serial port 2
IRQ4		Serial port 1
IRQ5		Parallel port 2
IRQ6		Disk controller
IRQ7		Parallel port 1

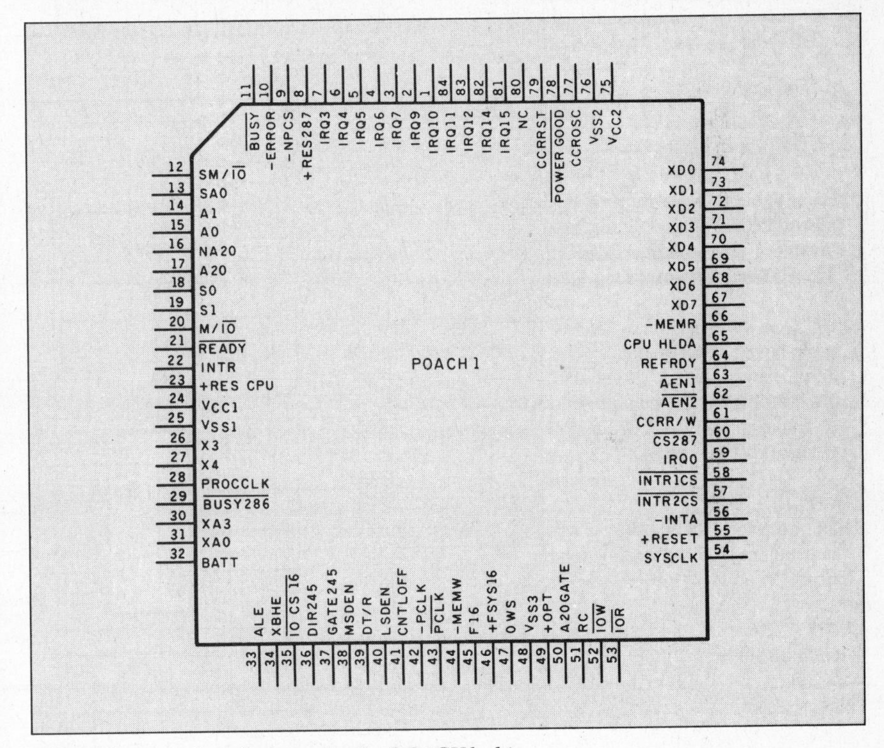

Figure 2: *A pin-out diagram for the POACH1 chip.*

The local address and data buses are tied directly to the 80286 and 80287. Twenty-four address lines and 16 data lines form the heart of the AT. The address lines are latched by three LS573s that buffer the local address bus from the system address bus. Because the 80286 can do word and byte data transfers, and word transfers need not be aligned, the AT data-bus interface has to differentiate between the high-bus byte and the low-bus byte. ("Aligned" refers to the fact that the word address is even, that is, A0 = 0).

If the system has to transfer a word over the bus to an odd address location, it requires two bus cycles—one to transfer the low byte and one to transfer the high byte (this is a nonaligned word transfer). The local data bus is separated from the system data bus by an LS245 buffer and an LS646, which not only buffers but also has a latch function.

The system address and data buses are the primary ones in the AT for both memory and I/O transfers, including the interface to the AT's expansion slots. The sys-

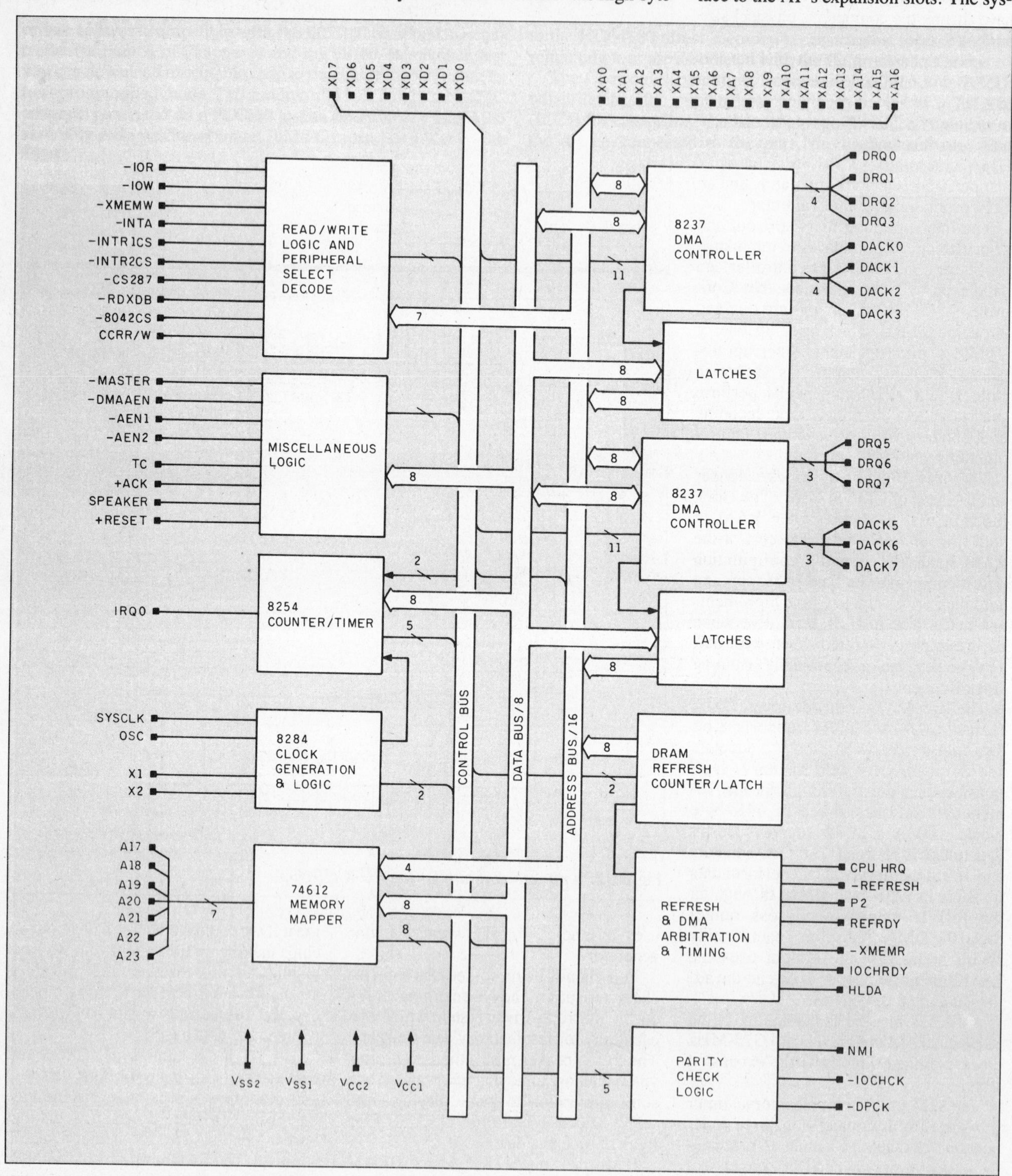

Figure 3: *The block diagram of POACH2.*

tem address bus is a latched 20-bit (SA0 through SA19) version of the 80286 local address bus. The system data bus maintains the low-byte and high-byte reference. The AT performs 16-bit transfers for memory and expansion data, but it does 8-bit data transfers for the X bus (which the AT motherboard uses to communicate with the 80286 peripheral chips).

The X address bus is a 17-bit bus that you can think of as private to the motherboard. The system uses this bus to address ROM (where the BIOS is kept) and motherboard I/O, as well as to generate addresses for DMA- and RAM-refresh operations. It is separated from the system address bus by LS245s.

The X data bus interfaces to functions like DMA controllers, interrupt controllers, the keyboard processor, and the clock/calendar/RAM hardware. Although the system uses the X address bus to select ROM data, this data is fed to the processor via the memory data bus, not the X data bus.

The memory address and data buses apply to DRAM on the motherboard. The 9 address lines (MA0 through MA8) of the memory address bus are a multiplexed version of 18 system-address lines. The memory data bus is a 16-bit motherboard bus that interfaces both DRAM and ROM.

The L address bus, hangs like an appendage off the local address bus. It is an unlatched 7-bit (LA17 through LA23) address that is always available except when an I/O processor gains control of the system. The L address bus gives the AT a 16-megabyte address range.

Complicated? You bet! We've just added 83 ICs for buffering, additional logic, and glue to tie the system together. This brings the total IC count for a 512K-byte system to 126 packages. Now you can see why the motherboard is approximately 14 inches square!

Microcomputers did not always have big motherboards. Originally, when they were much simpler, they used passive backplanes, with the processor board plugged into one of the backplane slots. Chip count and board size have an obvious effect on the cost of the system. That's where the new high-integration chips play such an important role.

The POACH Set

The engineers at ZyMOS partitioned the motherboard's logic into a set of 84-pin ICs. Two chips, POACH1 and POACH2, contain *all* the 82xxx-series peripherals and also some key logic. POACH1 is primarily responsible for system clocking and bus control (see figures 1 and 2). It contains the 8259A interrupt controllers, the 82284 clock generator and ready interface, the 82288 bus controller, and the 6818 clock/calendar/RAM. All the buffers and latches in the system are controlled by POACH1.

POACH2, shown in figure 3, controls the X address bus for DMA and refresh operations. It contains the two DMAs, the 74LS612 memory mapper, refresh logic, the 8284 clock generator, and the 8254 programmable interval timer. Figure 4 shows POACH2's pin-out diagram.

The buffers, latches, and remaining logic are in POACH3, the buffer chip. (I haven't provided a block diagram for POACH3 since it's simply a mass of uninteresting discrete logic.) POACH3 is mode-programmable, and we've used two in the CCAT, one for address buffering and the other for data buffering. I'll refer to the device set for address-buffering mode as POACH3-A and the device set for a data buffer as POACH3-D.

The AT POACH set brings the total IC count for a 512K-byte AT-compatible motherboard equivalent down to 23 IC packages and two SIMMs (single in-line memory modules).

Ultimately, our design is intended to be totally IBM compatible, with certain key advantages. Using the POACH chips, we can squeeze the whole motherboard into an AT expansion board with room to spare and build a system on a passive backplane. Since all the usual power-hungry ICs are now incorporated in the POACH devices (the four-chip set needs less than 100 milliamperes), we will also be able to construct a low-power AT.

The POACH set brings the total chip count for a 512K-byte AT compatible down to 23 IC packages and two SIMMs.

In Conclusion

In part 2 of Chapter 11, I'll complete this project with a full schematic and a detailed description of the Circuit Cellar AT computer.

The CCAT was a joint venture, and I'd like to note the contributions and help from ZyMOS, Micromint Inc., Award Software Inc., and the Circuit Cellar research staff. In addition, I'd like to personally thank Bob Andrews, Jeff Bachiochi, and Jeff Remmers for their efforts.

Editor's Note: Steve often refers to previous Circuit Cellar articles. Most of these past articles are available in book form from BYTE Books, McGraw-Hill Publishing Company, P.O. Box 400, Hightstown, NJ 08250, (1-800-2-MCGRAW).

Ciarcia's Circuit Cellar, Volume I covers

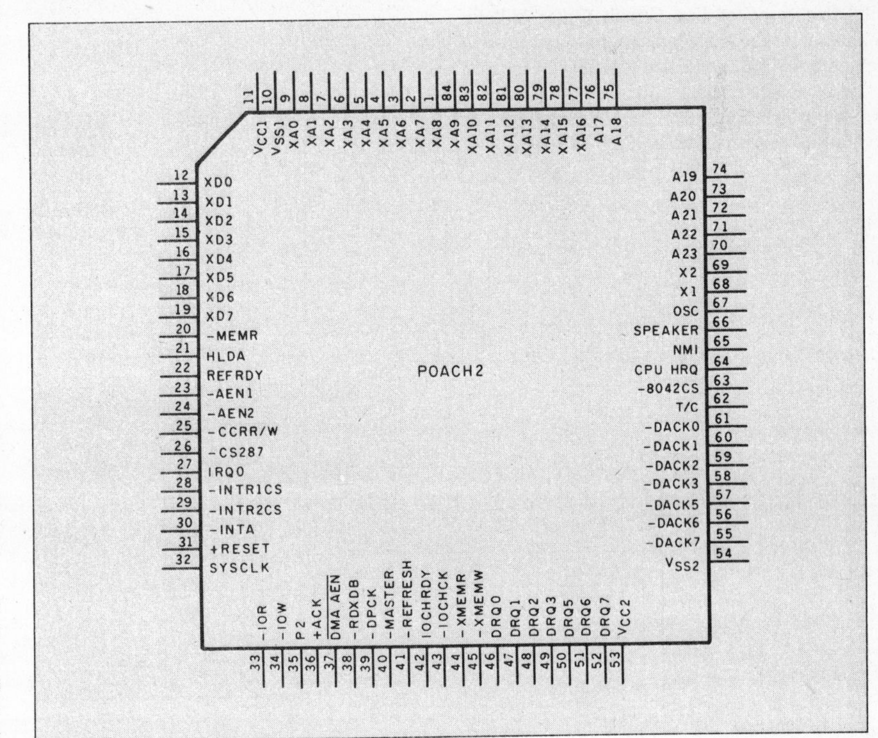

Figure 4: *POACH2's pin-out diagrm.*

articles in BYTE from September 1977 through November 1978. *Volume II* covers December 1978 through June 1980. *Volume III* covers July 1980 through December 1981. *Volume IV* covers January 1982 through June 1983. *Volume V* covers July 1983 through December 1984.

The following items are available from

Micromint Inc.
4 Park St.
Vernon, CT 06066
(800) 635-3355 (orders)
(203) 871-6170 (information)
Telex: 643331

1. Assembled and tested CCAT 80286 PC AT compatible CPU board with 512K bytes of RAM, Award BIOS in ROM, and user's manual. Available in 10 MHz.
   10 MHz, order OEM286LP/10 ....$620
2. Assembled eight-slot AT-style passive backplane with all connectors.
   order MBAT/8 ....$115

POACH chips and BIOS ROMs are available. Call for price and delivery information.

For either source above, all payments should be made in U.S. dollars by check, money order, MasterCard, Visa, or American Express. Surface delivery (U.S. and Canada only): add $5 for U.S., $8 for Canada. For delivery to Europe via U.S. airmail, add $14. Three-day air freight delivery: add $10 for U.S. (UPS Blue), $25 for Canada (Purolator overnight), $45 for Europe (Federal Express), or $60 for Asia and elsewhere in the world (Federal Express).

There is an on-line Circuit Cellar bulletin board system that supports past and present projects. You are invited to call and exchange ideas and comments with other Circuit Cellar supporters. The 300/1200/2400-bps BBS is on-line 24 hours a day at (203) 871-1988.

11

BUILD THE CIRCUIT CELLAR AT COMPUTER

PART 2: SCHEMATIC

Steve reveals the circuitry for his AT on a board

In part 1 of this chapter, we took a close look at the structure of the IBM PC AT computer, and I introduced the highly integrated POACH (PC on a chip) set from ZyMOS that makes an AT on a board a feasible project. In part 2, you'll see that, aside from the choice of board layout and perhaps memory configuration, the task is done. [Editor's note: *You'll want to refer to Chapter 11, part 1. Many AT circuit details Steve talks about here were described in part 1.*] Since most functions are part of the POACH set, the remaining circuitry must follow strict guidelines to remain 100 percent compatible. In essence, the rest of the design is just "cookbook" stuff and—as you can see from the circuit schematic for CCAT in figure 1—is fairly simple.

Inside CCAT

The CCAT is designed to run with either an 8- or 10-megahertz system clock, depending upon which 80286 chip is used (−8 or −10). POACH1 generates both PROCCLK and the system clock (SYSCLK) from a 16- or 20-MHz crystal connected across pins 26 and 27.

POACH1 requires a 32.768-kilohertz time base (CCROSC) for the 6818 clock/calendar/RAM. A CMOS 74HC04 inverter is used as the oscillator amplifier. The POACH1 6818 requires only 10 microamperes of standby current. Battery power for the 6818 is connected to pin 32 of POACH1.

The local data bus (D0 through D15) runs between the 80286, 80287, and POACH3-D. Local address lines A1 through A23 go directly to POACH3-A, but since the POACH2 memory mapper also generates A17 through A23 during direct-memory-access operations, those address lines are also connected to POACH2.

The system address bus lines SA1 through SA19 are generated by POACH3-A. Normally, CPUHLDA and ALE control system bus activity. +ACK from POACH2 gates SA17 through SA19 during DMA operations. SA0 is generated directly from A0 by POACH1. POACH1 also produces +CNTL OFF, XA0, LSDOE, MSDOE, and DT/R, which are required gate and direction control for the system data bus.

Memory-address and data-bus generation is a convoluted affair if you try to handle it with discrete logic, but it becomes straightforward for the CCAT project. POACH3-A generates the multiplexed MA0 through MA8 address lines that are controlled by +REFRESH and GA-2RAS. The memory data bus (MD0 through MD15) connects POACH3-D directly to the dynamic RAMs and ROM. XA0 gates the least significant byte; XBHE gates the most significant byte. -XMEMR and DIRMS control direction; when both signals are high, data flow is from memory to system.

Primarily, the X address bus runs between POACH2, POACH3-A, and ROM. POACH2 generates -DMAAEN, which POACH3-A uses to control direction on the X bus during DMA operations. X data flows between POACH1, POACH2, the 8742 keyboard processor, and POACH3-D. DIR245 and -RDXDB control direction on the bus, and GATE245 and +ACK gate control the data flow.

The overall effect of integration on the AT is obvious when you view design complexity (including chip count) and board space. But some very subtle benefits don't immediately come to mind.

For instance, EMI (electromagnetic interference) and RFI (radio frequency interference) decrease dramatically. Every trace running across the circuit board is an antenna that radiates more ef-

Photo 1: *The Circuit Cellar AT computer: an AT clone on a board.*

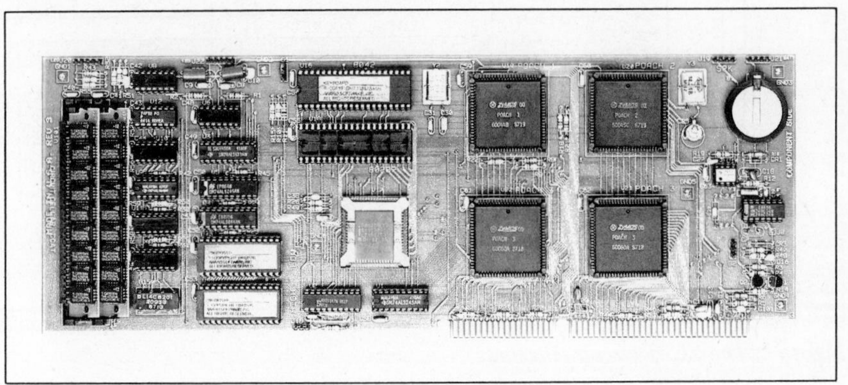

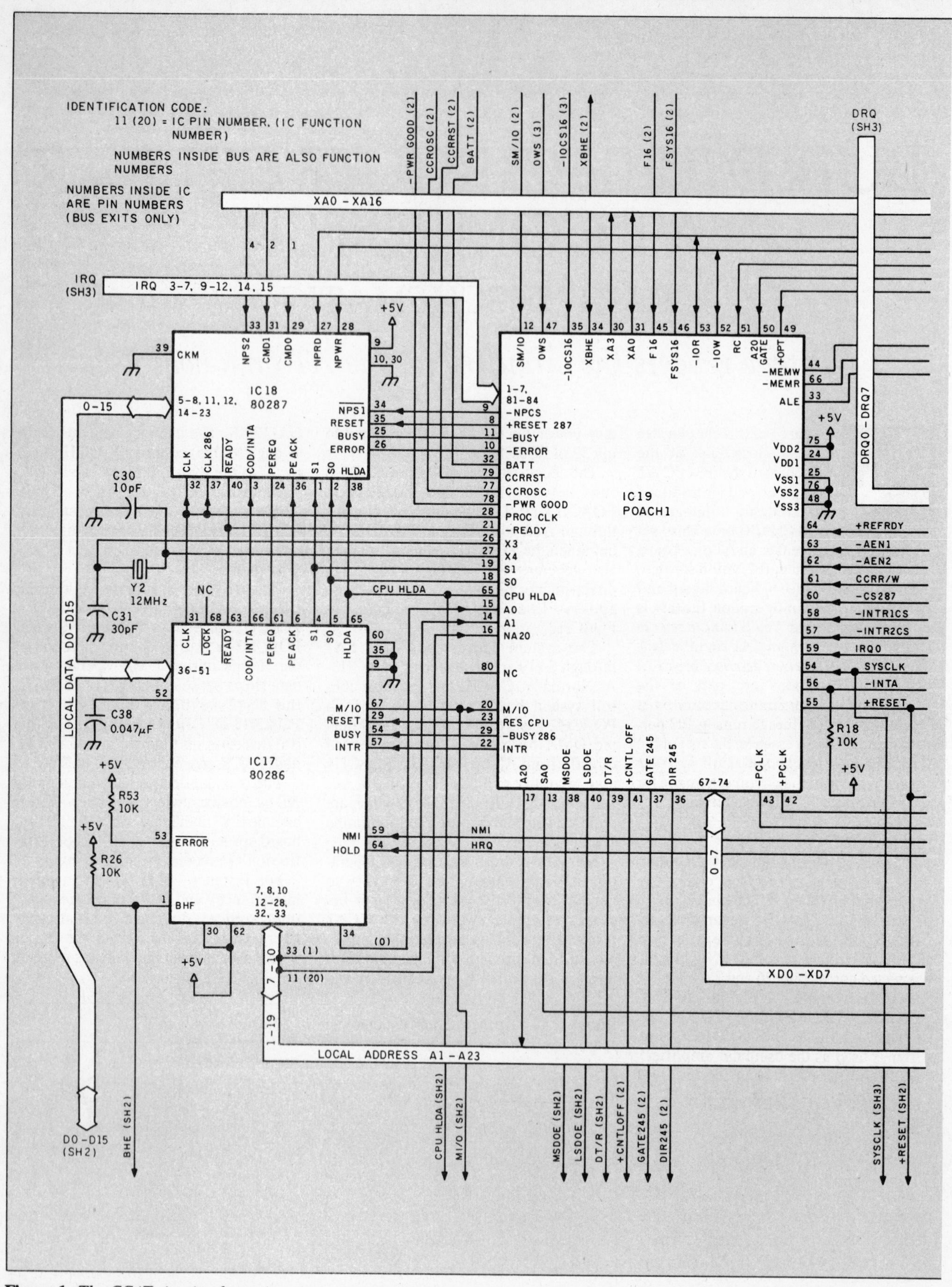

Figure 1: *The CCAT circuit schematic.*

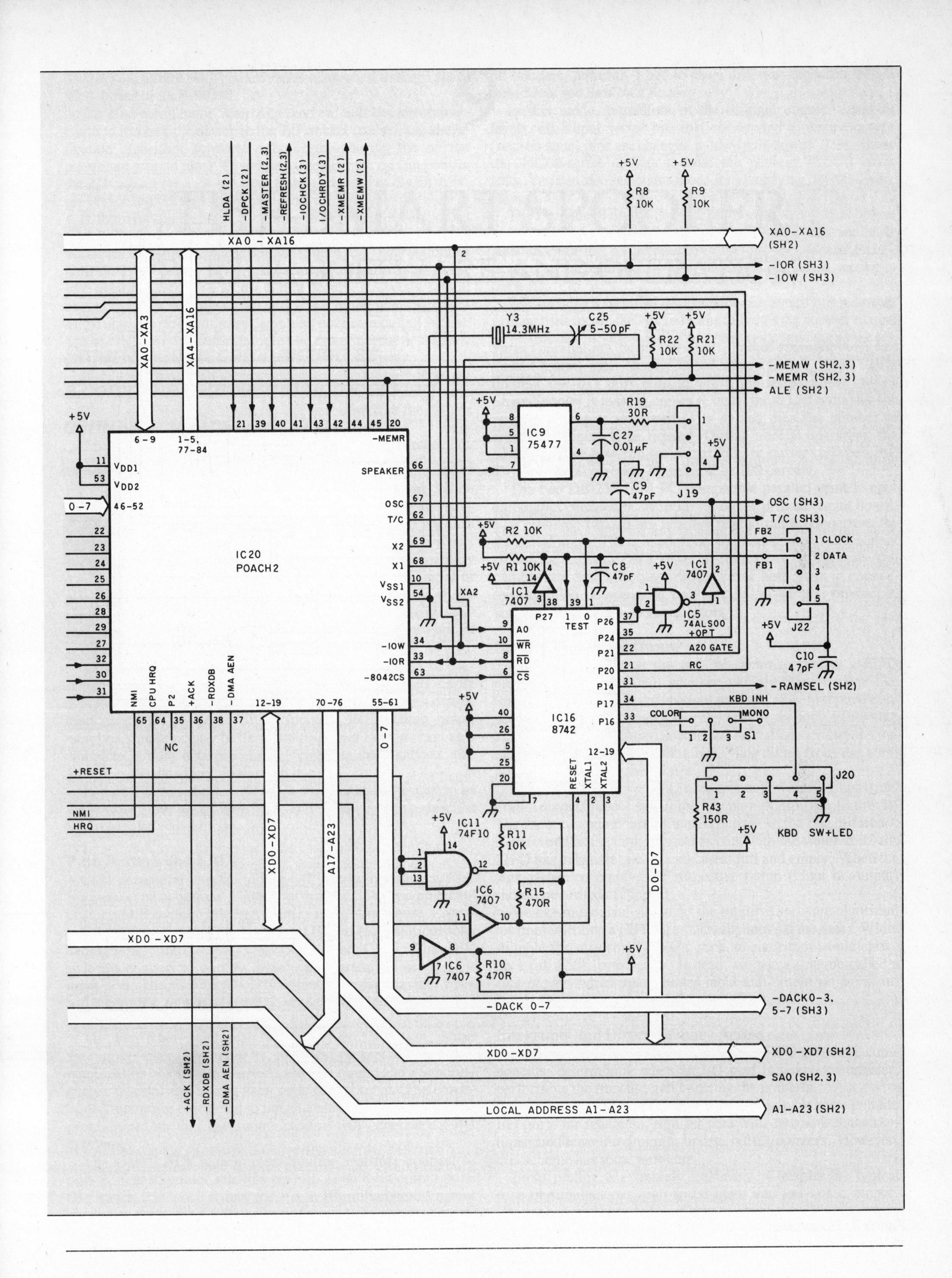

HLDA (2)
-DPCK (2)
-MASTER (2,3)
-REFRESH (2,3)
-IOCHCK (3)
I/OCHRDY (3)
-XMEMR (2)
-XMEMW (2)
XA0 - XA16
XA0-XA16 (SH2)
-IOR (SH3)
-IOW (SH3)
XA0-XA3
XA4-XA16
Y3 14.3MHz
C25 5-50pF
R8 10K
R9 10K
R22 10K
R21 10K
-MEMW (SH2, 3)
-MEMR (SH2, 3)
ALE (SH2)
IC9 75477
R19 30R
C27 0.01μF
C9 47pF
J19
IC20 POACH2
OSC (SH3)
T/C (SH3)
R2 10K
R1 10K
C8 47pF
IC1 7407
IC5 74ALS00 +OPT
1 CLOCK
2 DATA
J22
C10 47pF
IC16 8742
A20 GATE
RC
-RAMSEL (SH2)
KBD INH
COLOR
MONO
S1
J20
R43 150R
KBD
SW+LED
IC11 74F10
R11 10K
IC6 7407
R15 470R
R10 470R
XD0-XD7
A17-A23
D0-D7
-DACK 0-7
-DACK0-3, 5-7 (SH3)
XD0-XD7 (SH2)
SA0 (SH2, 3)
LOCAL ADDRESS A1-A23
A1-A23 (SH2)
+ACK (SH2)
-RDXDB (SH2)
-DMA AEN (SH2)
+RESET
NMI
HRQ

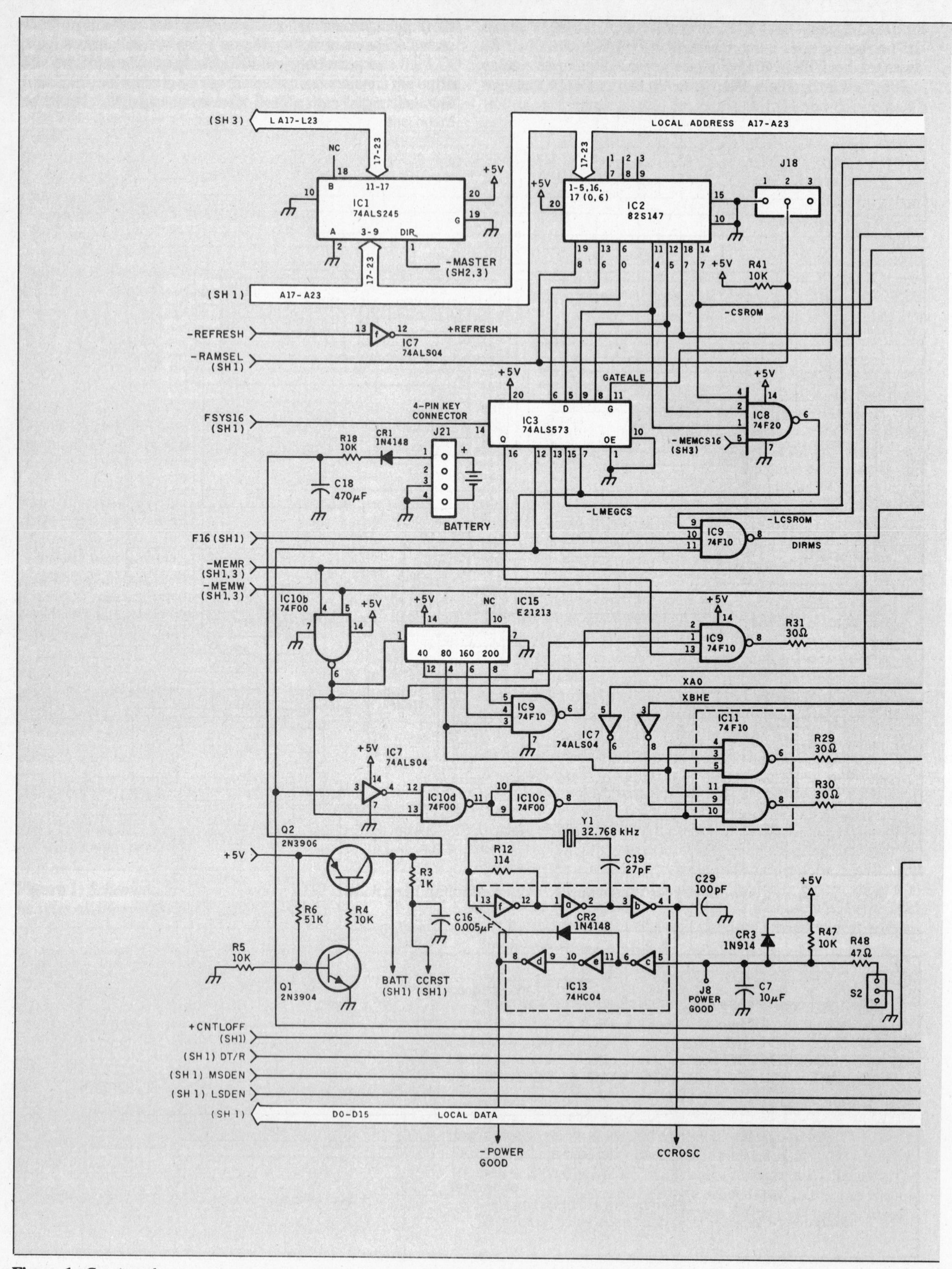

Figure 1: *Continued.*

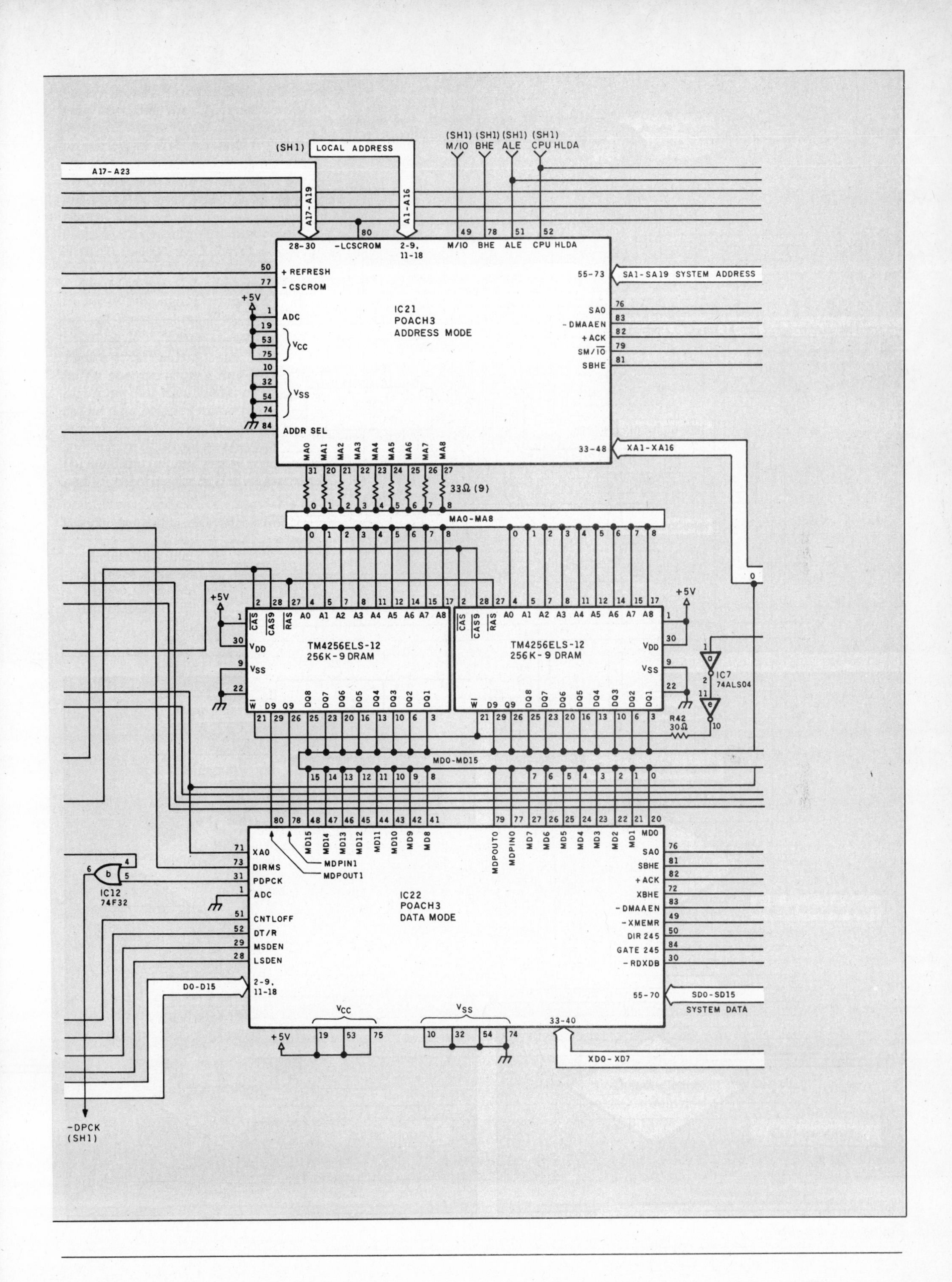
(SH1) LOCAL ADDRESS
(SH1) M/IO
(SH1) BHE
(SH1) ALE
(SH1) CPU HLDA
A17-A23
A17-A19
A1-A16
IC21
POACH3
ADDRESS MODE
+ REFRESH
- CSCROM
ADC
ADDR SEL
-LCSCROM
SA1-SA19 SYSTEM ADDRESS
SA0
- DMAAEN
+ ACK
SM/IO
SBHE
XA1-XA16
33Ω (9)
MA0-MA8
TM4256ELS-12
256K-9 DRAM
74ALS04
R42
30Ω
MD0-MD15
IC22
POACH3
DATA MODE
XA0
DIRMS
PDPCK
CNTLOFF
DT/R
MSDEN
LSDEN
MDPIN1
MDPOUT1
IC12
74F32
D0-D15
XBHE
- XMEMR
DIR 245
GATE 245
- RDXDB
SD0-SD15
SYSTEM DATA
XD0-XD7
-DPCK
(SH1)

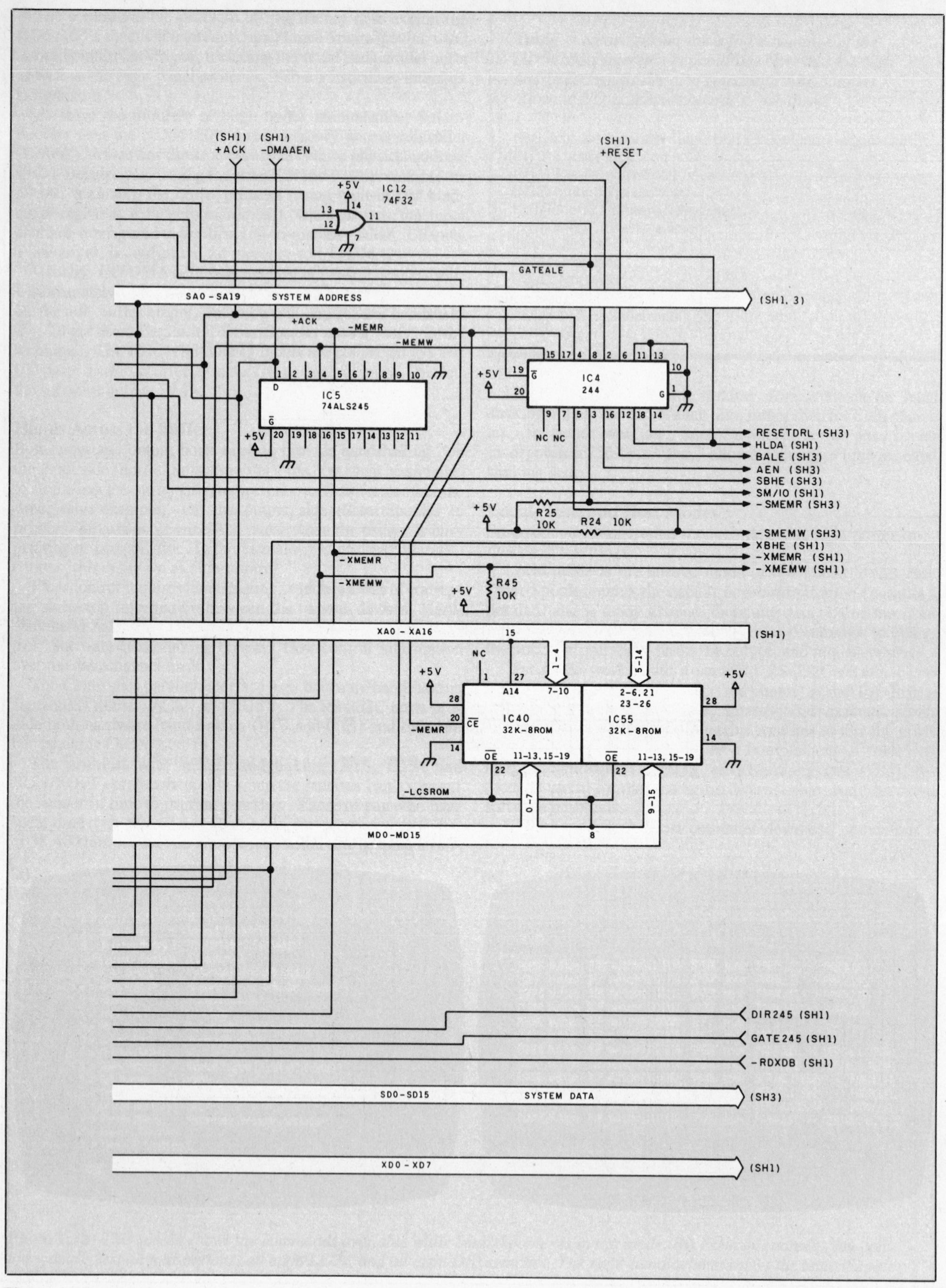

Figure 1: *Continued.*

fectively as the length increases, particularly if it becomes tuned to the carrying frequency. Integration collapses trace lengths, subsequently reducing emissions. Also, since the POACH set is CMOS, power requirements are substantially reduced. The original IBM PC AT motherboard drained 5 amperes. The CCAT requires less than 1 A (typically, about 0.8) at 5 volts.

The final configuration agreed upon for the CCAT was a six-layer printed circuit board with an AT expansion-board form factor: 4.8 inches high by 13.12 inches long (see photo 1). The four 84-pin packages in the center right portion of the board are the POACH chips. To their left are the three processors in the system: the 8742 keyboard controller, the 80287 math coprocessor, and the 80286 16-bit microprocessor. The Award Software ROM BIOS is located in the two sockets on the lower left portion of the board, labeled Low and High for low byte and high byte, respectively. The 16-MHz crystal, which provides clocking for the whole system, lies between the 8742 and POACH1.

Power, ground, system-address, data, and control signals are available at the edge connectors (see figure 2). These edge connectors follow the standard pin-out of the IBM PC AT bus. The speaker and external backup-battery connections are located on the upper right edge of the card, immediately above the on-board battery. Both are 4-pin Berg strips with pin assignments, as shown in table 1.

A hardware clone of an AT is not considered compatible unless its operating system and application software also function in a manner equivalent to the way they would on an IBM PC AT. This compatibility is facilitated through the ROM and keyboard-controller BIOSes provided by Award Software for the CCAT. This licensed software is among the most efficient available to IBM PC-compatible developers.

One handy feature of the Award BIOS on the CCAT is that the Setup program usually provided on disk is built into the ROM. Invoking the Ctrl-Alt-Esc key sequence enters Setup where you can set the following options: date, time, diskette1, diskette2, disk1, disk2, video, base memory, extended memory, and error halt.

Putting the CCAT to Use

Even though it might be obvious at this point, I want to emphasize that the CCAT is the equivalent of the IBM PC AT motherboard. Like any AT motherboard, it needs other peripheral cards to function as a computer system. To create a system, you plug the CCAT and all the peripheral cards into a passive backplane (often called a passive motherboard) that carries all the signals from the CCAT to the other peripheral cards. (A passive backplane contains no circuitry—only connectors and connecting wires.)

Neglecting some of the new super-multifunction boards, a minimum CCAT system would require a display-driver card (e.g., EGA, CGA, and monochrome), an AT floppy disk or AT floppy/hard disk controller card, and a keyboard. For a complete system, you would add one more combo card containing memory expansion and serial and parallel ports. Thus, a full-function CCAT computer is made up of four cards. Visualize four expansion cards plugged into your present XT or AT, and you will see that it takes relatively little volume. That giant motherboard and power supply were hogging all the space!

I've already stated that the CCAT is both smaller and more power-efficient than a standard AT configuration. Being more efficient, it needs no power-supply fan or monster power supply (power requirements beyond the CCAT, of course, depend upon the specific peripherals you plan to use). Newly introduced VLSI display and disk-controller cards are also more energy-efficient as well.

In Conclusion

While the greatest audience for the CCAT design will eventually be OEMs looking for a better 80286-based computer, such testaments are relatively boring to an end user reading this chapter. Instead, to provide a suitable demonstration, it was only natural for me to consider making a briefcase-size, battery-operated portable computer as the conclusion to this project.

The CCAT is both smaller and more power-efficient than a standard IBM PC AT configuration. It needs no power-supply fan or monster power supply.

Unfortunately, like most computer systems these days, objective is not necessarily reality. I briefly considered building something to rival one of the commercially produced portable computers, but that was like trying to fit 10 pounds in a 5-pound bag. The CCAT does indeed reduce the size of the AT electronics to a point where such a project is conceivable, but standard form-factor peripheral cards, disk drives, batteries, a

Table 1: *Speaker and external backup-battery pin connections.*

Pin	Speaker	Battery
1	Data out	Ground
2	Key	Not used
3	Ground	Not used
4	+5 V DC	6 V DC

Photo 2: *CCAT in a box.*

With the CCAT board, the only task becomes that of mechanically fitting off-the-shelf peripheral hardware in the smallest box.

display, and an AT keyboard just wouldn't fit in a briefcase.

Short of redesigning everything and making this portable computer a bigger project than the CCAT, I had to resort to using a larger case. Fortunately, I had this nice camcorder case sitting around holding an infrequently used camcorder. Out went the camcorder, and in went the CCAT and a bunch of other junk. Ten pounds eventually evolved into 30 pounds, with my minimal but efficient configuration losing out to enclosing everything but the kitchen sink. Equipment cases are like mass storage—the more room you have, the faster it seems to be filled.

My CCAT portable uses a backlit LCD, as most portables do. This $1395 display from Axonix Corp. (2257 South 1100 East, Suite 2C, Salt Lake City, UT 84106) has 640- by 200-pixel resolution and connects to the RGB output of a CGA display card, making it suitable for homebrew applications.

This display—and everything else in the unit—is powered from a pair of 4-A-hr 12-V gel-cel batteries (read that as *heavy*). The 12-V battery output is also converted to +5 V for the computer section and −12 V for RS-232.

The brain is a four-card AT system mounted on its side to reduce space. It consists of the CCAT board, a standard AT-style keyboard, an IBM CGA board, and clones of standard AT floppy/hard disk controllers and AST SixPakPlus expansion boards. Since we didn't know any better and nobody lifted the case until we finished, we added both a half-height floppy disk drive and a 40-megabyte hard disk drive.

There is nothing more I can add about this portable, beyond telling you what is in the case. With the CCAT board, the only task becomes that of mechanically fitting all this off-the-shelf peripheral hardware in the smallest box. We succeeded, and it works well indeed (see photo 2).

It operates for about 2 hours on the internal batteries, but it is better to plug it in the wall with the rest of the computers.

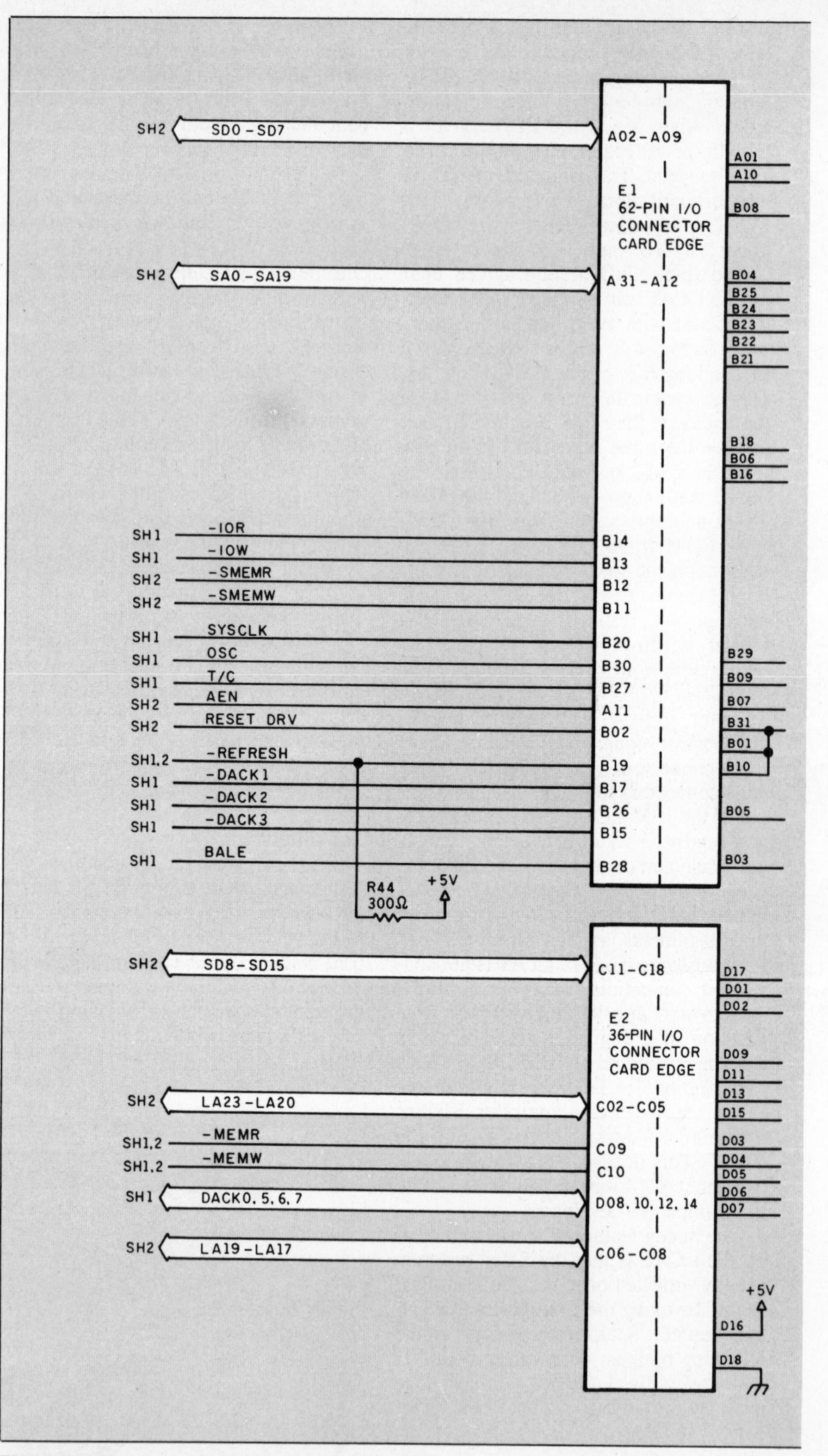

Figure 2: *Edge-connector pin-outs for the CCAT.*

And, while technically portable, it is a bit too large to use on your lap in an airplane. Still, it serves well as a transportable, and these 8088 portables are put to shame when you turn on the switch and crank up the CCAT's 10-MHz 80286.

The CCAT was a joint venture, and I'd like to note the contributions and help

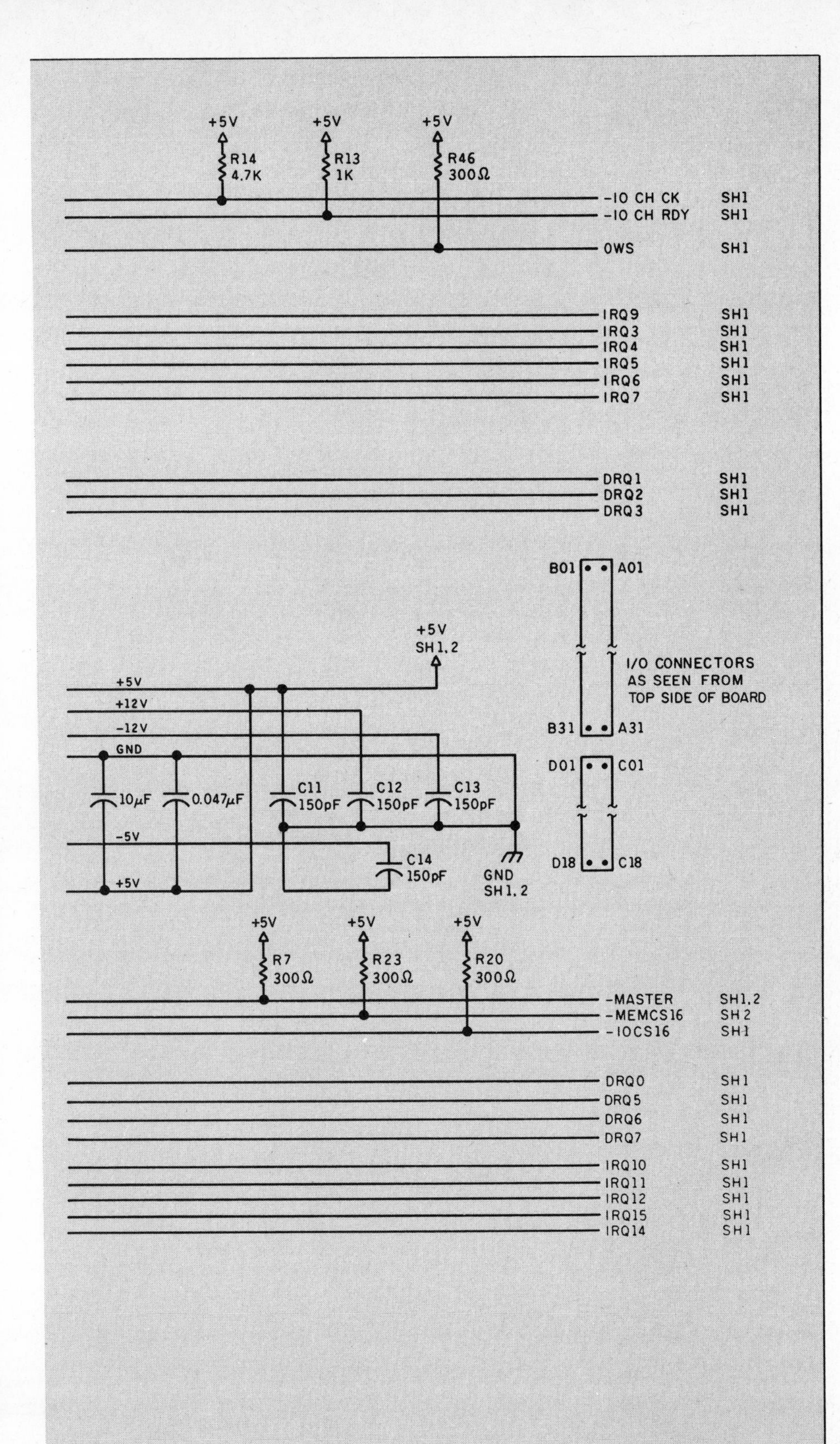

from ZyMOS, Micromint, Award Software, and the Circuit Cellar research staff. In addition, I'd like to personally thank Bob Andrews, Jeff Bachiochi, Jeff Remmers, Steve Smith, and Charles Skyles for their efforts.

Editor's Note: Steve often refers to previous Circuit Cellar articles. Most of these past articles are available in book form from BYTE Books, McGraw-Hill Publishing Company, P.O. Box 400, Hightstown, NJ 08250, (1-800-2-MCGRAW).

Ciarcia's Circuit Cellar, Volume I covers articles in BYTE from September 1977 through November 1978. *Volume II* covers December 1978 through June 1980. *Volume III* covers July 1980 through December 1981. *Volume IV* covers January 1982 through June 1983. *Volume V* covers July 1983 through December 1984.

The following items are available from

Micromint Inc.
4 Park St.
Vernon, CT 06066
(800) 635-3355 (orders)
(203) 871-6170 (information)
Telex: 643331

1. Assembled and tested CCAT 80286 PC AT-compatible CPU board with 512K bytes of RAM, Award BIOS in ROM, and user's manual. Available in 10 MHz.
 10 MHz. order OEM286LP/10........$620
2. Assembled eight-slot AT-style passive backplane with all connectors.
 order MBAT/8 ....$115

For either source above, all payments should be made in U.S. dollars by check, money order, MasterCard, Visa, or American Express. Surface delivery (U.S. and Canada only): add $5 for U.S., $8 for Canada. For delivery to Europe via U.S. airmail, add $14. Three-day air freight delivery: add $10 for U.S. (UPS Blue), $25 for Canada (Purolator overnight), $45 for Europe (Federal Express), or $60 for Asia and elsewhere in the world (Federal Express).

There is an on-line Circuit Cellar bulletin board system that supports past and present projects. You are invited to call and exchange ideas and comments with other Circuit Cellar supporters. The 300/1200/2400-bps BBS is on-line 24 hours a day at (203) 871-1988.

12

A SUPERCOMPUTER

PART 1

Steve begins a supercomputer project by looking at multiprocessing basics

Every month, I get several hundred letters from readers. Many of them ask for help with specific hardware or software problems, but there are always a few letters chiding me for not building a *real* computer. They imply that even 80386 and 68030 machines are simply uninteresting, and that I should design a supercomputer of one sort or another.

Unfortunately, the problem with a supercomputer is that it takes super software to drive it. Remember, my favorite programming language is solder, and that doesn't make me particularly fond of introducing "Yet Another Computer System" with "Yet Another Programming Language."

But around the Circuit Cellar we like to do things that are out of the ordinary. I thus decided to see what it would take to build a supercomputer, and I thought you would be interested in how I determined the proper architecture. The result is a three-part chapter describing multiprocessing that starts with the theory and ends with reality.

The Circuit Cellar supercomputer is a multiprocessor computer specialized to evaluate the iterative formula describing the Mandelbrot set, so a more accurate name is the Circuit Cellar Mandelbrot engine. A driver program running on an IBM PC AT presents the results in real-time color on an EGA or a VGA, with smooth panning and scrolling so you can examine the results on the fly. System performance increases as you install more processors. You can start with a single processor, graduate to 64 processors (as in my example), or work up to a monster system with 256 processors.

The Circuit Cellar Mandelbrot engine starts at roughly twice the performance of an 8-MHz AT with just one card of 8 processors, increases smoothly past a 16-MHz 80386 with three cards (24 processors), and zooms far beyond it with 64 processors (eight cards). I haven't found anything (under $500,000) to compare with a full-bore system of 32 cards. Not bad, considering that the Mandelbrot engine is based on the lowly Intel 8051.

Here in part 1 of Chapter 12, I'll describe how to increase the performance of single-processor systems and show why there's a definite upper limit to processing speed. The solution seems to be using multiple computers on the same problem, so I'll explore some of the different ways to connect multiple processors in arrays, and the troubles that arise from these connections.

Before launching into a discussion of multiprocessing, I'll review some of the performance problems and solutions for single-processor systems. I'll start with the simplest possible system and work up to pipelining and caching.

Building Performance

In comparing performance, you must be careful what you're measuring. A convenient unit is the number of instructions per second, which you get by dividing the total number of instructions executed by the elapsed time from start to finish. A processor that executes twice as many instructions in a given time has twice the performance, for an increase of 100 percent. (Some of the examples I'll give will focus on the number of cycles per instruction, which is the reciprocal of the number of instructions per cycle. Be careful not to compute the performance ratio upside down.)

Figure 1 shows the two essential components of a computer: a CPU and memory to hold the program and data. The fat arrow between the two represents the address, data, and control lines running between them. For these discussions, I'll ignore the necessary I/O hardware and presume that the program and data are already loaded into memory.

This CPU is so simple that it doesn't include any registers; all operations must refer to memory locations. For example, an ADD instruction must specify three memory locations: one each for the two numbers to be added and where to put the result. Although your favorite microcomputer may not have such an ADD instruction, the earliest computers (back in the Good Old Days of relays and vacuum tubes) actually worked this way. Figure 2 shows the execution sequence for the ADD instruction I've described. Each vertical line marks a single CPU clock cycle or memory access.

The first step, of course, is to fetch the ADD instruction from memory. After the instruction arrives in the CPU, it is decoded to determine the addresses of the operands. The CPU then fetches the operands and performs the addition. Finally, the CPU stores the result back into memory. This sequence repeats for each instruction, with some variation.

What's of interest is that a single instruction requires four memory accesses: an instruction fetch, two data fetches, and one data store. During two more cycles, memory is idle while the CPU decodes and executes the instruction. Other instructions have different sequences, but the overall pattern is similar.

The memory in this example must be able to return data within a single CPU cycle and also be ready for another access at the start of the next cycle. Dynamic RAMs need some time after an access to get ready for the next operation; the minimum time between accesses is the DRAM's cycle time. The memory's cycle time is necessarily longer than the access time required to return data. Typically, DRAMs have a cycle time that's about twice the access time.

For example, premium DRAMs with a 100-nanosecond access time have a cycle time of about 200 ns, so each CPU cycle must be 200 ns. The ADD instruction I've presented will therefore take 6 × 200 ns, or 1200 ns. Some instructions may be faster and some slower, so the CPU will run at about 800,000 instructions per second.

Increasing the performance of this machine by a factor of 2 sounds simple enough: Double the clock frequency and reduce the CPU and memory cycle times to 100 ns. Unfortunately, DRAMs with a 50-ns access time and a 100-ns cycle time are on the forefront of technology right now, and more expensive than you can imagine. But all is not lost.

Figure 1: *The essential parts of a computer system.*

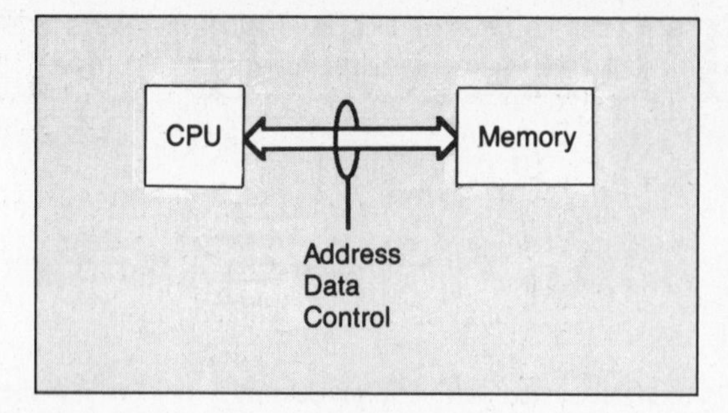

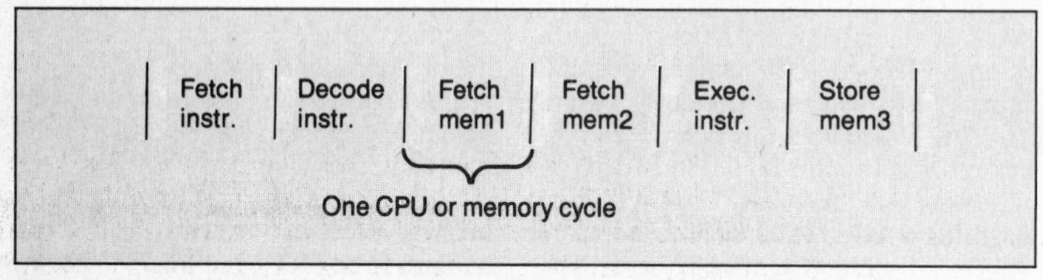

Figure 2: *Execution sequence for the instruction* ADD mem1,mem2,mem3.

Figure 3: *Adding registers to the CPU of figure 1.*

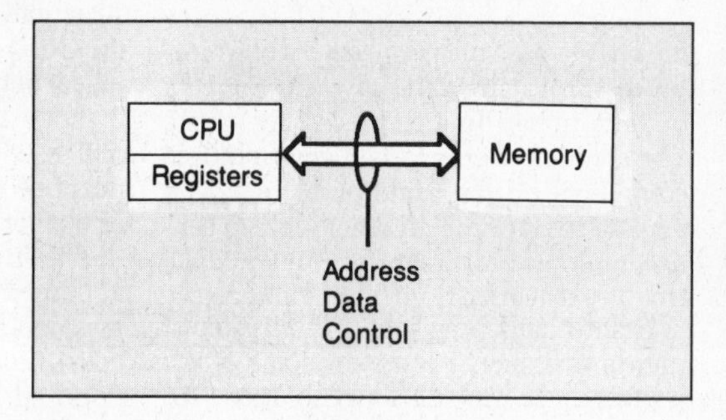

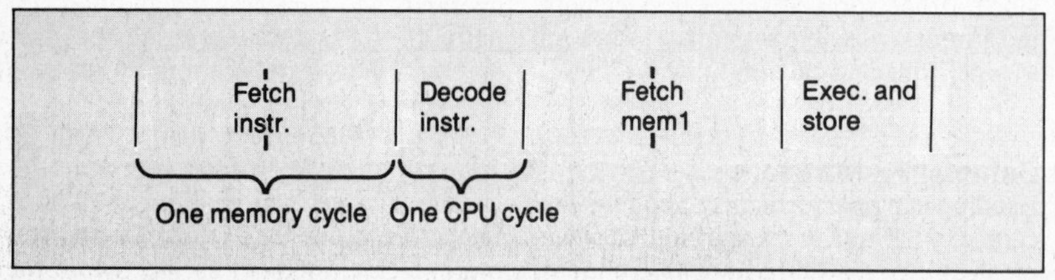

Figure 4: *Execution sequence for* ADD mem1,reg. *Note that memory access is now two CPU cycles (see text).*

Registered Speedup

Figure 3 shows a more complex CPU with internal data registers. Each register can hold the same amount of data as one memory location, so an instruction can refer to either memory or registers. Because the registers are on the CPU chip, register accesses are faster than memory accesses. To take advantage of this, the definition of an ADD instruction changes so that it now adds the contents of a memory location to an internal register and puts the result back into the same register. This reduces the number of memory accesses to two: one instruction fetch and one data fetch.

The reason for these changes is to let the CPU run with a faster clock rate than the memory could otherwise tolerate. The CPU cycle time can now be 100 ns, half the memory cycle time of 200 ns. Any memory access must include one extra cycle, but operations within the CPU can now proceed twice as fast as before.

Figure 4 shows the execution of the new ADD instruction. Fetching the instruction takes two clock cycles because of the memory access, but decoding it takes only one. Fetching the operand from memory takes two more cycles, but the result is computed and stored in a register in a single cycle. The ADD instruction takes six cycles from start to finish, but the faster CPU clock rate reduces the total elapsed time to only 6 × 100 ns, or 600 ns, half that of the processor in figure 1.

Both processors use the same type of memory, but the second system has twice the performance of the first. Bearing in mind that a typical system has only one CPU and several megabytes of memory, a more complex CPU is a good way to improve the overall system performance without increasing the overall cost beyond reason.

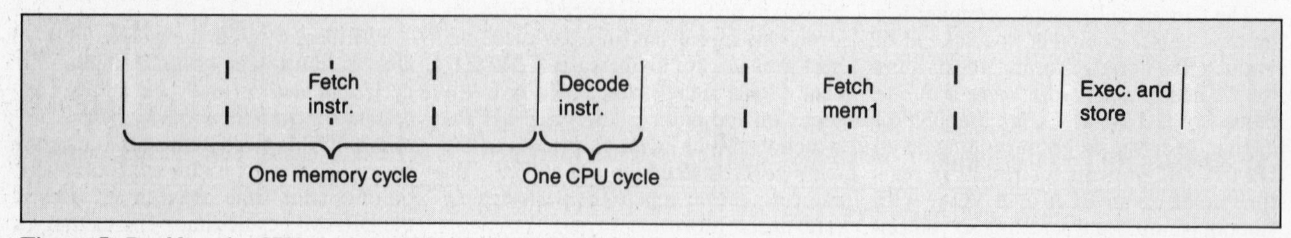

Figure 5: *Doubling the CPU clock speed for the instruction shown in figure 4 yields this sequence. Since the CPU is running twice as fast, memory access now requires four cycles.*

The extra cycle for each memory access is commonly called a wait state. Many of the newer AT clones run with zero-wait-state memory, which simply means that the memory can keep up with the processor. The complete details are a little more complicated than I've shown here, because the AT's memory is actually measured by access time rather than cycle time, but the principle is similar.

If doubling the CPU clock rate helped so much, how about doubling it again? Memory accesses now require 4 cycles (4 × 50 ns, or 200 ns), and figure 5 shows what happens. The ADD instruction now takes 10 cycles, for an elapsed time of 10 × 50 ns, or 500 ns. Doubling the clock rate improves performance by only about 20 percent because the CPU now spends most of its time waiting for memory accesses.

But if 100-ns-cycle-time memory was too expensive, you can imagine what 50-ns memory will cost. At some point, the system will outrun the fastest DRAMs, so static RAMs are the only choice. SRAMs have about 25 percent the density of DRAMs, so the chip area that can hold a 1-megabyte DRAM will hold only 256K bytes of SRAM. Prices are driven by chip area, so the memory cost increases by a factor of 4, even without considering the additional cost of the faster memory.

Access Caching

There's another trick we can use: memory-access caching.

Although the system may have megabytes of memory, most program instructions are clustered in small groups. For example, a loop may execute a dozen instructions hundreds of times. Data accesses can be clustered in the same way, as with a word processor updating successive characters in a buffer.

A cache takes advantage of this typical program behavior by storing the most recently accessed instructions and data in a local memory that's much faster than the main memory. Figure 6 shows a cache inserted between the CPU and the memory unit. Instructions and data in this cache can be returned in one cycle, just like the CPU registers. But if there's a cache miss and the cache must access the main memory, the access will take five CPU cycles.

Assuming that the CPU is running at 50 ns, figure 7a shows that an ADD instruction with all cache hits takes only 4 × 50 ns, or 200 ns. If those hits turn into misses, the ADD instruction takes 12 × 50 ns, or 600 ns (see figure 7b). It's obvious that the cache hit ratio determines the overall system performance.

More complex caches guess where the processor will need data and prefetch from those locations so that the CPU's accesses will be hits. Some systems have separate instruction and data caches with different updating strategies to take advantage of the differences in access patterns. In fact, a cache is one of the trickiest parts of a system, and it can harbor the most obscure bugs for the longest times.

The Last Drop: Pipelining

We can squeeze more performance from the processor by introducing instruction pipelining. Pipelining, also known as overlapped execution, takes advantage of the fact that each instruction breaks down into several distinct phases. The ADD instruction I've been using has four phases, which I'll call I-fetch, I-decode, D-fetch, and D-store. By adding CPU hardware to handle each phase separately, we arrive at figure 8.

Figure 9 shows the sequence of events as the CPU begins executing a series of ADD instructions, each with different memory and register operands. A new ADD instruction starts every clock cycle, so, after the pipeline fills, the throughput is one ADD instruction every clock cycle. Although an ADD instruction (with cache hits) still takes four cycles, one ADD instruction finishes every cycle, so the overall performance is 50 ns per ADD instruction.

In this example, the pipeline hardware improved performance by a factor of 4 at the same clock rate. As with the other tricks, pipelining doesn't always provide that much improvement. For example, if an instruction needs a register set by a previous ADD instruction, the CPU must ensure that the instructions complete in the right order.

Not all instructions have the same number of phases, so the pipeline may not always be full. Branch instructions are a particular problem, because the system can't determine the next instruc-

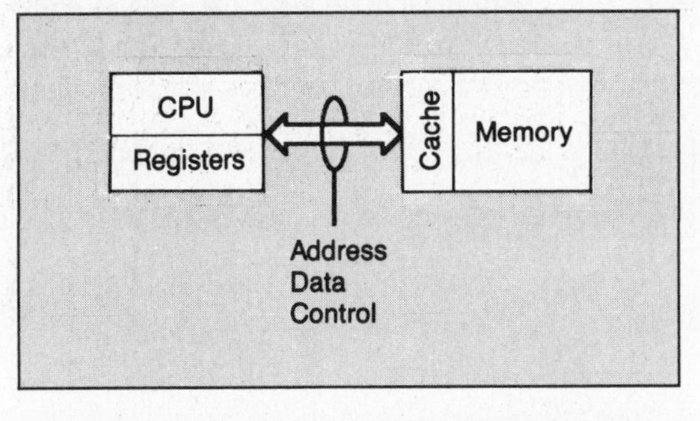

Figure 6: *The modified system from figure 3 is further modified by adding a cache between the CPU and memory.*

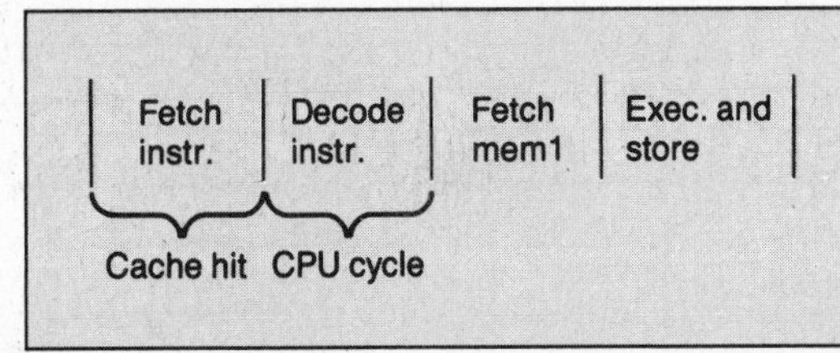

Figure 7a: *Executing the instruction* `ADD mem1,reg` *on the cached system, assuming one-cycle cache hits.*

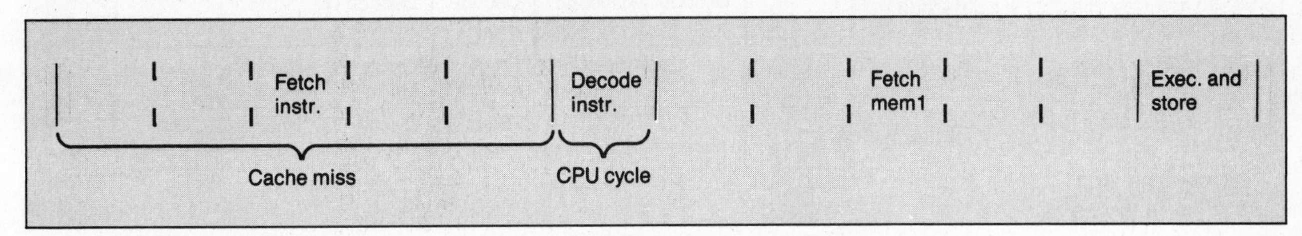

Figure 7b: *Executing the same instruction as in figure 7a with five-cycle cache misses.*

tion address until after the branch is decoded. As with caches, pipelines require some very subtle logic to take account of all the possible combinations.

It's worthwhile to remember that these tricks do not always improve the processor's performance. For example, if a program doesn't use a small set of data that fits into the cache, every data reference will be a cache miss. Similarly, if the program has many branches that flush the pipeline, the throughput will drop off to the level of a processor that hasn't been pipelined. No matter what the hardware assumes, someone can always write a program to bring the system to its knees.

Performance Limits

From these examples, you can see that the ultimate limit to a processor's performance is one instruction per clock cycle. Although several instructions may be in the pipeline, each clock cycle will produce only one result at a time. That's the upper, theoretical, ideal limit, so cache misses and pipeline flushes can only detract from it.

Because a single processor can produce only one result in a clock cycle, the only way to boost performance (for a given processor design) is to increase the clock frequency. This explains the "clock race" pushing 80386 machines beyond 20 MHz. Unfortunately, performance increases by the same factor as the clock rate. Therefore, a 24-MHz 80286 is, at best, three times faster than an 8-MHz 80286.

What the block diagrams do not show is the incredible amount of hardware required to add caches, pipelining, and all the other sorcery to support the higher clock rates. All this circuitry must run at the increased speed, and designers find out that tricks that worked well at lower speeds aren't as effective at higher speeds.

Regardless of your cleverness, the hardware circuit technology will set an upper limit to the clock rate. Mainframe computers, with custom LSI chips and optimized interconnections, run at about 60 MHz. In round numbers, 30 MHz is the fastest clock rate you'll see on your desktop computer for quite a while, because the price of the technology goes up dramatically beyond that point.

A 30-MHz clock rate means that a processor can produce one result every 33 ns at best. If the program and hardware can support one instruction per clock cycle, the processor will hit 33 million instructions per second. In real life, of course, your mileage may vary. . . .

To put this into perspective, a standard 4.77-MHz IBM PC runs at about 0.1 MIPS. A stock 8-MHz AT runs at about 0.5 MIPS, and a 20-MHz PS/2 Model 80 does about 3 MIPS. The best we can expect in the near future is only another order of magnitude faster.

Obviously, something has to give if we want still more performance from a computer system.

Architectural Choices

If one processor can deliver one answer in one clock cycle, how many answers will two processors deliver in the same time? Although this sounds like a child's riddle, the answer is profound: two!

The work needed to solve some problems can be divided more or less equally among multiple processors. If each processor can run simultaneously, the total performance increases directly with the number of processors. Two processors will produce the result in half the time of a single processor, and so on.

There are several different types of multiprocessor systems, each suited for a different class of problem. I'll discuss some of the main architectural choices and describe what sort of problem each is best suited to solve; after that, the description of the Circuit Cellar supercomputer will make more sense.

A *scalar* is a mathematical term indicating a value that can be expressed as a single number. A scalar computer, therefore, can work on a single number at a time. You are probably most familiar with scalar processors, although you may never have thought of your computer in quite that way before. All the examples in the previous sections have dealt with scalar processor design.

A *vector* is a value that must be expressed with two or more numbers. For example, the coordinates for a point in space consist of three numbers: the distance from the origin along the *x*, *y*, and *z* axes. The notion of a vector is more flexible than that, though, and can describe

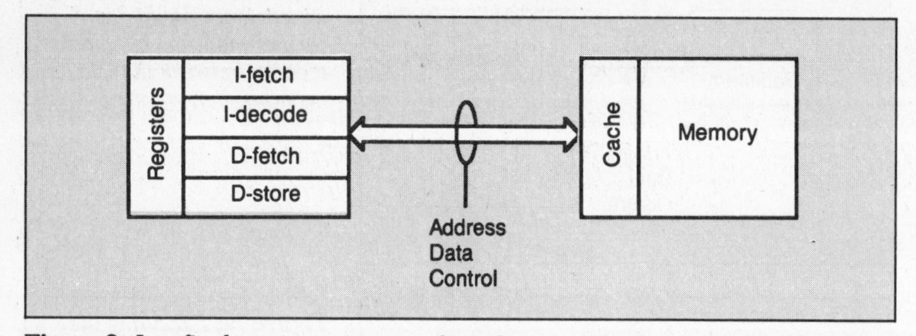

Figure 8: *In a final attempt to improve throughput, pipelining is added to the CPU.*

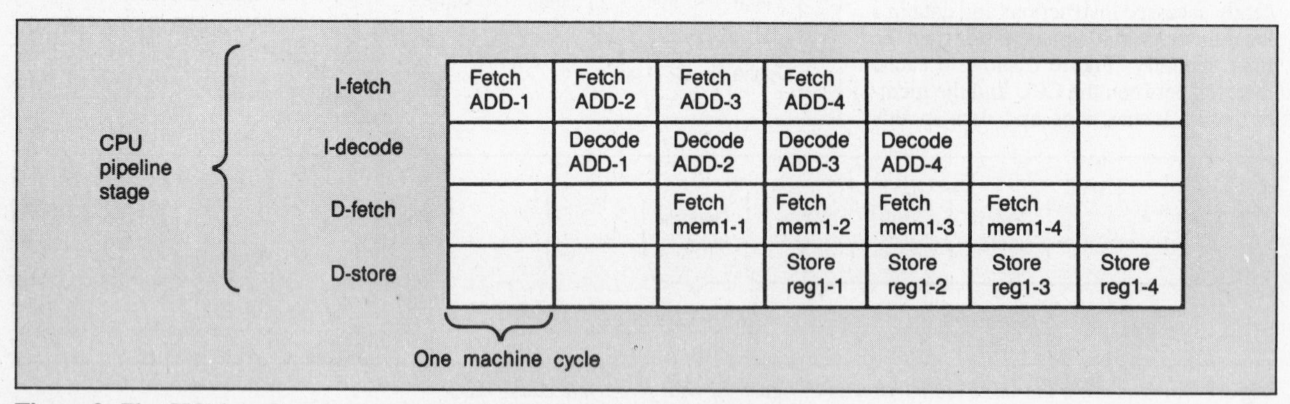

Figure 9: *The CPU from figure 8 executes a series of* `ADD mem1,reg` *instructions.*

Figure 10: *Two processors connected to a common memory. Access to the memory is regulated by a memory control.*

Figure 11: *Both CPUs of the system described in figure 10 execute* ADD *instructions.* ▼

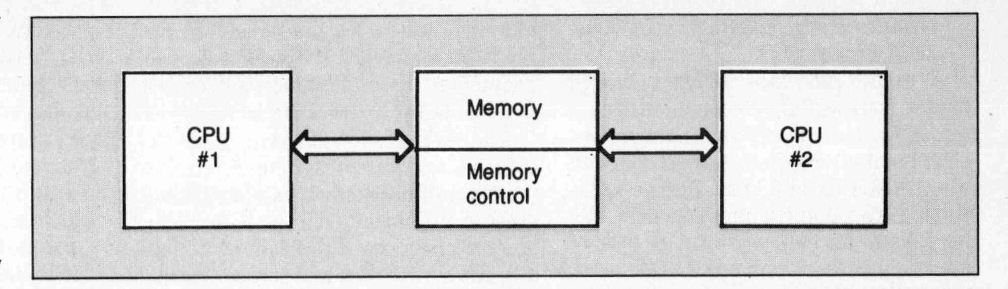

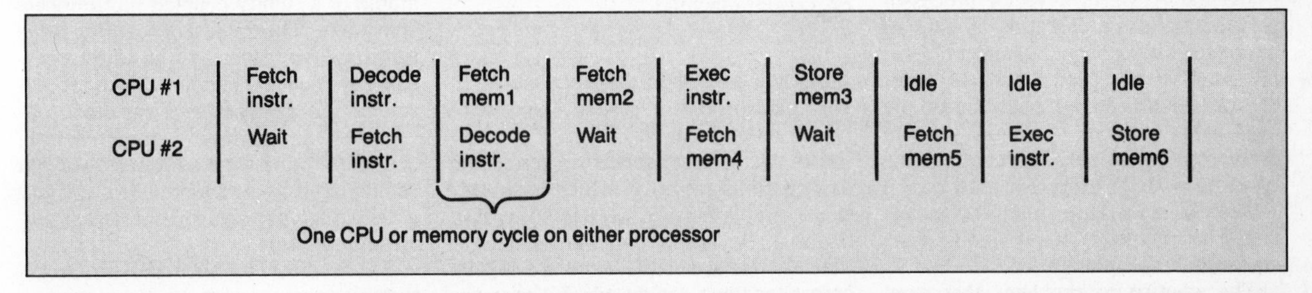

Figure 12: *A four-way tightly coupled multiprocessor system.*

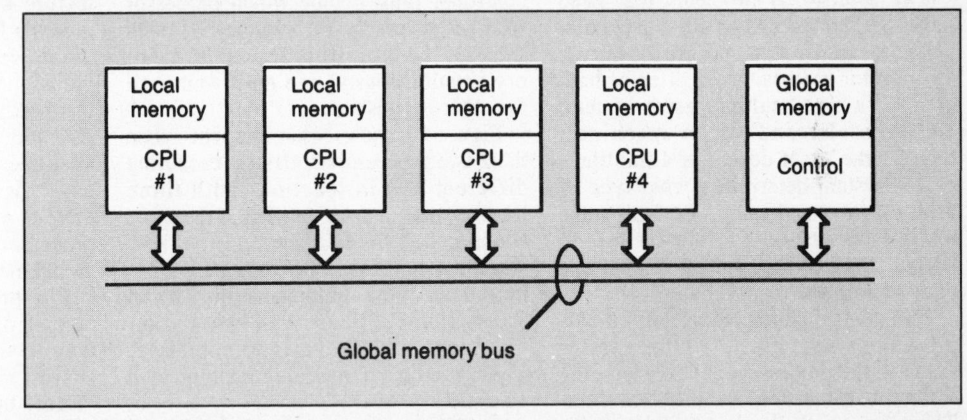

Figure 13: *A four-way direct-connected multiprocessor system.*

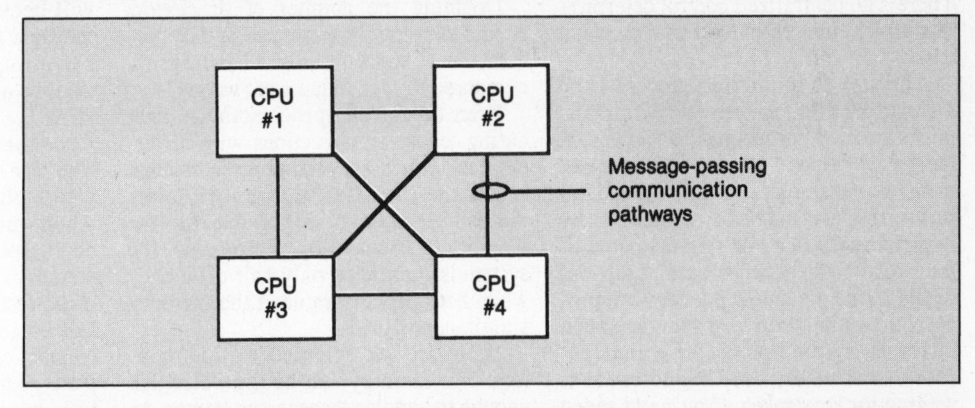

Figure 14: *A multiprocessor system consisting of six CPUs connected via a common message bus.*

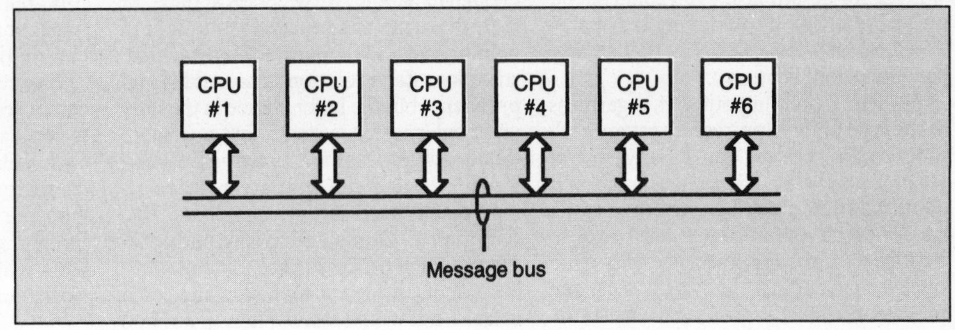

points with hundreds or thousands of associated numbers.

A vector processor includes enough hardware to perform computations on all (or at least many of) the numbers for each point at one time. A machine with three ALUs could add all three coordinates of two points simultaneously, delivering the result in one-third the time of a scalar processor with one ALU.

All the ALUs in a vector processor perform the same operation on each of the vector's components. Many problems involving vectors need exactly this sort of lock-step processing, but some simply need more freedom. Although you can create some variations, in general, all the ALUs in a vector processor must do the same thing at the same time. This means that some problems simply don't fit the vector processor pattern.

The solution to problems that need more performance than the best scalar processor can provide, but are too unruly for a vector processor, can often be handled by a true multiprocessor system. Unlike scalar and vector machines, though, the exact design of a multiprocessor system determines what type of problems it can handle. In fact, some multiprocessors on the market are so specialized that they can solve only one class of problem.

The distinguishing feature of a multiprocessor system is that—unlike a vector system—the processors are all executing different instructions on different data. There's no centralized control determining which instruction to use on what data.

Although using multiple processors on a single problem can provide a dramatic performance improvement, not all problems will respond to this sort of treatment. For example, a program that computes the factorial of a number by recursive calls can't be split up on multiple processors, because each result depends on the preceding one. A word processor won't attain a dramatic speed increase on a multiprocessor system, because most of the time the software is waiting for keystrokes. (You could speed up reformatting by assigning one paragraph to each processor, but that's not a convincing application.)

It's worth pointing out that, regardless of the type of multiprocessor system, each individual processor can be any sort of scalar processor you'd like to use. Any and all of the tricks described earlier to crank up scalar performance are fair game in multiprocessor applications. The only catch is that, because the system has many processors, the cost goes up dramatically as each processor becomes more complex. Sometimes, as we'll see in part 2 of this chapter, many simple processors can outperform fewer complex ones.

And, as I mentioned above, the design of a multiprocessor system determines the types of problems it can handle with greatest efficiency. Just as with caches and pipelines, some problems will actually run slower on a multiprocessor than on a scalar machine.

Tightly Coupled Multiprocessors

Any multiprocessor system starts with two or more scalar machines. Figure 10 shows two simple processors connected to a common memory, which is similar to the simple DRAM we started with in figure 1. Because the memory can handle only one access at a time, a memory controller must decide which processor will get access to the memory on each cycle. If the processors access the memory simultaneously, one must wait until the other is finished.

Figure 11 shows what happens when these two processors both start executing different `ADD` instructions with three memory operands each. This is the same situation described in figure 2, but now you can see the lost time when CPU #2 is locked out of the common memory by an access from CPU #1. The two `ADD` instructions take 9 cycles to complete, where two on a pure scalar machine take 12 cycles.

Doubling the number of processors should increase performance by 100 percent, but it went up only 33 percent ($\frac{2}{9}$ compared to $\frac{2}{12}$). What went wrong?

Even though the processors are executing separate instructions with different data addresses, both processors must access the shared memory to get information. Because it is possible for the memory to handle only one request, the system is running at only half efficiency when both processors need the memory simultaneously.

There are two solutions to this problem: The memory can become complex enough to handle two accesses at once, or the processors can reduce the number of memory accesses required for each instruction. Each solution raises additional problems, but the latter choice is the only practical one for systems that use more than a few processors. Imagine building a memory that can support a dozen simultaneous accesses.

Figure 12 shows a four-way multiprocessor. Each processor has a local memory for its program and working variables. Results and status flags are stored in common memory, which is accessed over the global memory bus connected to each processor. The memory controller decides which of the four processors will gain access during each memory cycle. Processors that lose the battle for access to common memory must wait for the next free memory cycle.

Because each processor has direct access to the global memory, this is an example of a tightly coupled multiprocessor system. The tightest of coupling is the limiting case occurring when the processors have no local memory. Each processor can change the state of any other processor's computation by simply writing new data in the right addresses. As you can imagine, this may not be an entirely good thing, particularly for debugging errant programs.

The hardware problem with this architecture is that the global memory bus requires a large number of signal lines. If the processors use 32-bit words and the global memory has only 1 megabyte, the bus needs over 50 lines for just the data and address. High-frequency transmission-line techniques are required to extend this bus more than a few feet, so there is a distinct limit to the number of processors that can connect to a single global memory bus.

For problems that demand a large number of processors, there is no feasible way to connect each processor to a shared high-bandwidth memory bus. Worse, the contention for that memory will begin to wipe out the advantage of multiple processors (remember the simple example in figure 11). Again, there must be a better way.

Loosening the Bonds

The best performance for a tightly coupled multiprocessor architecture occurs when you are running programs that don't need much access to the global memory. If that is indeed the case, the wide, fast, expensive global memory bus can be replaced by a relatively narrow connection between processors. The ultimate reduction is a bidirectional serial link, but it could be a byte-wide or word-wide channel with some handshaking control lines.

Figure 13 shows a four-way multiprocessor connected by narrow ports between each pair of processors. Because each link has a relatively low bandwidth compared to the previous global memory bus, the processors can exchange only limited amounts of data. But for problems with fairly strict partitioning, this works reasonably well.

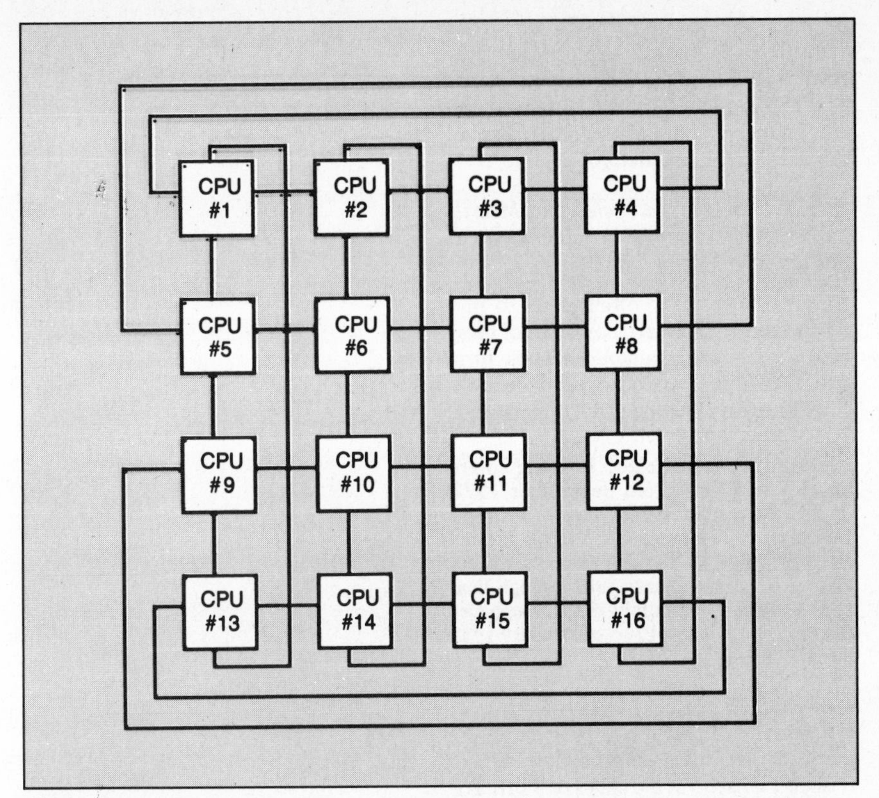

Figure 15: *A 16-processor system. Each processor can communicate with its four closest neighbors.*

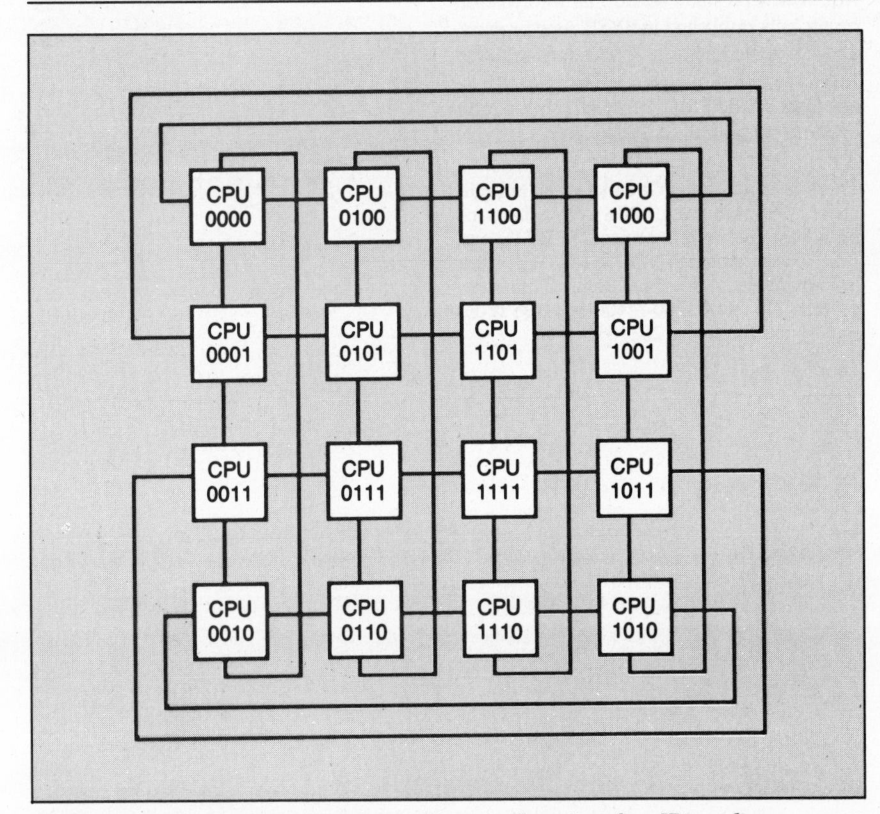

Figure 16: *The processors from figure 15 are each assigned an ID number to simplify message passing.*

Connecting each processor to all the others simplifies the task of exchanging data, but it requires $n-1$ ports on each processor. For four processors this is feasible, but I challenge you to draw the connections for a 16-processor system. For lots of small processors, there's a problem fitting all the connections into the available physical space.

One way around this is to connect all the processors to a common "message bus," as shown in figure 14. Although the figure looks much like figure 12, the difference is that the common bus is relatively narrow and doesn't connect to a global memory. Any processor can send a message to any group of the others, but only one transmitter can be active at any one time.

You'll notice a striking resemblance between figure 14 and the standard block diagram for a local-area network. In fact, although LANs are usually thought of as a way to share peripherals, they're also useful for coordinating the work of many processors. I've seen some work that uses otherwise idle computers on a LAN to perform "background" computations on a complex problem, shifting the calculations around the network to take best advantage of the available hardware.

The problem with a LAN, of course, is that there can be only one message active at a time. Regardless of the LAN bandwidth, there will be some lost time when the processors queue up to use that single resource.

Figure 15 shows one way around this problem. Each processor can send messages to its four closest neighbors, with processors on the edges of the array wrapped around to the other side. Depending on how the processors are programmed, each can support up to four messages at once. If the code is particularly clever, any processor can send a message to any other one by routing it along the shortest distance between the two.

That layout will work if most of the messages are to adjacent processors. Sending a message across the entire array will involve all the processors between the two nodes, and the overhead involved in figuring out the proper path can be significant. A slightly different way of connecting the processors can improve message passing, at the cost of greater wiring complexity.

Figure 16 looks almost the same as figure 15, with each processor linked to four others. If you look closely, you'll see that the processor numbers differ by only 1 bit across each link. That change makes the message-routing problem almost trivial.

When a processor receives a message, it compares the destination address in the message header with its own address by performing an exclusive-OR. If the result is 0, the two addresses are the same and it can process the message. If the result isn't 0, the message must be forwarded to another processor. The processor sends the message to the neighboring processor that has an address differing in that same bit position.

Using this method, you can interconnect 65,536 processors with only 16 links per processor and a maximum message delay of 16 transmissions. This seems to be the best way to connect truly large arrays to minimize transmission delay without unduly complicating the backplane wiring.

Communication Bottlenecks

What should be obvious from these sample systems is the need to figure out just what level of communication will occur between the processors. A large communication volume requires high-bandwidth connections, with the attendant complexity of common memory design. Programs with shorter, less frequent messages between fewer processors can run effectively on processors with "narrower" links, perhaps using message passing between links to reduce the number of distinct connections.

A multiprocessor system must be designed to solve a particular class of problems. Ideally, it will handle that class with particular efficiency (although every now and again the designers find that there's a skeleton in the closet that hampers performance). Attempting to fit a problem from a different class onto that machine will result in poor performance, perhaps even lower than on a scalar machine.

If you're at all familiar with the calculations behind the Mandelbrot set, you can probably tell why I elected to use it as the foundation for a supercomputer. A single image requires massive amounts of computation, but it can be easily divided between an arbitrary number of processors. The results of the computations can be summarized by 1 or 2 bytes, so the output data transmission can use a relatively slow link. Even better, the algorithm doesn't need any communication between neighboring processors, so the interprocessor communication isn't a critical issue.

In Chapter 12, part 2, I'll describe the system architecture of the Circuit Cellar Mandelbrot engine and explore the mathematics and algorithms behind its operation. In Chapter 12, part 3, I'll cover the array hardware and the control/display software for the Mandelbrot engine.

Special thanks to Ed Nisley and Merrill Lathers for their expert contributions to this project.

Editor's Note: Steve often refers to previous Circuit Cellar articles. Most of these past articles are available in book form from BYTE Books, McGraw-Hill Publishing Company, P.O. Box 400, Hightstown, NJ 08250, (1-800-2-MCGRAW).

Ciarcia's Circuit Cellar, Volume I covers articles in BYTE from September 1977 through November 1978. *Volume II* covers December 1978 through June 1980. *Volume III* covers July 1980 through December 1981. *Volume IV* covers January 1982 through June 1983. *Volume V* covers July 1983 through December 1984. *Volume VI* covers January 1985 through June 1986.

Circuit Cellar Ink

It's virtually impossible to provide all the pertinent details of a project or cover all the designs I'd like to in the pages of BYTE. For that reason, I have started a bimonthly supplemental publication called Circuit Cellar Ink, which presents additional information on projects published in BYTE, new projects, and supplemental applications-oriented materials. For a one-year subscription (6 issues), send $14.95 to Circuit Cellar Ink. Credit card orders can call (203) 875-2199.

There is a multiline Circuit Cellar bulletin board system (running TBBS 2.0M) that supports past and present projects in BYTE and Ink. You are invited to call and exchange ideas and comments with other Circuit Cellar supporters. The 300-/1200-/2400-bps BBS is online 24 hours a day at (203) 871-1988.

12

A SUPERCOMPUTER

PART 2

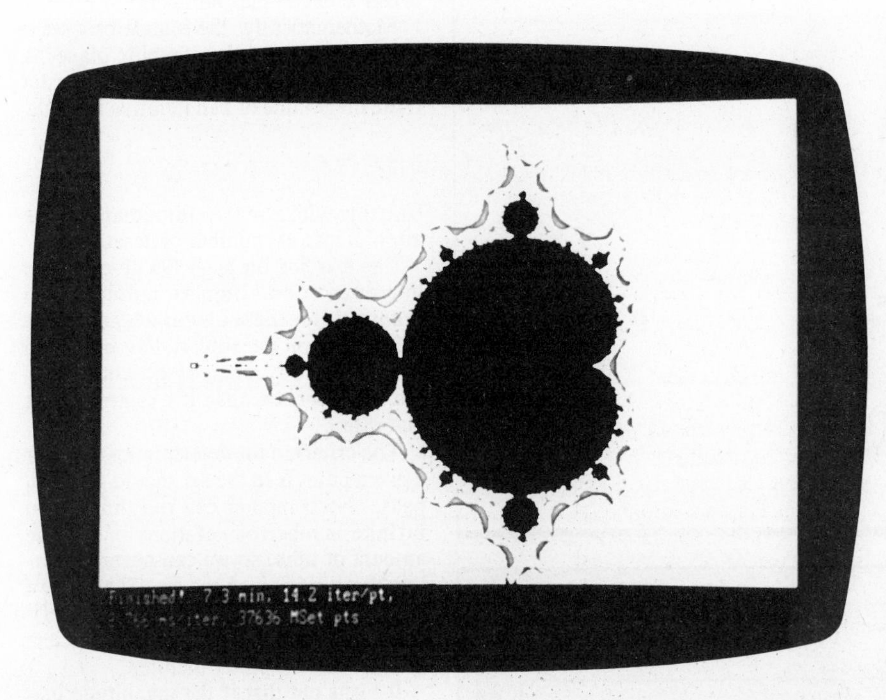

Photo 1: *The complete Mandelbrot set, as displayed by an EGA monitor (no magnification).*

Steve continues the supercomputer project with a look at the Mandelbrot set

In part 1 of Chapter 12, I explored the limits of computer performance and described several multiprocessor architectures. In parts 2 and 3, I will describe the Circuit Cellar Mandelbrot engine, a small-scale supercomputer designed to produce the familiar Mandelbrot set images much faster than is possible with an affordable scalar processor.

This installment deals with overall design issues and the algorithms and methods used in the software.

Picking the Problem

Many of the letters that prompted this project detailed the letter writer's favorite fantasy computer. Some of the projects were awesome in scope, featuring multitudes of bit-slice processors, ganged up with megabytes of static RAM and kilobytes of custom microcode!

As I explained in part 1 of this chapter, the Mandelbrot set calculations are nearly ideal for a multiprocessor because the problem can be divided among any number of processors and there is no need to communicate between processors during the solution. (I could have tackled North American weather forecasting, but the resulting project would have been out of the reach of nearly everyone.)

But a Circuit Cellar project is a blend of the challenging and the commonplace. It must be challenging enough to be interesting and informative, but it must use relatively commonplace components so that interested readers will be able to afford it. Building a supercomputer out of the latest custom LSI components would certainly be challenging, but few could afford the resulting hardware.

The cost of a multiprocessor is controlled by two major factors: the cost of each individual processor element and the level of communication between them. Element cost is obvious: If a single element costs $500, how many people would be willing to buy 64 of them? The level of communication determines how many ports each element must have and how complex those ports must be.

The Circuit Cellar Mandelbrot engine uses Intel 8751 processors rather than the 8088 or 80286 chips found in IBM PCs, ATs, and compatibles. The 8751 is similar to the 8031 used in many of my Circuit Cellar projects during the last few years, except that it contains 4K bytes of on-chip EPROM for program storage. Although processors in the 8051 family are often thought of as simple controllers, you are about to see what happens when a bunch of them concentrate on a single task: The engine becomes a beehive of activity!

There are several compelling reasons for choosing a processor from the 8051 family: performance, capability, and price. The performance with a standard 12-MHz crystal is just under 1 million instructions per second because most instructions take one or two cycles of 12 clock periods each. The chip includes a bidirectional serial port and extensive interrupt support, as well as a reasonably competent instruction set. Finally, a processor element can be a single-chip 8751 or an 8031 with an external EPROM, depending on whether cost or board space is more important; the cost per element ranges from under $10 to $40.

Figure 1 shows a block diagram of the Mandelbrot engine and the connection to the AT controller. The engine can include any number of processor elements between 1 and 255, and performance increases smoothly with the number of elements. The AT connection uses a standard serial port for data transfer, with the port control lines handling handshaking and array resets.

Listing 1: (a) *Pseudocode for the Mandelbrot set program kernel.* **(b)** *The same code without the* SQRT *function.*

(a)

```
(Given a complex point c and an
 iteration limit K)
 z = 0+0i
 count = 0
 while (count <= K) and
        (SQRT(Re(z)^2 + Im(z)^2) < 2.0)
    z = z2 + c
    count = count + 1
 end while
```

(b)

```
(Given a complex point c and an
 iteration limit K)
 z = 0+0i
 count = 0
 while (count <= K) and
        ((Re(z)2 + Im(z)2) < 4.0)
   z = z2 + c
   count = count + 1
 end while
```

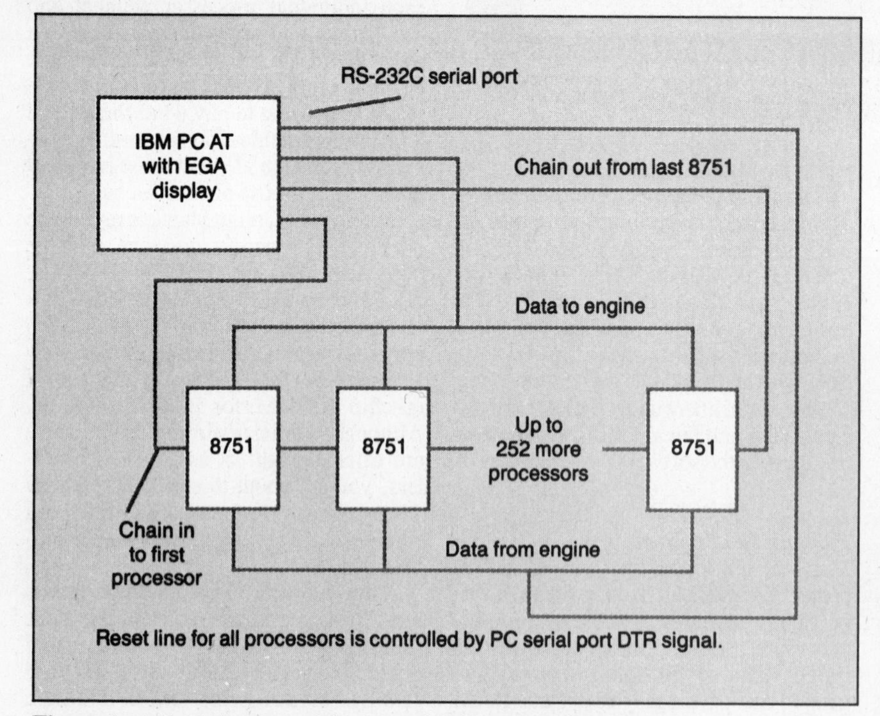

Figure 1: *A block diagram of the Circuit Cellar Mandelbrot engine.*

The Complex Plane

Photo 1 shows the entire Mandelbrot set displayed on an EGA monitor. You can look at the image as either a mathematical object or the output of a computer program, but you've got to understand both views to know how the Mandelbrot engine works.

Discovered by Benoit Mandelbrot, an IBM Fellow at the Thomas J. Watson Research Center, the Mandelbrot set is probably the best-known example of a fractal figure. Indeed, for many people, "Mandelbrot set" is synonymous with "fractal" and constitutes their only brush with complex numbers.

Mathematically, the Mandelbrot set is a set of points in the complex plane. A complex point c is in the Mandelbrot set if the magnitude of z in iterative formula

$$z = z^2 + c$$

(starting with $z = 0 + 0i$) remains finite after an infinite number of iterations.

The text box on page 194 gives a brief introduction to complex numbers and arithmetic for those of you who may be a little shaky on the subject. Pay particular attention to the process of squaring a complex number, because it's central to the algorithm.

The criterion for determining whether a given point is in the set should give you pause. No computer can run through an infinite number of iterations in a finite amount of time, so we can never be certain that a point is in the set. The best we can do is show that a given point is definitely not in the set, but that is sufficient to create all the pretty pictures.

It turns out that if the magnitude of z exceeds 2.0 after any iteration, it will eventually "blow up" to infinity after some additional number of iterations; therefore, that point is definitely not a member of the Mandelbrot set. If it remains below 2.0 for all the iterations we attempt, the best we can say is that it hasn't blown up yet and that the point may be a member of the set.

The test becomes more reliable with more iterations. For example, if one point blows up after 2 iterations and its neighbor survives 1000 iterations, the second one is probably a Mandelbrot set point. The iteration limit must be the same for all the points in a given image, although the particular value depends on where that image lies in the complex plane.

The algorithm behind the Mandelbrot set is shown in listing 1a. As you can see, despite the foregoing discussion, it's simple enough to fit on only a few lines—the sort of program I like.

Figuring the magnitude of a complex number requires a square root, but that is a difficult and time-consuming operation for a microprocessor. Fortunately, we can avoid extracting a square root by simply changing the comparison: Whenever the magnitude exceeds 2.0, the square of the magnitude will exceed 4.0. Listing 1b shows the revised loop.

Determining whether a point is in the Mandelbrot set is simply a matter of evaluating the algorithm in listing 1b until it terminates. If the final count is less than the iteration limit, the point is definitely not in the Mandelbrot set. If the count equals the iteration limit, the point is either in the set or the limit is too low to rule it out; in either case, we assume that the point is in the set.

By convention, all the points in the Mandelbrot set are colored black. The iteration count for the remaining points selects one of the colors available on the display, so the rainbow hues are a direct indication of the size of the iteration count. The blue pixels in photo 1 had the fewest iterations, the green ones had more, and so on up to the bright white pixels adjacent to the central black area marking the set proper.

Floating Points

The center of the image in photo 1 lies at -0.5 + 0i in the complex plane, and the horizontal axis is about 3.6 units long. The EGA was in 640- by 350-pixel mode, so 224,000 separate pixels are shown on the screen. You have to evaluate the algorithm in listing 1 at each one of those points, so you can see why Mandelbrot set images take so long to create.

Because the entire area of the Mandelbrot set must lie within a circle 2.0 units in radius centered on the origin (think about it for a while), the calculations must be done with numbers that can represent fractions. The familiar integer variables used for most purposes in programs simply do not have the range of values needed for the Mandelbrot set.

Because IBM designed the original PC with a socket for a math coprocessor, nearly everyone is familiar with floating-point numbers. In fact, that single design decision may have had more to do with the spread of PCs in the engineering domain than any other; for the first time, floating-point math on a small computer wasn't prohibitively slow.

But it turns out that floating-point numbers are not well suited for the Mandelbrot set calculations, despite the fact that they are quite easy to use. The Mandelbrot engine uses real numbers, to be sure, but they have a fixed-point representation rather than floating-point. To explain why, I must first detour into floating-point numbers and return via fixed points.

All numeric values within a computer are represented by a fixed number of bits. Remember that a single byte of 8 bits can take on 256 different values, and that 2 bytes can count 65,536 values. The C programming language (and others, as well) allows long integers that have 4 bytes and about 4×10^9 different values. The key idea is that a given number of bits can take on only a fixed number of different values.

The mathematical definition of a real number poses a problem for computer implementations. Between any two real numbers, no matter how close together, are an infinite number of other real numbers. Representing an infinite quantity requires an infinite number of bits, so computers simply can't represent all the real numbers correctly.

Floating-point numbers represent "lots" of real numbers in a "few" bits. The trick is similar to the scientific notation I used just above. Instead of writing 4,000,000,000, I used 4×10^9 and saved quite a bit of space. Each floating-point number has three parts: the mantissa, the sign, and the exponent. The mantissa encodes the significant figures of the real number, the sign tells you if the mantissa is positive or negative, and the exponent represents the magnitude.

Figure 2 shows the floating-point equivalents of some common numbers. Note that the exponent is actually an exponent of 2 rather than the 10 used in scientific notation. The exponent is selected so that the mantissa is always between 0.5 and 1.0, which you can see by comparing the values for 1, 0.5, and 0.25. Because the decimal point (or, more exactly, the binary point) can be located anywhere in the mantissa, the representation is called floating-point.

Figure 3 shows the IEEE standard format for the floating-point numbers

$0.0 = -1^0 \times 0 \times 2^0$
sign = 0
mantissa = 0
exponent = 0 (by convention)
$1.0 = -1^0 \times 1 \times 2^0$
sign = 0
mantissa = 1
exponent = 0
$2.0 = -1^0 \times 1 \times 2^1$
sign = 0
mantissa = 1
exponent = 1
$-2.0 = -1^1 \times 1 \times 2^1$
sign = 1
mantissa = 1
exponent = 1
$0.5 = -1^0 \times 1 \times 2^{-1}$
sign = 0
mantissa = 1
exponent = −1
$-0.25 = -1^0 \times 1 \times 2^{-2}$
sign = −1
mantissa = 1
exponent = −2

Figure 2: *Some common numbers written in the floating-point format.*

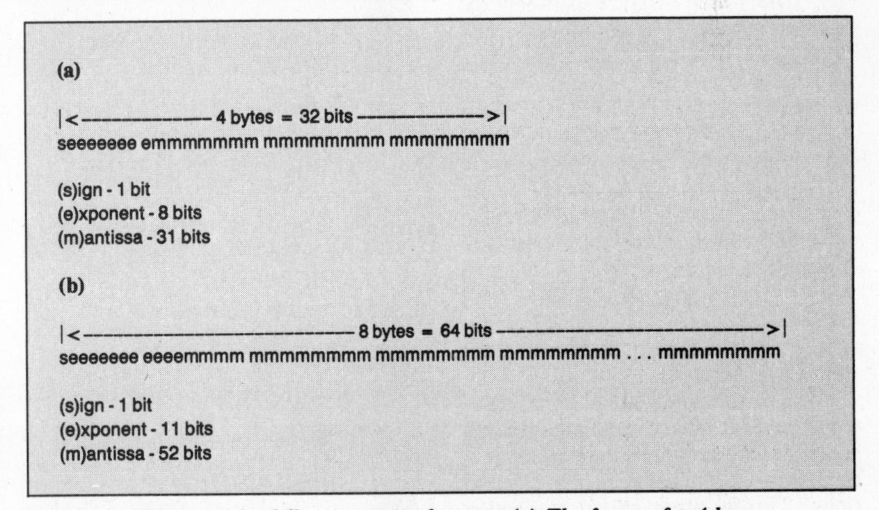

Figure 3: *IEEE standard floating-point formats.* **(a)** *The format for 4-byte floating-point numbers* (`float`). *This representation has a range of* $+/- 3.4 \times 10^{-38}$ *to* $+/- 3.4 \times 10^{38}$ *and a precision of six to seven decimal digits.* **(b)** *The 8-byte* `double` *format, which has a range of* $+/- 1.7 \times 100^{-308}$ *to* $+/- 1.7 \times 10^{308}$ *and a precision of 15 to 16 decimal digits.*

known to Microsoft C programmers as `float` and `double` variables. C programs typically use `double` variables whenever floating-point numbers are required, but there's a trap for the unwary in this practice.

Many common real-world programming problems are well behaved, in that the real numbers are all within a few orders of magnitude of each other. For example, in electrical engineering, the range of values extends from about 10^{-15} to 10^{12} or so (1 femtofarad to 1 gigahertz, for example). Calculations tend not to exceed these bounds, so programmers can get away with double variables most of the time.

Mandelbrot set calculations, on the other hand, require more precision than even `doubles` can provide. Photos 2a through 2c show the spike on the left-hand edge of the Mandelbrot set in increasing magnification. As you can see, there's not much change in structure as you zoom in on the set. This is one of the defining characteristics of a fractal figure. The level of detail and overall structure are similar at all magnifications. In some sense, the picture is the same regardless of the magnification.

To contrast this with ordinary objects, try magnifying the period at the end of this sentence by a factor of 10,000. The period is about 0.015 inch across, so the disk would be 150 inches in diameter. The difference in structure between the page and the period should be obvious.

Photo 2c magnifies the tip of the spike by 10^{12}. This is a magnification almost beyond comparison, but here's an analogy: The mean distance between the earth and sun is 92×10^6 miles or 5.83×10^{12} inches. One inch magnified by a factor of 10^{12} is one-fifth of the way to the sun.

Pixels in photo 1 are 0.0056 (3.6/640) units across, while pixels in photo 2c are 5.6×10^{-15} units across. The horizontal coordinates of the left edge of the spike are about 2.0 in both pictures, so you are looking at a single picture that spans 15 orders of magnitude. Recall that `float` variables have only six digits of precision, and you will begin to see why they would be inadequate.

Because the Mandelbrot set calculations use the results of one iteration as the source for the next pass, there is an inevitable loss of precision in the calculations. This is often called truncation error, because the true value of a real number must be truncated to fit within the available precision. The size of the truncation error increases with the number of iterations and is roughly the logarithm of the iteration count.

The calculations in photo 1 were carried out with an iteration limit of 64, so about two digits (log 64 = 1.8) of precision were lost to truncation near the black areas marking the Mandelbrot set. Many interesting images require hundreds of iterations, so the precision loss can exceed three or four digits.

If the image requires 15 digits of precision just to represent the coordinates and 4 digits to contain the truncation error, what happens when the program uses `float` variables having only 6 digits? The answer is painfully obvious: See photo 3. Some of you have seen this effect in your programs, and now you know what caused the failure.

Using `double` variables doesn't solve the problem. Photo 2a needs about 18

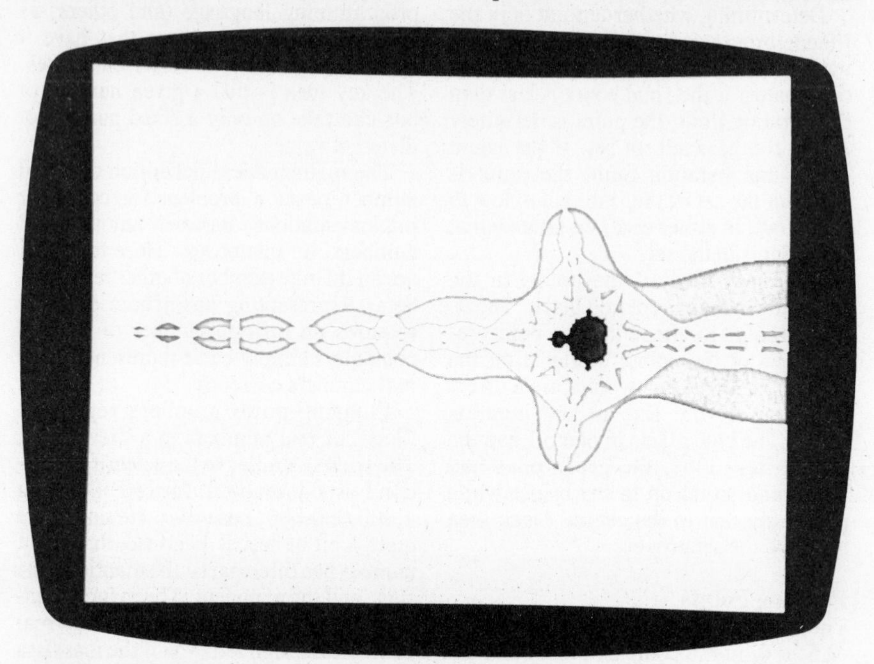

Photo 2a: *The "spike" at the left end of the set magnified 10 times.*

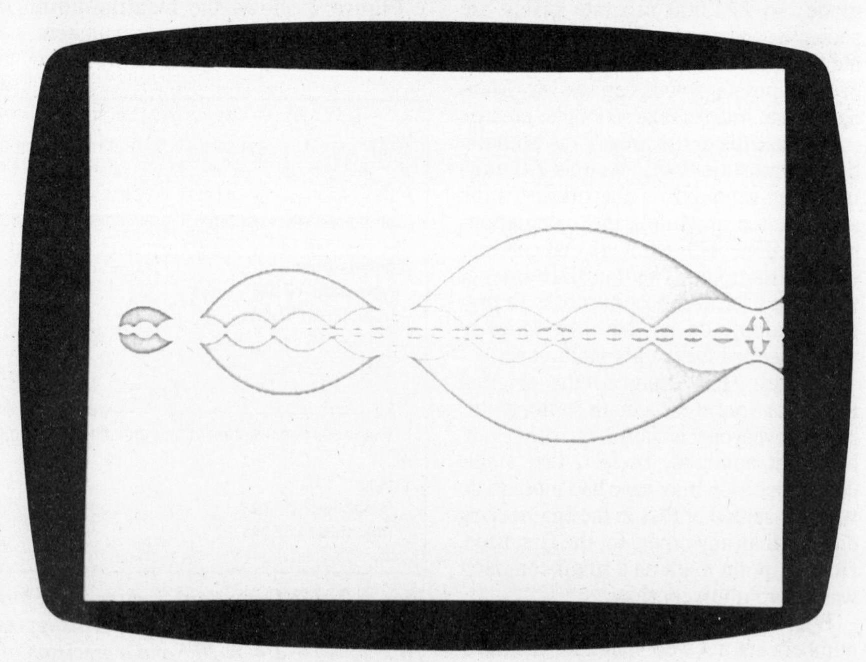

Photo 2b: *The "spike" at the left end of the set magnified 100 times.*

digits of precision, which is 2 or 3 more than `double` numbers can offer. What can you do?

Fixed Precision

Although the Mandelbrot set calculations require high precision, they don't require a large dynamic range. The entire Mandelbrot set lies within coordinates of +2.0 to −2.0, and the largest useful magnitude is only 4.0; some of the exponent bits in `float` and `double` numbers are wasted. Being able to represent numbers as large as 10^{308} is no advantage in these calculations.

Figure 4 shows how real numbers are encoded in the Mandelbrot engine. There is no exponent, so there is no way to "float" the binary point in the mantissa. With the binary point in a predetermined location, these numbers are *fixed-point* numbers. The values are stored in two's complement notation, rather than the sign-magnitude used in IEEE floating-point numbers. The complete value occupies the same number of bytes as a `double` number, but it has at least three more digits of precision.

It turns out that fixed-point number programs are both simpler and faster than floating-point code, simply because there's no need to work with the exponent. Simplicity is particularly important in the Mandelbrot engine, because the code is written in 8051 assembly language.

Fixed-point addition is also easy to understand. Listing 2 shows the 8051 code needed to add two fixed-point numbers, given two pointers to the start of each number. The loop simply adds corresponding bytes and propagates any carries to the left.

It's worth noting that the 8051, despite its Intel heritage, stores 2-byte integers with the high-order byte in the low-order address, so the addition proceeds downward from the high addresses. This is the convention used by Motorola processors and is exactly opposite from the Intel 8088 family. As you might expect, there are conversion routines in the AT driver code to convert from the 8088 convention (low byte first) to the 8051 convention (high byte first).

Although addition is simple, multiplication and division are not. I was able to skip division because the Mandelbrot set calculations didn't need it, but multiplication was essential. The 8051 instruction set includes a multiply that produces a 16-bit product from two 8-bit inputs in only 4 microseconds (μs), but combining the partial products into the final answer took some tricky coding.

Multiplying two 8-byte fixed-point numbers produces a 16-byte product. A little pencil work will show you that there are 64 multiplies and 56 2-byte additions to combine the partial products. Because each addition can generate a carry into the next byte, there are a considerable number of additional steps to propagate the carries throughout the partial products. Finally, the product has to be aligned so that the binary point is in the right location (if the point can't float, the number must).

Normal extended-precision multiplication routines use a series of nested loops to handle the repetitive calculations. Because this routine is used four times in each iteration, I coded it as a monster macro that creates about 2.5K

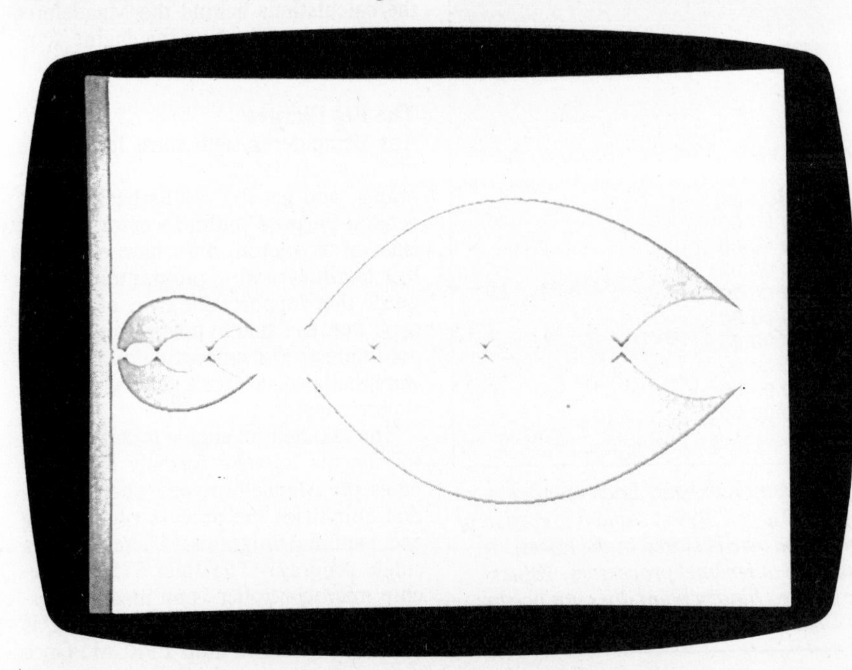

Photo 2c: *The "spike" at the left end of the set magnified 10^{12} times.*

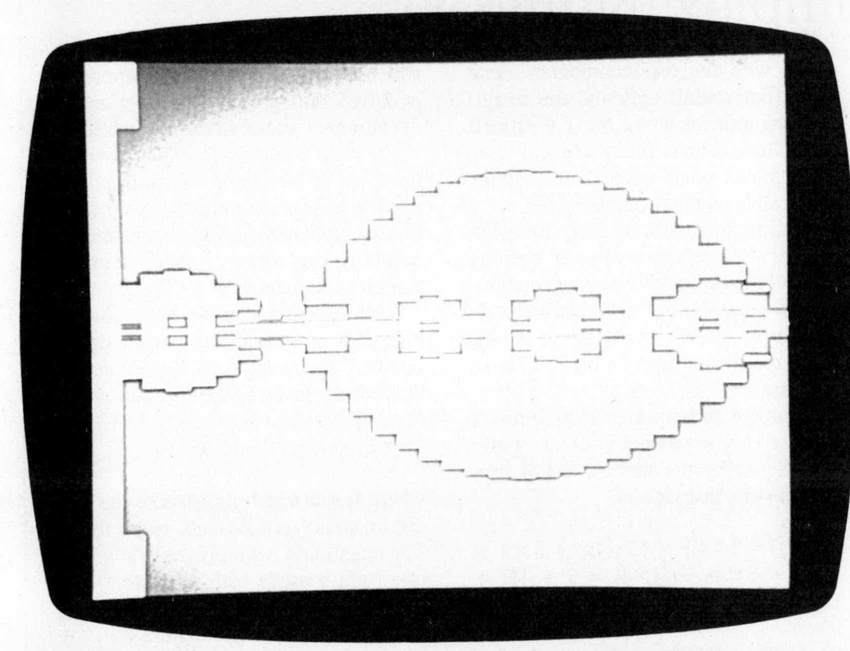

Photo 3: *Using* `float` *variables while performing the same magnification as in photo 2c (10^{12} times) makes precision loss due to truncation obvious.*

bytes of straight-line code. Space considerations prohibit printing such an incredible bulk here, but it's available for your inspection on the Circuit Cellar BBS.

Because the dynamic range is so low, many of the math routines check their inputs to prevent overflows. For example, the squaring routine will flag an attempt to square a number greater than 2.0. Unlike most math routines (which signal an error), that overflow simply indicates that the magnitude of a complex number will exceed 4.0, so the Mandelbrot calculations for that case are finished.

```
|<-------- 8 bytes = 64 bits -------->|
siiiffff ffffffff ffffffff ffffffff ffffffff . . . ffffffff

sign - 1 bit
integer - 3 bits
fraction - 60 bits

Sample values, with only the first 4 bytes shown:
```

Value	Binary	Hexadecimal
0.0	0000.0000000000000000000	00 00 00 00
1.0	0001.0000000000000000000	10 00 00 00
2.0	0010.0000000000000000000	20 00 00 00
-1.0	1111.0000000000000000000	F0 00 00 00
-2.0	1110.0000000000000000000	E0 00 00 00
0.5	0000.1000000000000000000	08 00 00 00
0.25	0000.0100000000000000000	04 00 00 00
-0.5	1111.1000000000000000000	F8 00 00 00
-0.25	1111.1100000000000000000	FC 00 00 00

Figure 4: *Circuit Cellar Mandelbrot engine fixed-point numbers. Each number occupies 8 bytes of storage and has a range of −8.0 to +7.999. . . and a precision of 19 decimal digits. In Intel 8051 code, the high-order byte is stored in the lowest address; this is the reverse of the convention used in other Intel processors. Values are stored in two's complement notation. The implied binary point does not occupy a bit location.*

Back in the bad old days, when computers filled rooms with racks of vacuum tubes, there were no math coprocessors. A great deal of work went into creating algorithms that could use fixed-point numbers to give useful results. Having to deal with a little of that analysis for this project made me appreciate the early pioneers a whole lot more.

Now that you have an appreciation for the calculations behind the Mandelbrot set, I can describe how the engine actually works.

The Big Picture

Any computer system must have some means to get data in, select and run programs, and get the results back out. A general-purpose multiprocessor (if that isn't an oxymoron) must have extensive I/O facilities and a programming language that supports the array of processors. But, as I said in the beginning, I'm not interested in supporting all the software that's needed for a general-purpose supercomputer.

The Mandelbrot engine is dedicated to solving the iterative formula that computes the Mandelbrot set. That dedication simplifies the process of selecting and running programs: There is but a single program! The Intel 8751 single-chip microcontroller is an ideal processor element, because that program is burned into the on-chip EPROM. Once

Complex Numbers

A complex number is made up of two separate numbers, which, for historic reasons, are called the real and imaginary components. The imaginary component is distinguished by a lowercase i either before or after the number. For example,

1+2i or 1+i2

is a complex number with a real part of 1 and an imaginary part of 2. The real and imaginary components are rarely integers, so you'll often see complex numbers written with decimal fractions:

1.2345 + 0.4321i

Electrical engineers often use j instead of i, because the symbol for electric current is i.

Because complex numbers have two components, they can be plotted on a plane, with the real component along the *x* (horizontal) axis and the imaginary component along the *y* (vertical) axis. This plane is often referred to as the complex plane because it contains all possible complex numbers.

Adding and subtracting complex numbers is a simple matter of keeping track of the real and imaginary components separately. For example, the sum (1+2i) + (3+4i) evaluates to 4+6i, while the difference (1+2i) − (3+4i) becomes −2−2i.

Forming the square of a complex number is a somewhat trickier operation. Squaring the number 3+4i proceeds in this fashion:

$$(3+4i)^2=(3+4i) \times (3+4i) = 3\times3 + 3\times4i + 3\times4i + 4i\times4i = 9 + 12i + 12i + 16i^2 = 9 + 24i - 16 = -7+24i$$

Simplifying $(16i)^2$ to −16 uses the fact that i denotes the square root of −1, which is why the imaginary component is called "imaginary"—it isn't one of the numbers we use in the real world.

I'll avoid describing complex division because the Mandelbrot formulas don't need it. If you're interested, your local library probably has a book on complex numbers that shows how division of complex numbers works.

Each complex number has a magnitude that represents the "size" of the number. The magnitude of the complex number z is given by the formula

$$Mag(z) = \sqrt{(Re(z)^2 + Im(z)^2)}$$

where Re(z) and Im(z) denote the real and imaginary components, respectively. The magnitude is simply the hypotenuse of a right triangle with sides Re(z) and Im(z). For example, the magnitude of 3+4i is $Mag(3+4i) = \sqrt{(Re(3+4i)^2 + Im(3+4i)^2)} = \sqrt{(3^2 + 4^2)} = \sqrt{(9 + 16)} = \sqrt{(25)} = 5$.

programmed, the chip doesn't need a disk drive or tape drive to get started; simply turn on the power and release the reset line, and the program is up and running immediately.

The issues of data communication are not so simply resolved. Once again, though, because we are dealing only with the Mandelbrot set, there is little need for interprocessor communication during the computations. Although the 8751 includes an on-chip serial port, there isn't enough room for a general-purpose network, and there isn't enough space in the EPROM for much of a network operating system.

Refer back to figure 1 to see how the communications are handled. All the 8751 serial inputs are connected together and driven by the AT's serial output. Similarly, all the 8751 serial outputs are connected together to drive the AT's serial input. Of course, there are communication buffers along the way to ensure that the ports are not overloaded.

Each processor has a unique ID number assigned during the initialization sequence, which I'll describe in Chapter 12, part 3. The AT can address a message to any number of processors in the array. If the message goes to a single processor, all other elements ignore the message and continue with their computations.

It's easy enough to control transmissions from the AT to the engine's processors, because there is only one sender on the line. Getting information back without collisions between processors could be very complicated, but another characteristic of the Mandelbrot set calculations came to our rescue.

Despite the exquisite precision required during the enormous number of calculations in each engine, the results can be summed up in 1 or 2 bytes. Polling each processor for its result didn't make any sense, because the necessary handshaking would reduce the effective data rate by at least a factor of 4. Some coordination is required, though, because the processors must return their results in a known sequence.

The "chain-in" line shown in figure 1 organizes the output communications. It daisy chains through all the processors in the engine, so the AT is connected to the first processor, which drives the second, which drives the third, and so on to the last processor, which is connected back to the AT. The key to this is the 8751 program, which will output its result only when its chain-in line is active, then toggle its chain-out line when transmission is complete.

The AT driver program gets ready to receive a set of results, then toggles its chain-out line. The first processor sends its result, then the second, and so forth through the array. The last processor sends its result and toggles the AT's chain-in line, which tells the AT that all results are complete. The cycle repeats when the AT is ready for the next set.

Because the driver program knows how the processors were assigned to pixels, it can put the results into the right places. The actual screen painting occurs between sets of pixels to reduce the loop overhead, so the results are metered out of the engine in bursts. The 8751 program will send only 1 byte if the iteration limit is less than 256, which reduces the

Listing 2: *The 8051 code to add two fixed-point numbers.*

```
;-----------
; Add two long integers
; R0 points to the high order byte of the target
; R1 points to the high order byte of the source
; Mashes A, B, and R1
; Returns R0 unchanged
NUMLEN      EQU 8    ; bytes per fixed point number
long_add    PROC
PUBLIC  long_add
  MOV       A,R0          ; point to end of target
  ADD       A,#NUMLEN-1
  MOV       R0,A
  MOV       A,R1          ; point to end of source
  ADD       A,#NUMLEN-1
  MOV       R1,A
  MOV       B,#NUMLEN     ; number to add
  CLR       C             ; set up for loop
  L?loop    EQU  $
  MOV       A,@R0         ; pick up target
  ADDC      A,@R1         ; tack on buffer
  MOV       @R0,A         ; drop into target
  DEC       R0            ; tick pointers
  DEC       R1
  DJNZ      B,L?loop      ; repeat for all bytes
  RET
long_add    ENDPROC
```

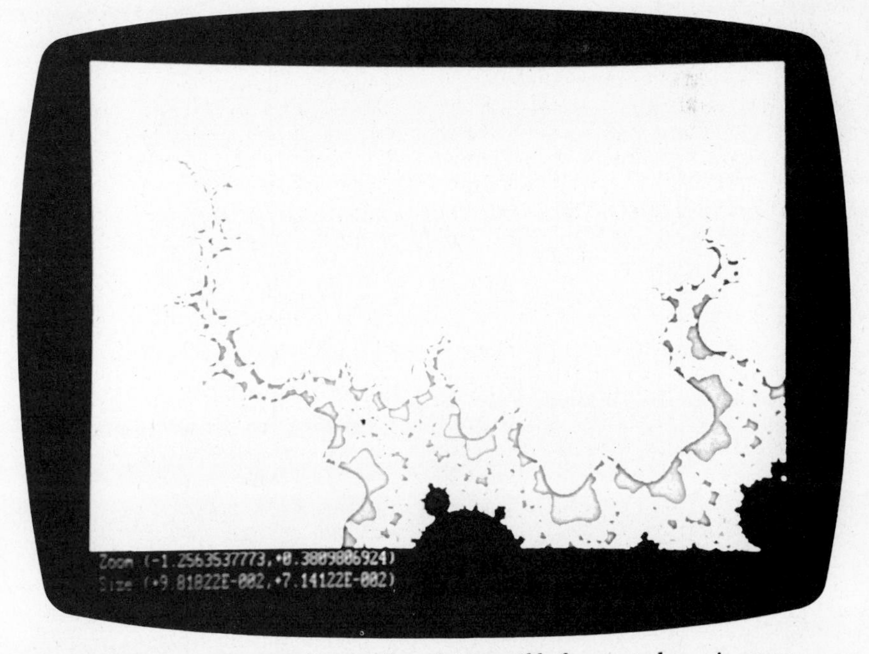

Photo 4: *The driver program's zoom box. It is possible for you to home in on an interesting part of the array and magnify it.*

data transmission time by a factor of 2 for most images.

You may wonder why I would pick a serial link instead of a parallel port. The answer is simple: It works! A standard RS-232C serial link can send 1 byte in about 380 μs at 28,800 bits per second. Compared to the time required to compute that byte, the transmission time is small. Only for trivial pictures does the transmission time for the full array approach the computation time.

You might also wonder what effect the daisy chain has on the result throughput. After all, all the down-chain processors must wait on any processor that's not ready when its chain-in signal becomes active. I'll discuss performance in part 3 of this chapter, but suffice it to say that it's not all that bad. The processors already polled have started on their next set of computations, and the remaining processors are probably still running through the current iterations. Only in the case where one processor hits a point in the Mandelbrot set while all others are outside does this delay become significant.

Because results from the Mandelbrot set calculations are shown on a graphics display, it's only natural that you should use a graphical means to select which part of the array to view. The driver program creates an overall view of the array, shown in photo 4.

In Conclusion

Now that you understand how the Circuit Cellar Mandelbrot engine works, I can describe the actual hardware used to build it. In the final part of Chapter 12, I'll cover the hardware construction and the AT driver program needed to get the images up on the screen.

Special thanks to Ed Nisley and Merrill Lathers for their expert contributions to this project.

Editor's Note: Steve often refers to previous Circuit Cellar articles. Most of these past articles are available in book form from BYTE Books, McGraw-Hill Publishing Company, P.O. Box 400, Hightstown, NJ 08250, (1–800–2–MCGRAW).

Ciarcia's Circuit Cellar, Volume I covers articles in BYTE from September 1977 through November 1978. *Volume II* covers December 1978 through June 1980. *Volume III* covers July 1980 through December 1981. *Volume IV* covers January 1982 through June 1983. *Volume V* covers July 1983 through December 1984. *Volume VI* covers January 1985 through June 1986.

There is a multiline Circuit Cellar bulletin board system (running TBBS 2.0M) that supports past and present projects in BYTE and Ink. You are invited to call and exchange ideas and comments with other Circuit Cellar supporters. The 300-/1200-/2400-bps BBS is online 24 hours a day at (203) 871-1988.

12

A SUPERCOMPUTER

PART 3

This final part looks at hardware nuts and bolts and also at the driver program

The first two parts of this chapter gave you a background in multiprocessor architectures and described the basic algorithms used in the Circuit Cellar Mandelbrot engine. In this part, I'll go into the nuts and bolts of the engine's hardware and explain how the driver program in the IBM PC AT controls the engine.

Many Processors

A multiprocessor system is made up of perhaps hundreds or thousands of separate computers. The overall cost of the system is largely determined by the cost of each individual computer, so there is a strong motivation to keep costs under control. If a system uses 256 processors that each cost $1000, few people can afford the full system.

As I explained in Part 1, the multiprocessor's hardware architecture must match the problems that it will solve. If the problems require extensive communication between processors, the hardware must provide high-bandwidth intercomputer channels. The fact that those links are expensive simply means that's the price of solving the problem in the fastest way possible.

The iterative formula that produces Mandelbrot-set images is ideal for a multiprocessor implementation. The calculations require high-precision arithmetic but need little communication to set up the processors and report the results. A simple processor with low-cost communications can handle the problem adequately.

The Mandelbrot engine uses the Intel 8751 single-chip microcontroller as its basic building block. Figure 1 shows all the circuitry associated with a single processor. The hardware is deceptively simple, because the 8751 is a complete computer on a chip—there's a lot of hardware behind those pins!

I've long believed that something was lost when computers stopped sporting front panels with blinking lights, so there are two LEDs for each processor. The program turns on one LED when it's computing and the other when it's done. A glance at the front panel indicates the state of every processor in the engine. The LEDs also assist in program and hardware debugging, because the program blinks them in specific patterns when it's waiting in various loops.

The 8751 receives data and setup information through its serial input port, which is driven by a buffer that is shared by all the chips on a single card. The results of the Mandelbrot set calculations, as well as ID and version information, are sent over the 8751's serial output port. Each chip has a separate output buffer that drives the single serial output line leaving the card.

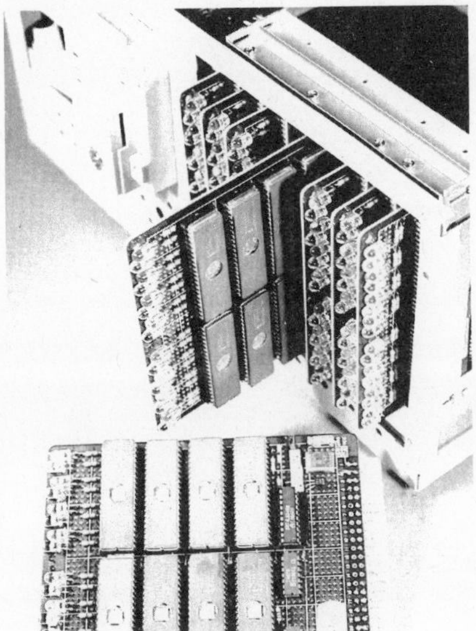

The chain-in and chain-out lines provide control and timing information during the calculations. When the program finishes evaluating the iterative formula for its current point, it waits for a pulse on the chain-in line before transmitting its result. After the final stop bit of the result message is finished, the chip pulses its chain-out line to signal the next processor in line that the serial output is available.

With those simple connections, each processor can receive data and send results in an orderly manner. The next step is to connect 256 of these processors into a single system.

It's in the Cards

Although I could have designed a single card with 256 40-pin DIP sockets and 512 LEDs, I figured that it made more sense to build the engine from smaller units. After all, a single-card computer would occupy about 5½ square feet and cost a small fortune.

Each 8751 is rated at no more than 250 milliamperes of current from the 5-volt power supply. While ¼ ampere doesn't sound like much, a full-bore engine with 256 processors will draw about 64 A! The two LEDs at each processor add another ampere or so to the bill. Obviously, there would be some problems getting enough power onto that single board.

After worrying about the problem for a while, I decided that a board with eight 8751s made a nice unit. Photo 1 on page 202 shows a hand-wired prototype with all eight processors installed and 16 LEDs peeking over the edge of the card. The card draws about 2 A of current, low enough not to pose a problem for the edge connectors.

Figure 2 shows the connections for a single card. To reduce the complexity of the drawing, I've shown the eight processors as blocks with I/O connections

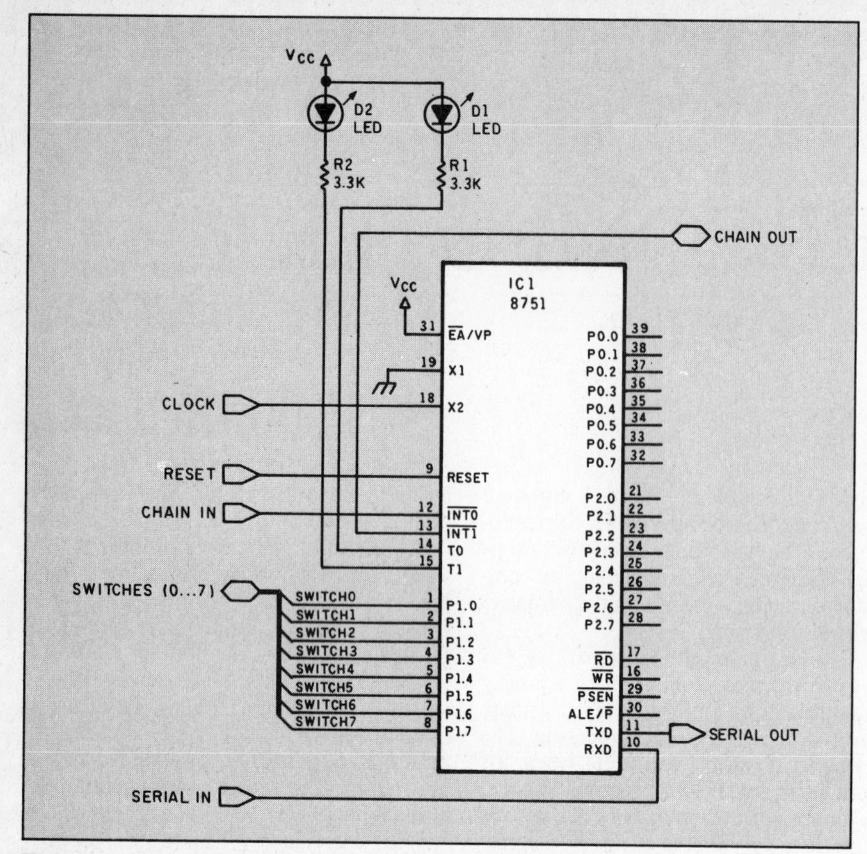

Figure 1: *The schematic for one processing element of the Mandelbrot engine.*

rather than as separate chips. Notice how the serial I/O ports are connected in common to all chips, while the chain-in and chain-out lines snake throughout the card.

The LEDs are connected to output-port pins on the 8751 through 3.3-kilohm resistors that limit the current to about 1 mA. Normally, a transistor buffer providing 10 mA or 20 mA drives the indicator LEDs, but I didn't think the additional hardware was worth it. Instead, I used high-efficiency red LEDs that are surprisingly bright at 1 mA.

The 8751, like all computers, requires a clock signal to sequence its internal registers and buses. Normally, a crystal connected directly to the 8751 provides that clock signal; each 8751 would then use a separate crystal. In this application, it didn't make much sense to have eight crystals on one board, so I included some circuitry to buffer the clock signal from one 8751 to the other seven. The few parts needed for the buffer cost much less than seven more crystals.

Because all the processors are driven by a common clock signal, their internal bus activity is synchronized. This can produce awesome current spikes on the 5-V supply, a situation that demands 100-nanofarad bypass capacitors across every IC on the board. These capacitors supply the transient current needed by each processor and reduce the spikes on the rest of the card. In addition, the cards have a ground plane on the top surface and heavy power-bus lines on the bottom, so there is a relatively low impedance connection to the 5-V supply.

One port of the 8751 is dedicated to switches that provide setup and test information. Rather than put a separate DIP switch on each card and risk having one switch in the wrong position, I used a single DIP switch for all the cards. Figure 2 shows port P1 connected to eight card-edge pins that provide the switch information.

Since the software can handle any number of processors, it made sense to design the hardware the same way. A jumper block on the card connects the chain-out line from each processor to the card's chain-out line. The first processor must be installed in socket IC1, and all processors must be contiguous, but by setting the jumper correctly, you can install any number of 8751s on the card.

Notice that, since I stored the Mandelbrot engine program in EPROM in each processor, there is no boot sequence before the program is ready to run. Unlike

IBM PCs and other common computers, the Mandelbrot engine is easily and quickly reset. In fact, all the 8751 reset lines are connected together and driven by the DTR line from the AT's serial port (through a level converter and buffer, of course), so that the AT can reset the engine at the start of calculations. This ensures that the engine is in a known state and ready to receive setup information.

Cards on the Rack

A complete Circuit Cellar Mandelbrot engine consists of 32 cards with 8 processors on each card (256 processors). Since that would require a card rack about 1¾ feet long and involve distributing 64 A of power supply equitably along the rack, I decided to take a more modest approach—only 64 processors.

Photo 2 shows a prototype rack that holds eight Mandelbrot engine cards. The power requirement for this rack is about 15 A, which is reasonable for a single power supply. (One point that bears mentioning is that a failure in the power supply can crisp 64 processors in one shot, so the rack also includes an overvoltage crowbar protector.)

Figure 3 shows the connections for one rack, with each card shown in place. One of the cards, the Rack Master, has a socket for the DIP switch needed to tell the 64 processors what serial-port rate to use, and so forth. The switches in each rack are identical.

A full Mandelbrot engine requires four racks. Figure 3 also shows the PC serial-port connections required to control the whole multiprocessor. As is indicated, you need relatively few wires for a computer with 256 processors. (The 64-processor engine we put together at the Circuit Cellar has about $6000 in hardware and 250 hours of wiring. Using a conservative $40-per-hour technician rate, our 64-processor Mandelbrot engine cost about $16,000.)

I've glossed over one small complexity up to this point: The serial-port connections from the PC use the standard RS-232C voltage levels, while the connections within the Mandelbrot engine are strictly TTL. One card, the Array Master, includes an MC145406 level converter (IC5) that handles the translations between the two standards. That card is also the Rack Master for the first rack, so it also has a DIP switch.

Skipping Jumpers

From reading the description of the Mandelbrot engine, you'd think that there are three different types of cards: a single Array Master, three Rack Masters, and 28 ordinary cards. While I could have designed three cards with three layouts, it made a lot more sense to design a single card that could serve in all applications.

The ordinary way to do that would use a set of jumpers on the card to select various I/O configurations and route signals to the appropriate places on the card. You could then select the "personality" of the card by the various jumper locations.

In this application, I decided that using personality jumpers was a bad idea. With up to 32 cards in a full Mandelbrot engine, it was almost certain that one or two jumpers would be misplaced, and it could take hours to track down the offending card. So the ideal goal was to eliminate jumpers entirely, except for the single jumper that designates a card with fewer than eight 8751s.

If you refer back to figure 2, you'll see the schematic for the Array Master card. The three Rack Master cards omit the MC145406 level converter and the LS21 chip that combines the serial data from the racks. All the cards can be identical, with the appropriate chips omitted from cards that don't need them.

It turns out that all the connections normally made through jumpers can be made by back-panel wiring between the cards. This explains the odd connections shown in figure 3.

For example, the AT's serial port connects to the serial data-in line on pin A7. The MC145406 converter translates the RS-232C signal into TTL voltages and drives four gates in the 74LS241 buffer. One of these connects to all the serial-port inputs on the card, and when the card operates as the Array Master, it also drives the seven other cards in the rack through pin B15.

The other three gates drive the remaining three racks through pins B14, B13, and B12. Because only the Array Master card has the 74LS241 buffer, each line from the Array Master to a rack connects to pin B15 on all the cards in that rack. If you trace the circuitry, you will see that pin connected to all the processors on the card; the pin doesn't care which way the electrons are flowing.

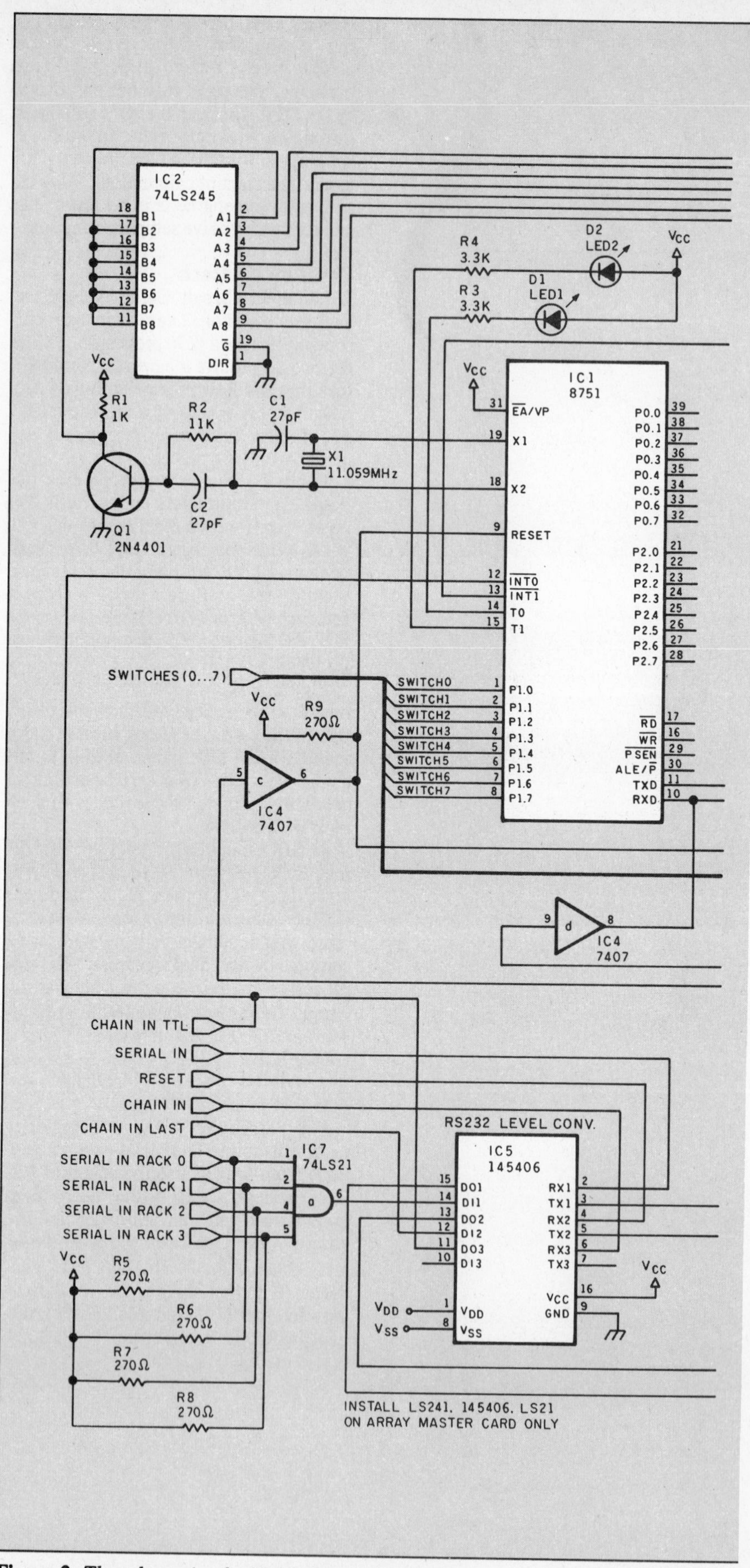

Figure 2: *The schematic of a Mandelbrot engine card.*

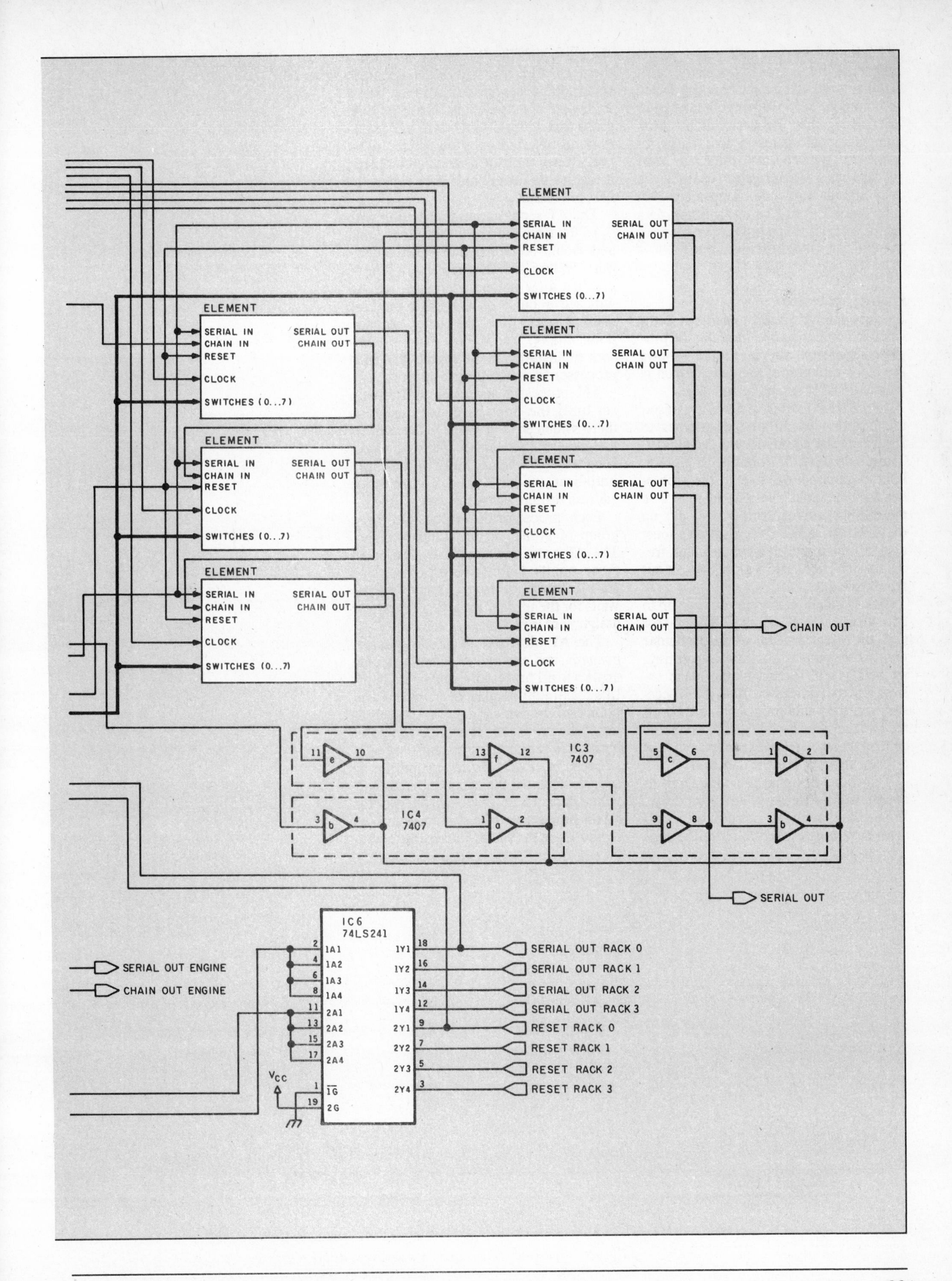
ELEMENT
SERIAL IN
CHAIN IN
RESET
CLOCK
SWITCHES (0...7)
SERIAL OUT
CHAIN OUT
CHAIN OUT
IC3
7407
IC4
7407
SERIAL OUT
IC6
74LS241
SERIAL OUT RACK 0
SERIAL OUT RACK 1
SERIAL OUT RACK 2
SERIAL OUT RACK 3
RESET RACK 0
RESET RACK 1
RESET RACK 2
RESET RACK 3
SERIAL OUT ENGINE
CHAIN OUT ENGINE
Vcc

Each card has a single serial output line from all the processors on the card. All eight outputs in a rack are connected in common, so 64 open collector gates are tied together. Because I didn't think connecting 256 gates in four racks together made much sense, the Array Master card has a 74LS21 gate to combine the four signals into one. Remember that only one 8751 will be transmitting data on the serial line at a time, so only one of the 256 gates will be active.

Engine Software

Because the Mandelbrot engine is dedicated to computing one formula, the program structure is simple. Figure 4 shows the block diagram of the code, which is written in 8051 assembly language.

Immediately after a hardware reset, the program examines the switches to figure out the bit rate for the serial port. Timer 1 in the 8751 controls the bit rate, but the timer period is set by the processor's clock input. Various members of the 8051 family can run with clocks up to 16 MHz, so the switch settings must also tell the program what the clock frequency is. There's no way to figure it out from within the 8751.

The Mandelbrot engine can have up to 256 identical processors. The data for each processor depends on the particular part of the complex plane under scrutiny, as well as the section of the plane assigned to that processor. If all the processors run the same program and they all receive the same serial data, how can the control program tell them apart, and how do they know which one they are?

The chain-in signal provides a way to assign each processor a unique address. When the AT begins assigning the addresses, it toggles the chain-in line connected to the first processor and sends the first address. The 8751 program waits until it sees an address immediately following a pulse on its chain-in line, grabs that address as its own, and transmits an inverted copy on its serial-out port. When the transmission is complete, it toggles its chain-out line to signal the next processor.

The AT sends addresses starting with 0, so the first address is 00 hexadecimal and should be returned as FF hexadecimal. When all the processors have grabbed their unique addresses, the AT will see a pulse on the chain-out line coming back from the array. At that point, the AT knows how many processors are installed in the engine, and each processor knows its unique address.

The processor addresses will remain set until the next hardware reset, so there's no need to go through the address assignment loop again. The AT must keep track of whether the array has been initialized, so that it doesn't perform the address assignment at the wrong time.

Each processor is responsible for a group of points in the complex plane. The AT must download the location of those points before the processor can start computing, so the entire engine waits for the initial data after the address assignment is finished.

The AT assigns data in two passes, *common* and *individual*. Common data applies to all processors, and it sets items like maximum number of iterations and point size. Individual data varies according to the processor; it controls the point location in the complex plane.

Each processor begins computing as soon as it has all the command and individual data. There's no reason to wait for all the processors to be started in unison, so you'll see the LEDs flicker into action

(text continues on page 206)

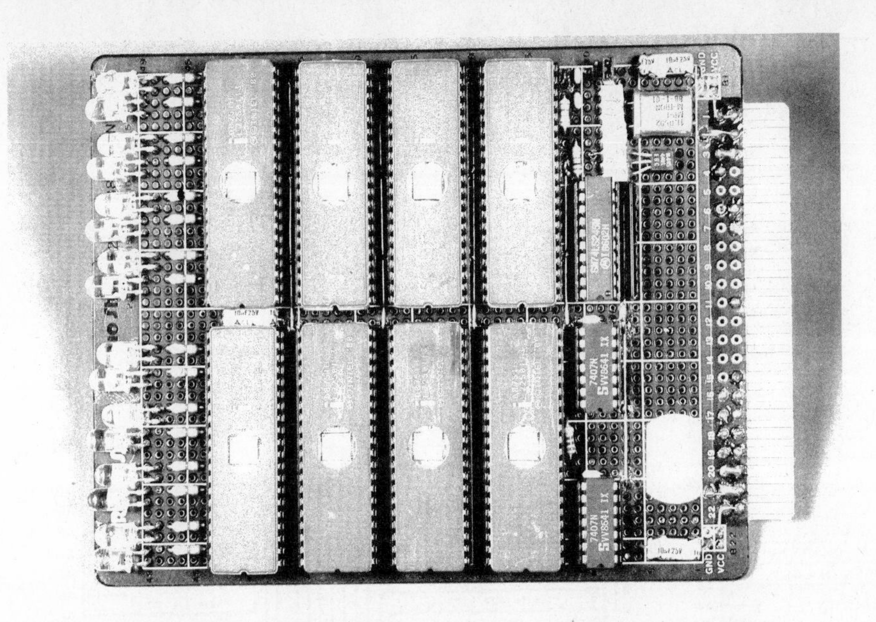

Photo 1: *A prototype of one eight-processor card from the Mandelbrot engine. High-efficiency LEDs are along the left of the board, and notice the windows on each CPU for programming the on-chip EPROM.*

Photo 2: *A prototype rack chassis, designed to hold eight cards (for a total of 64 processors). Note that this chassis also houses a Circuit Cellar AT, complete with a hard disk drive.*

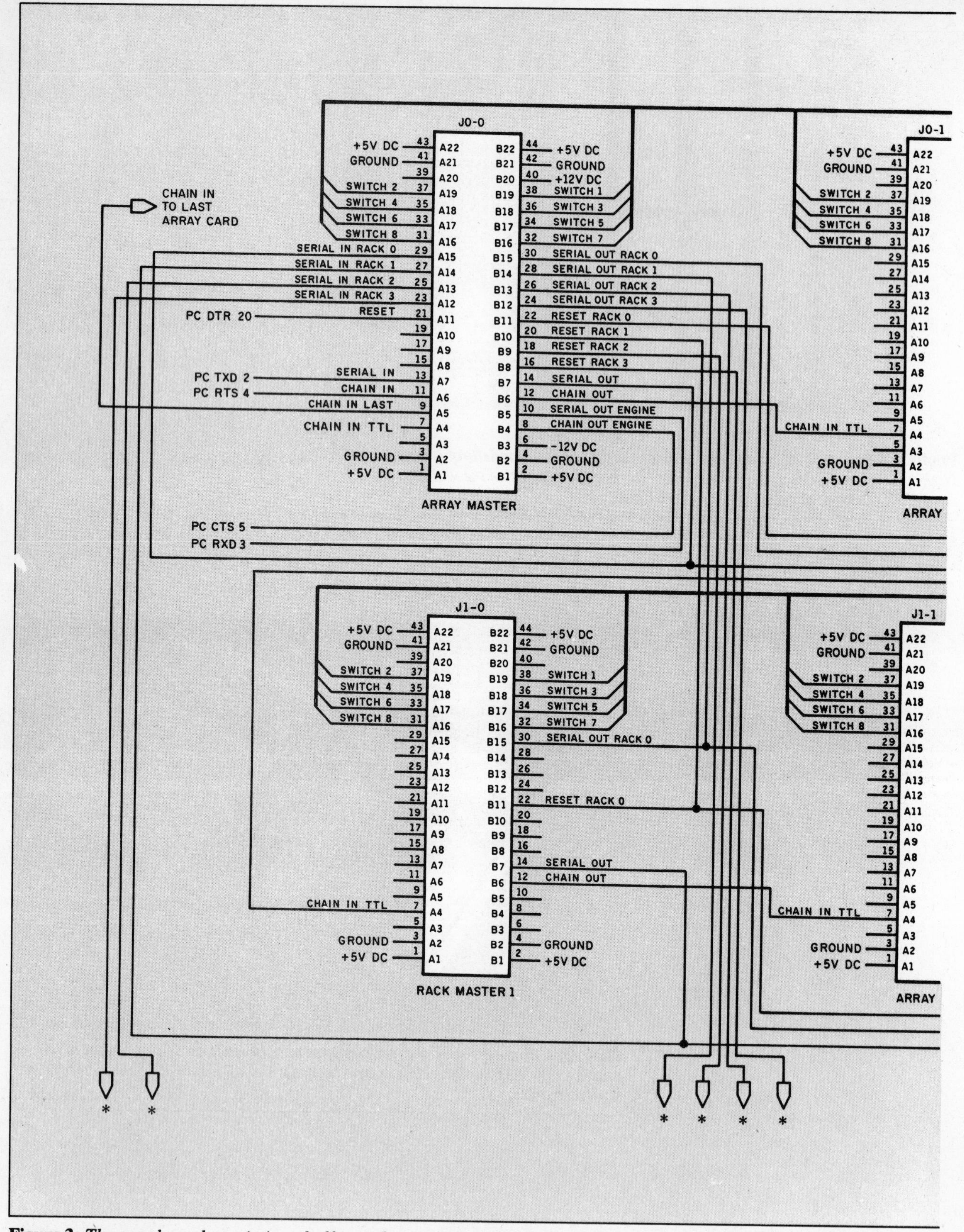

Figure 3: *The complete schematic (sans buffers and support chips) for the Mandelbrot engine, detailing how the processors are chained together.*

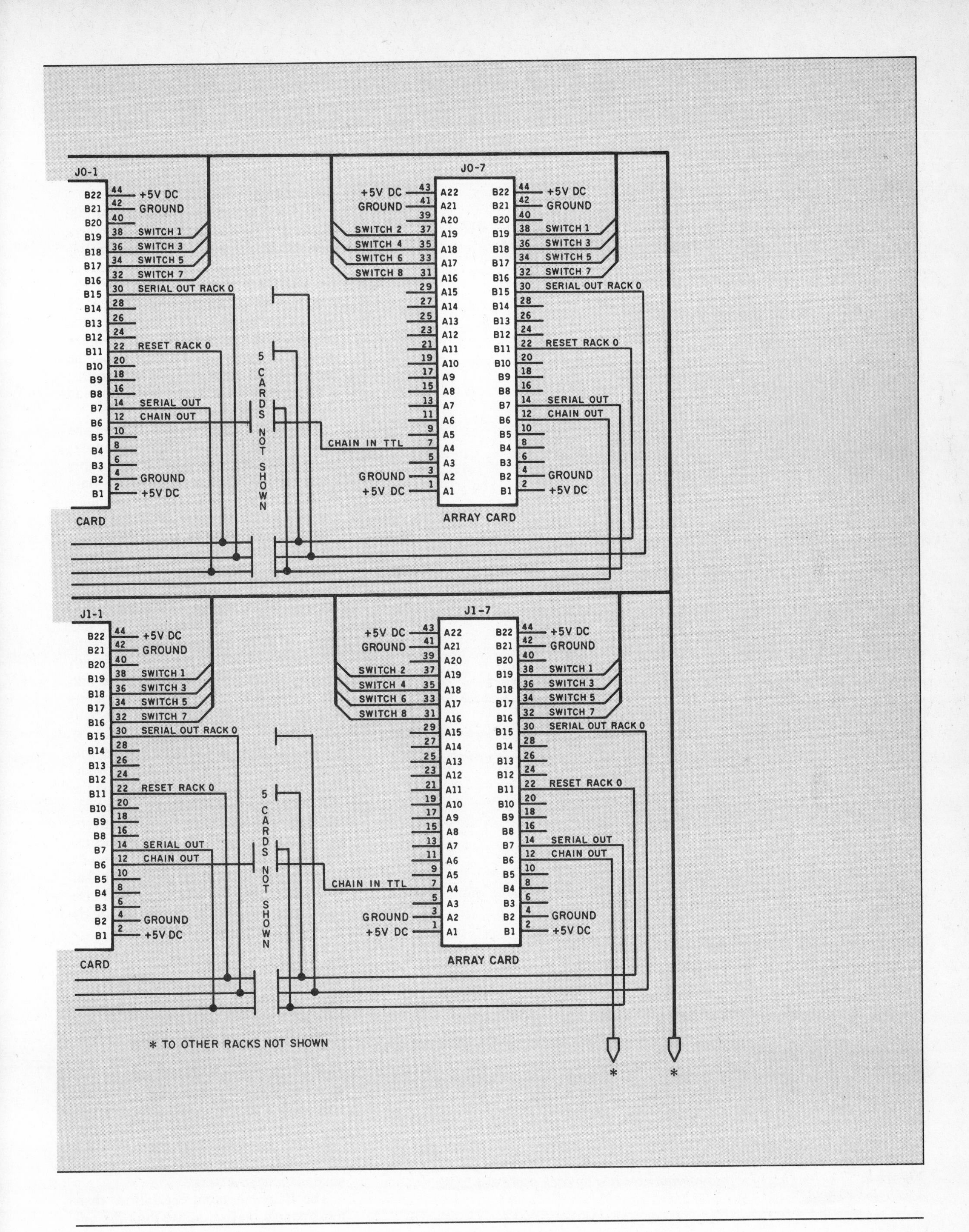
J0-1
J0-7
+5V DC
GROUND
SWITCH 1
SWITCH 2
SWITCH 3
SWITCH 4
SWITCH 5
SWITCH 6
SWITCH 7
SWITCH 8
SERIAL OUT RACK 0
RESET RACK 0
SERIAL OUT
CHAIN OUT
CHAIN IN TTL
5 CARDS NOT SHOWN
CARD
ARRAY CARD
J1-1
J1-7
* TO OTHER RACKS NOT SHOWN

in a wave as the AT assigns data to each processor.

I described in Chapter 12, part 2, how the AT reads data out of the array. To summarize, it toggles the chain-in line to the first processor, which responds with the results of its calculations on the first point and toggles the chain-in line for the next processor. Once each processor has responded, the AT formats the data into dots on the screen and then repeats the process.

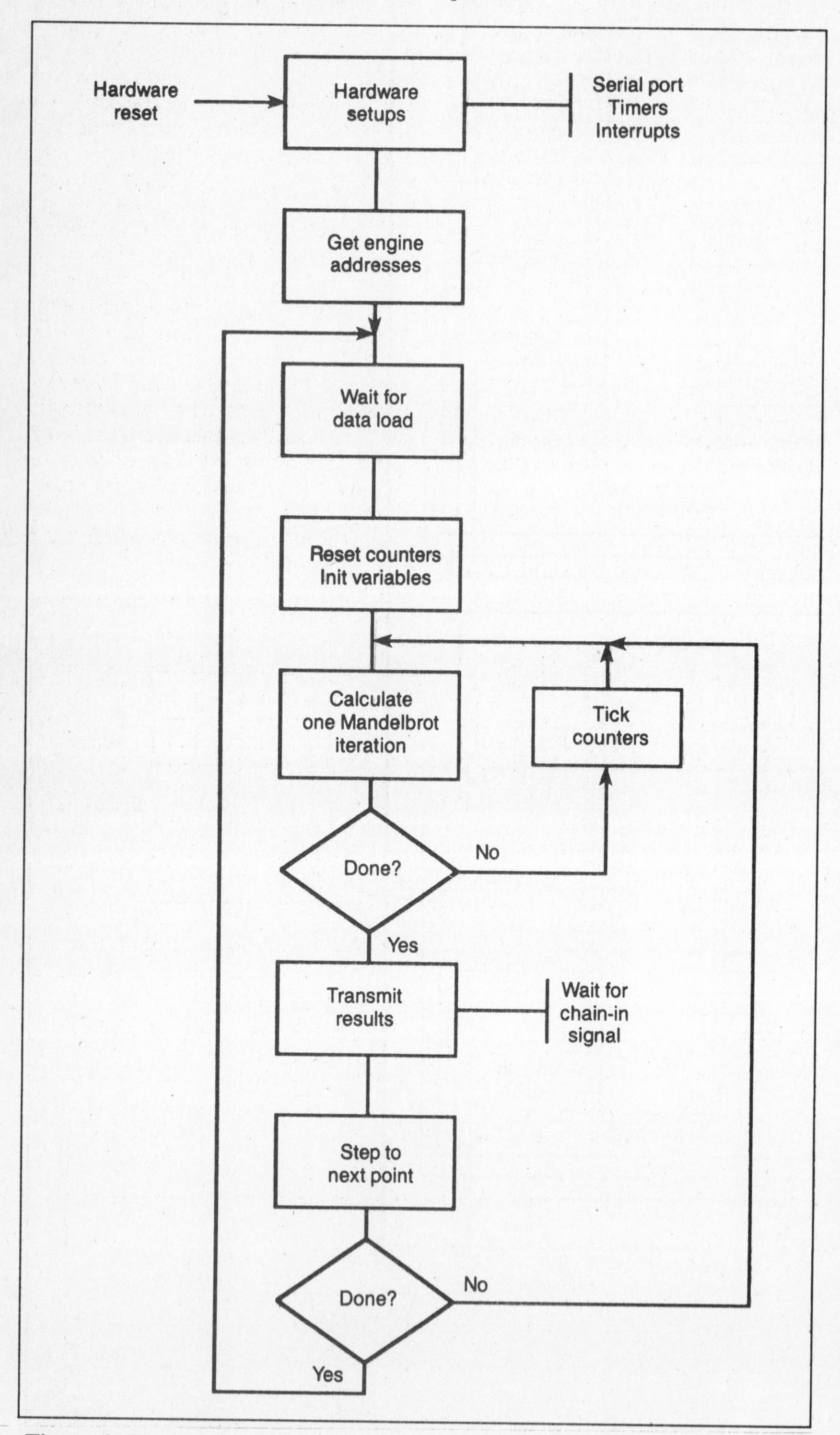

Figure 4: *The flowchart for the program executed by each processor of the Mandelbrot engine.*

The AT program divides the number of points on the screen by the number of processors in the engine, to decide how many points each processor must handle. For example, an EGA display has 224,000 pixels, so a four-processor engine will have 56,000 (224,000÷4) points assigned to each processor.

Figure 5 illustrates how those 224,000 pixels are divided among the four processors. Each processor is responsible for every fourth pixel across the line, and the assignments repeat on each line. The common setup data tells every processor that it must compute 160 pixels along the real axis before stepping the complex axis value. After 350 complex steps, the processor has finished its 56,000 points; it simply returns to the top of the loop and waits for the next set of data.

The sequence of events follows the classic "set up, read data, compute, write data, repeat forever" loop. In this case, though, the processors have only two LEDs and a serial port for communication, which is quite different from most computers you've seen before.

Although the AT program can handle engines of any size, it picks a line length that is the largest multiple of the actual number of processors. If it finds 9 processors installed (one full card and one lonely 8751 on another), the line length will be 639 (9 × 71). Six full cards with 48 processors will give a 624-pixel line (48 × 13), and so on. Installing a number of processors that "fits" into 640 pixels is best.

All the processors in the engine will cease computing after the complete Mandelbrot set is finished, at which time the AT can reload new common and individual data without having to reassign processor addresses. If the AT needs to interrupt a calculation in progress, it sends a single command to all processors that forces them into the idle state. If all else fails, of course, the AT can simply toggle the reset line and start from scratch.

Images On-Screen

The results of the Mandelbrot engine's calculations could be presented as a printed listing of the 224,000 counts needed to fill an EGA display, but it makes a lot more sense to view them directly. The program should also let you specify coordinates and areas by simply indicating a section of the screen. After all, typing in 19-digit numbers is no fun! Function and cursor keys control all the AT program's actions, so little typing is needed to make it work.

The EGA has more capabilities than most people realize, so we took the opportunity to show off a few of them in the Mandelbrot engine driver program (known simply as DRIVER). It's written in Microsoft C 5.1, with a few routines in Macro Assembler 5.1 to handle the video interrupts and some ugly machine details. (This code is available on the Circuit Cellar bulletin board system.)

All current EGA cards come equipped with 256K bytes of video memory, and DRIVER uses it all. To run DRIVER, you will need a 256K-byte EGA card and a display similar to an IBM Enhanced Color Display or an NEC MultiSync that can show 640 by 350 pixels in 16 colors. It does not support CGA cards, Hercules adapters of any sort, or (of course) the nongraphics IBM monochrome display.

At this point, DRIVER treats a VGA card just like an EGA, but we may at some time in the future add more capability to get better images using the VGA 320- by 200-pixel 256-color mode. We did a few sample VGA screens (sorry, the code isn't in distributable form yet), and once you've seen Mandelbrot-set images in full color, you'll never want to go back to 16 colors again. (Those of you with old true-blue 64K-byte IBM EGAs are out of luck. You are overdue to add the memory daughterboard and RAM chips. Given the current price of memory, you may need to shop around, but the expansion will be worth the price.)

Figure 6 shows how DRIVER uses the 256K bytes of video buffer. The EGA hardware supports smooth horizontal panning and vertical scrolling, so the buffer is laid out with 546 scan lines of 960 pixels each. The first 21 scan lines are dedicated to a hardware split screen that shows status messages and prompts, so the area available for graphics is actually 960 by 525 pixels. The 640 by 350 pixels visible on the screen can be scanned smoothly over the background using the cursor keys.

The results from the Mandelbrot engine can fill either the screen area or the whole video buffer. If DRIVER is in full-buffer mode, you can tour the complex plane while the image is being drawn. The scrolling and panning is controlled by an interrupt routine driven by the vertical interrupt from the EGA card.

Because that video interrupt is so critical, DRIVER will not work if you have a true-blue IBM VGA Display Adapter card for your PC, XT, or AT. For whatever reason, IBM did not implement the vertical interrupt signal on the PC version of the VGA card. DRIVER cannot update the screen at the right times without the vertical interrupt signal. Real VGAs work just fine, however, because the PS/2 system board includes the interrupt signal. Nice move, IBM.

I've had trouble with "super EGA" boards that automatically switch between various graphics modes. DRIVER is badly behaved in that it directly manipulates the hardware, so make sure that your board switches are set to disable all the super features if it doesn't run on your hardware. Successfully running DRIVER is a good test of your board's compatibility with a genuine IBM EGA.

Messages and prompts appear on a split-screen line that appears only when the messages are needed. At all other times, the full 350 scan lines are filled with Mandelbrot-set graphics. The vertical interrupt routine scrolls the split screen up and down, while the rest of the program continues about its business.

DRIVER can save images on disk, but it dumps the complete contents of the buffer with no compression. A single Mandelbrot image requires 252,512 bytes of disk space, so a hard disk drive is essential. The prototype engine with the AT controller that I built sports a 32-megabyte hard disk drive, which filled up surprisingly quickly.

DRIVER's main purpose is to control the engine, but it also includes functions that get the engine running and that measure its performance. Various function keys pop up statistics describing the current image, and a summary line appears after an image is finished.

The Problem with Benchmarks

Benchmarks can be misleading under the best of circumstances. When it comes to evaluating a multiprocessor system, benchmarks can be downright dangerous. There's more to it than just running a program between stopwatch clicks.

The ordinary benchmark sequence for a new computer seems to involve running the Sieve program at least once, along with a few other programs that exercise memory, disk accesses, and floating-point math. The problem, of course, is that the Mandelbrot engine can't run any of those programs: It's designed to solve one problem very well, and it can't handle any other task.

So a more appropriate benchmark

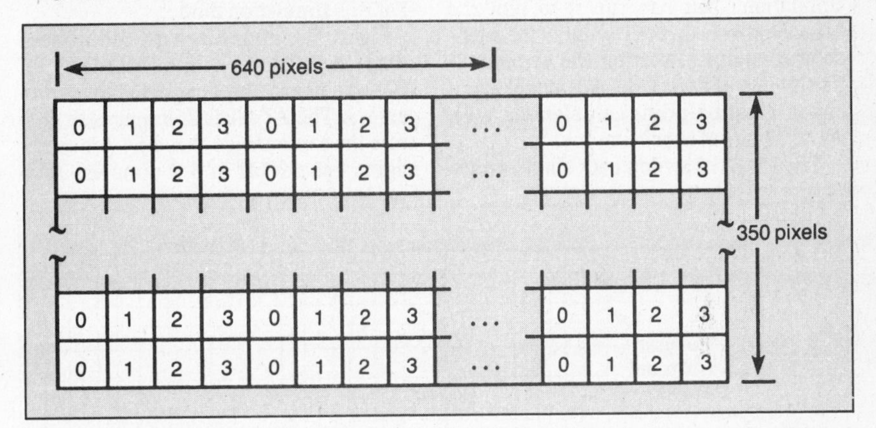

Figure 5: *Assigning pixels to processors. If you're running a four-processor engine, the control program assigns pixels to successive processors by moving through the screen buffer horizontally left to right, then down.*

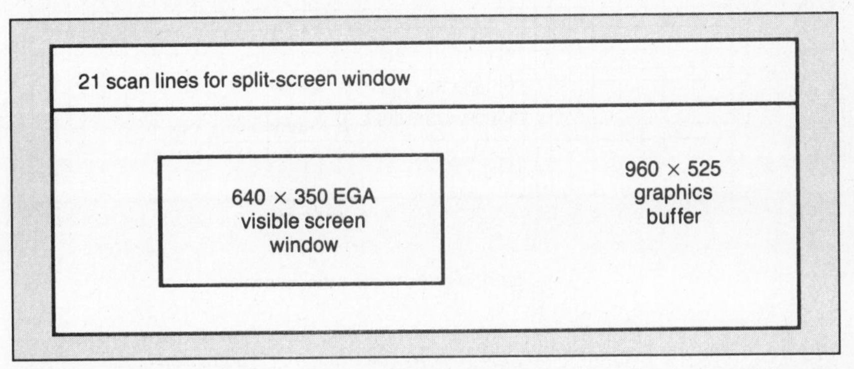

Figure 6: *A map of the EGA video buffer (256K bytes), as used by the Mandelbrot driver program.*

might compare the Mandelbrot engine against other Mandelbrot-set programs that run strictly on a PC. The problem here is that the programs on the two machines are totally different, so it's not clear what "performance" the benchmarks would measure: hardware, software, or firmware.

Worse, the Mandelbrot engine requires a driver program running on a PC to handle all its setup and I/O. Obviously, a faster PC will run the driver program faster, so the performance will increase, even though the engine itself remains unchanged. Should the benchmarks include testing on various PCs to isolate the effects of the driver?

I decided that the only reasonable way to benchmark the Mandelbrot engine was to add code to DRIVER that simulated the engine's calculations. The comparisons could then take DRIVER as a constant and concentrate on the differences between the engine and the PC simulation. This also ensured that any hardware differences on the PC would affect the engine and the PC simulation equally.

DRIVER can handle three levels of simulation. You can run it to plot the Mandelbrot set directly without the Mandelbrot engine. Entering the command-line switch `-FAKE:1` invokes double-precision floating-point calculations with about 52 bits of precision.

The `-FAKE:2` switch uses single-precision floating-point math with about 24 bits of precision. The `-FAKE:3` switch triggers a precise emulation with fixed-point numbers identical to those used in the engine.

The two floating-point simulations (as well as the rest of the code) take advantage of a math coprocessor if the PC has one available. The fixed-point simulation is written in C rather than tightly optimized assembly language, so the performance is excruciatingly slow. The comparisons shown below use only the `-FAKE:1` simulation mode.

The AT is slower than the Mandelbrot engine by a factor of 22.

Figure 7 summarizes a head-to-head competition between an 8-MHz IBM PC AT and a Mandelbrot engine with 64 processors. The AT takes 85 minutes to generate the same image that a 64-processor engine knocks off in 3.9 minutes. Adding a 10-MHz 80287 to the AT improves its performance to 9.7 minutes, while the engine drops to 2.8 minutes because of the improved speed in DRIVER.

Those numbers are worth examining in more detail. The AT is slower than the engine by a factor of 22! Adding an 80287 improves the AT by a factor of 8.8, but the engine is 3.5 times faster than that. Not bad for a bunch of single-chip microcontrollers.

Remember that engine calculations use more precise numbers to derive the results, so the Mandelbrot-set images will be slightly different. I suspect that the engine results are more "correct," but I don't have any way to prove it (I'm certainly not going to run the calculations by hand).

The engine's performance depends on how many processors are sharing the work load, as with any multiprocessor system. Figure 7 plots the total elapsed time for each calculation against the number of processors, using log-log axes to show equal percentage changes as equal distances (I ran the tests with a maximum iteration limit of 64 to provide a better load on the processors, but the image area is the same as on the previous test). With the exception of the 64-processor engine, the results fall on a nearly perfectly straight line, which indicates that the additional processors are sharing the work load equally—the more processors, the better it runs.

The AT (with an 80287) takes 21.7 minutes to generate the test image, which is equivalent to a engine with only 13 processors. Although I didn't try it without the 80287, the factor of 8.8 found in the previous test indicates that it would take about 191 minutes. A Mandelbrot engine with 2 or 3 processors will do better than that!

The 64-processor engine is slower than the trend line for the other six configurations would indicate. Recall that each processor delivers its results in strict order, so any processors that get done calculating "early" must wait their turn on the serial link. This contention will increase as more processors share the link. We considered changing the 8751 program to buffer the results of one or two calculations so that the program can continue even if the link isn't ready, but there's no room left on the 8751 for even a tiny buffer.

It's worth noting that serial-link transmission time doesn't slow down the results by a significant amount. A byte on the link takes about 450 microseconds at 28,800 bits per second, so a complete Mandelbrot set with 224,000 points will

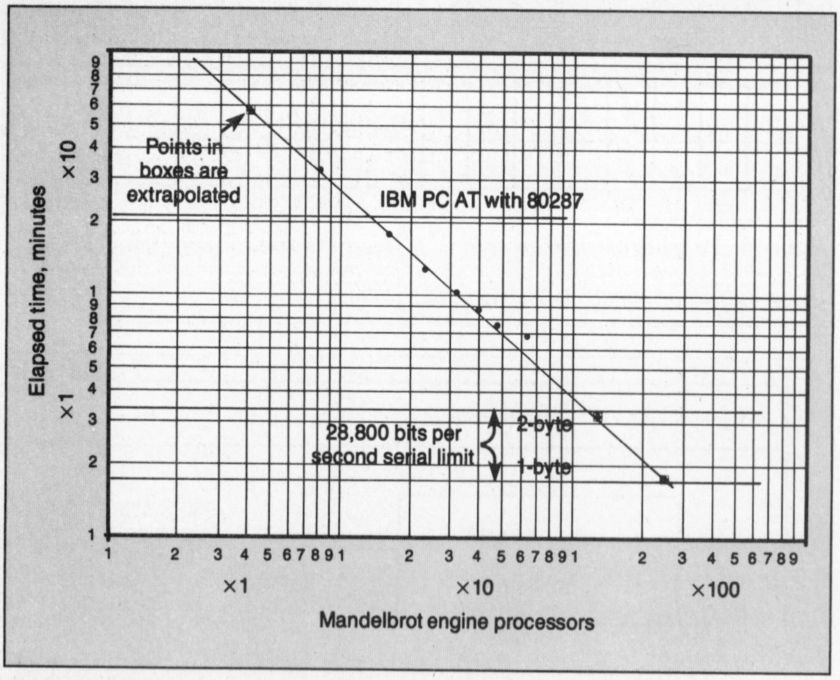

Elapsed time, minutes
Points in boxes are extrapolated
IBM PC AT with 80287
28,800 bits per second serial limit
2-byte
1-byte
Mandelbrot engine processors

Figure 7: *The log-log plot of elapsed time to calculate a given portion of the set versus number of processors. Also shown is the time for an AT to plot the same image.*

take 1.7 minutes to transfer if every processor is ready on time. Iteration limits above 255 require 2-byte counts, so the transfer time would double to 3.4 minutes. You can see that the serial link is not the limiting factor for "serious" calculations.

Conclusion

During this three-part chapter, I've explained why single-processor computer systems suffer from performance limitations that can't be wished away and how multiprocessor systems are able to "get around" some of those limits. The Circuit Cellar Mandelbrot engine demonstrates some of the design issues that arise in a multiprocessor computer and shows how you can get high performance from very simple computing elements.

Special thanks to Ed Nisley and Merrill Lathers for their expert contributions to this project.

Editor's Note: Steve often refers to previous Circuit Cellar articles. Most of these past articles are available in book form from BYTE Books, McGraw-Hill Publishing Company, P.O. Box 400, Hightstown, NJ 08250, (1–800–2–MCGRAW).

Ciarcia's Circuit Cellar, Volume I covers articles in BYTE from September 1977 through November 1978. *Volume II* covers December 1978 through June 1980. *Volume III* covers July 1980 through December 1981. *Volume IV* covers January 1982 through June 1983. *Volume V* covers July 1983 through December 1984. *Volume VI* covers January 1985 through June 1986.

There is a multiline Circuit Cellar bulletin board system (running TBBS 2.0M) that supports past and present projects in BYTE and Ink. You are invited to call and exchange ideas and comments with other Circuit Cellar supporters. The 300-/1200-/2400-bps BBS is online 24 hours a day at (203) 871-1988.

Index